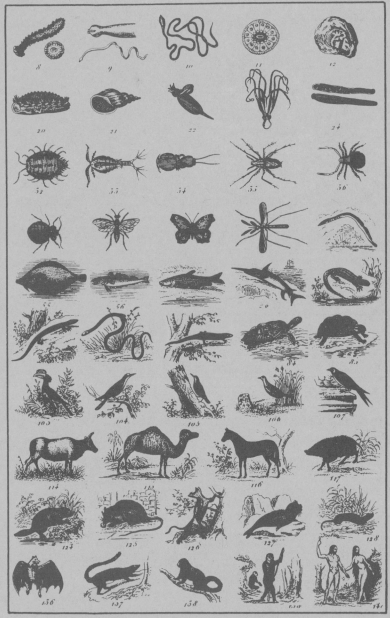

SHADOWS OF FORGOTTEN ANCESTORS

*A carving from the
Sepik River, central highlands
of Papua New Guinea.*

SHADOWS
OF FORGOTTEN
ANCESTORS

A Search for Who We Are

CARL SAGAN
ANN DRUYAN

RANDOM HOUSE

NEW YORK

Permissions acknowledgments for previously published
material can be found on pages 479–480.

Library of Congress Cataloging-in-Publication Data
Sagan, Carl
Shadows of forgotten ancestors / Carl Sagan and Ann Druyan.—1st ed.
p.   cm.
Includes bibliographical references and index.
ISBN 0-394-53481-6
1. Human Evolution.   2. Evolution (Biology)   I. Druyan, Ann, 1949–   .   II. Title.
QH325.S24   1992   304.4—dc20   92-50155

Manufactured in the United States of America
Book design by J. K. Lambert
24689753
Printed on recycled paper
First Edition

TO
LESTER GRINSPOON,
WHOSE EXAMPLE REASSURES US
THAT OUR SPECIES
MAY HAVE
WHAT IT TAKES

Thus she spoke; and I longed
to embrace my dead mother's ghost.
Thrice I tried to clasp her
image, and thrice it slipped
through my hands, like a
shadow, like a dream.

HOMER
The Odyssey[1]

Contents

Introduction

We were very lucky. We were raised by parents who took seriously their responsibility to be strong links in the chain of generations. The search that informs this book may be said to have begun in childhood, when we were given unconditional love and protection in the face of real adversity. It's an ancient practice of the mammals. It was never easy. In modern human society, it's even harder. There are so many dangers now, so many of them unprecedented.

The book itself began in the early 1980's when the rivalry between the United States and the Soviet Union was making a potentially fateful intersection with 60,000 nuclear weapons that had been accumulated for reasons of deterrence, coercion, pride, and fear. Each nation praised itself and vilified its adversaries, who were sometimes portrayed as less than human. The United States spent ten trillion dollars on the Cold War—enough to buy everything in the country except the land. Meanwhile, the infrastructure was collapsing, the environment was deteriorating, the democratic process was being subverted, injustice festered, and the nation was converted from the leading lender to the leading debtor on the planet. How did we get into this mess? we asked ourselves. How can we get out? *Can* we get out?

So we embarked on a study of the political and emotional roots of the nuclear arms race—which led us back to World War II, which of course had its origins in World War I, which was a consequence of the rise of the nation-state, which traces straight back to the very beginnings of civilization, which was a by-product of the invention of agriculture and the domestication of animals, which crystallized out of a very long period in which we humans were hunters and foragers. There was no sharp division along the way, no point at which we could say: *Here* are the roots of our predicament. Before we knew it, we were looking to the first humans and *their* predecessors. Events of remote ages, long before humans came to be, are critical, we concluded, for an understanding of the trap that our species seems to be setting for itself.

We resolved to look inside ourselves, to retrace as many of the important twists and turns of the evolution of our species as we were able. We made a compact with each other not to turn back, no matter where the search might lead. We had learned much from each other over the years, but our own politics are not identical. There was a chance that one or both of us might have to give up some of those beliefs we considered self-defining. But if we were successful, even in part, perhaps we could understand much more than just nationalism, the nuclear arms race, and the Cold War.

As we complete this book, the Cold War is over. But somehow we are not home free. New dangers edge their way onto center stage, and old familiar ones reassert themselves. We are confronted with a witches' brew of ethnic violence, resurgent nationalism, inept leaders, inadequate education, dysfunctional families, environmental decay, species extinctions, burgeoning population, and increasing millions with nothing to lose. The need to understand how we got into this mess and how to get out seems more urgent than ever.

This book addresses the deep past, the most formative steps in our origins. Later, we will gather up the threads laid down here. We have been led to the writings of those who preceded us in this search, to distant epochs and other worlds and across a multitude of disciplines. We tried to keep in mind the physicist Niels Bohr's aphorism, "Clarity through breadth." The breadth required can be a little daunting, though. Humans have erected high walls separating the branches of knowledge essential to this quest—the various sciences, politics, religions, ethics. We have searched for low doors in the walls, or sometimes tried to vault over or burrow under. We feel a need to apologize for our limitations. We are well aware of the inadequacies of our knowledge and of our discernment. And yet such a search has no chance of succeeding unless those walls are breached. We hope that where we have failed, others will be inspired (or provoked) to do better.

What we are about to say draws on the findings of many sciences. We urge the reader to bear in mind the imperfection of our current knowledge. Science is never finished. It proceeds by successive approximations, edging closer and closer to a complete and accurate understanding of Nature, but it is never fully there. From the fact that so many major discoveries have been made in the last century—even in the last decade—it is clear that we still have far to go. Science is always subject to debate, correction, refinement, agonizing reappraisals, and revolutionary insights. Nevertheless, there now seems to be enough known to reconstruct some of the key steps that led to us and helped to make us who we are.

On our journey we encountered many who were generous with their time, expertise, wisdom, and encouragement, many who carefully and critically read all or part of the manuscript. As a result, deficiencies were removed, and errors of fact or interpretation corrected. We particularly thank Diane Ackerman; Christopher Chyba, Ames Research Center, NASA; Jonathan Cott; James F. Crow, Department of Genetics, University of Wisconsin, Madison; Richard Dawkins, Department of Zoology, Oxford University; Irven de Vore, Department of Anthropology, Harvard University; Frans B. M. de Waal, Department of Psychology, Emory University, and Yerkes Primate Research Center; James M. Dabbs, Jr., Department of Psychology, Georgia State University; Stephen Emlen, Section of Neurobiology and Behavior, Cornell University; Morris Goodman, Department of Anatomy and Cell Biology, Wayne State University School of Medicine; Stephen Jay Gould, Museum of Comparative Zoology, Harvard University; James L. Gould and Carol Grant Gould, Department of Biology, Princeton University; Lester Grinspoon, Department of Psychiatry, Harvard Medical School; Howard E. Gruber, Department of Developmental Psychology, Columbia University; Jon Lomberg; Nancy Palmer, Shorenstein Barone Center on the Press and Politics, Kennedy School of Government, Harvard University; Lynda Obst; William Provine, Departments of Genetics and of the History of Science, Cornell University; Duane M. Rumbaugh and E. Sue Savage-Rumbaugh, Language Research Center, Georgia State University; Dorion, Jeremy, and Nicholas Sagan; J. William Schopf, Center for the Study of Evolution and the Origin of Life, University of California, Los Angeles; Morty Sills; Steven Soter, Smithsonian Institution; Jeremy Stone, Federation of American Scientists; and Paul West. Many scientists kindly sent us pre-publication copies of their work. C.S. also thanks his early teachers in the life sciences, H. J. Muller, Sewall Wright, and Joshua Lederberg. Of course none of these people are responsible for any remaining errors.

We are deeply grateful to those who ushered this work through its various drafts. For excellence in library research, transcription, file keeping, and much else we owe a special debt of gratitude to A.D.'s assistant, Karenn Gobrecht, and to C.S.'s long-time Administrative Assistant at Cornell, Eleanor York. We also thank Nancy Birn Struckman, Dolores Higareda, Michelle Lane, Loren Mooney, Graham Parks, Deborah Pearlstein, and John P. Wolff. The superb facilities of the Cornell University library system were a critical resource in the writing of this book. We also could not have written it without the help of Maria Farge, Julia Ford Diamond, Lisbeth Collacchi, Mamie Jones, and Leona Cummings.

We are indebted to Scott Meredith and Jack Scovil of the Scott Meredith Literary Agency for unstinting encouragement and support. We are happy that *Shadows* has come to fruition during Ann Godoff's tenure as our editor; and also thank Harry Evans, Joni Evans, Nancy Inglis, Jim Lambert, Carol Schneider, and Sam Vaughan at Random House.

Walter Anderson, the editor-in-chief of *Parade* magazine, has made it possible for us to present our ideas to the broadest possible audience. Working with him and Senior Editor David Currier has been an unalloyed pleasure.

This book is written for a wide readership. For clarity, we have sometimes stressed the same point more than once, or in more than one context. We have tried to indicate qualifications and exceptions. The pronoun "we" is used sometimes to mean the authors of this book, but usually to mean the human species; the context should make clear which is meant. For those who wish to dig deeper, references to other works, popular and technical—keyed to superscripts in the text—are in the back of the book. Also to be found there are additional comments, notes, and clarifications. Although the two works have little else in common, the haunting 1964 film by Sergei Parajanov gave us our title.

As for essential inspiration and a heightened sense of urgency, it was during the years of preparation of this book that we became the parents of Alexandra Rachel and Samuel Democritus—beloved namesakes of unforgettable ancestors.

CARL SAGAN
ANN DRUYAN
June 1, 1992
Ithaca, N.Y.

THE ORPHAN'S FILE

Having seen a small part of life, swift to die,
men rise and fly away like smoke, persuaded
only of what each has met with . . . Who then
claims to find the whole?

EMPEDOCLES
On Nature[1]

Who are we? The answer to this question is not
only one of the tasks, but *the* task of science.

ERWIN SCHRÖDINGER
Science and Humanism[2]

The immense, overpowering blackness is relieved here and there by a faint point of light—which, upon closer approach, is revealed to be a mighty sun, blazing with thermonuclear fire and warming a small surrounding volume of space. The Universe is, almost entirely, black emptiness, and yet the number of suns is staggering. The neighborhoods immediately encompassing these suns represent an insignificant fraction of the vastness of the Cosmos, but many, perhaps most, of those cheerful, bright, clement circumstellar regions are occupied by worlds. In the Milky Way galaxy alone there may be a hundred billion of them—neither too close by, nor too distant from, the local sun, around which they orbit in silent gravitational homage.

This is a story about one such world, perhaps not very different from many others—a story, especially, about the beings that evolved upon it, and one kind in particular.

Just to be alive billions of years after the origin of life, a being must be tough, resourceful, and lucky: There have been so many hazards along the way. Lifeforms endure by being patient, say, or ravenous, or solitary and camouflaged, or profligate with offspring, or fearsome hunters, or able to fly away to safety, or sleek swimmers, or burrowers, or sprayers of noxious, disorienting liquids, or masters at infiltrating into the very genetic material of other, unsuspecting, beings; or by accidentally being elsewhere when the predators stalk or the river is poisoned or the food supply dwindles. The creatures with which we are particularly concerned were, not so long ago, gregarious to a fault, noisy, quarrelsome, arboreal, bossy, sexy, clever, tool-using, with prolonged childhoods and tender regard for their young. One thing led to another, and in a twinkling their descendants had multiplied all over the planet, killed off all their rivals, devised world-transforming technologies, and posed a mortal danger to themselves and the many other beings with whom they shared their small home. At the same time, they set off to visit the planets and the stars.

———

Who are we? Where do we come from? Why are we *this* way and not some other? What does it mean to be human? Are we capable, if need be, of fundamental change, or do the dead hands of forgotten ancestors impel us in some direction, indiscriminately for good or ill, and beyond our control? Can we alter our character? Can we improve our societies? Can we leave our children a world better than the one that was left to us? Can we free them from the demons that torment us and haunt our civilization? In the long run, are we wise enough to know what changes to make? Can we be trusted with our own future?

Many thoughtful people fear that our problems have become too big for us, that we are for reasons at the heart of human nature unable to deal with them, that we have lost our way, that the dominant political and religious ideologies are unable to halt an ominous, long-term drift in human affairs—indeed, that they have helped cause that drift through rigidity, incompetence, and the inevitable corruption of power. Is this true, and if it is, can we do anything about it?

In attempting to understand who we are, every human culture has invented a corpus of myth. The contradictions within us are ascribed to a struggle between contending but equally matched deities; or to an imperfect Creator; or, paradoxically, to a rebellious angel and the Almighty; or to the even more unequal struggle between an omnipotent being and disobedient humans. There have also been those who hold that the gods have nothing to do with it. One of them, Nanrei Kobori, late Abbot of the Temple of the Shining Dragon, a Buddhist sanctuary in Kyoto, said to us

> God is an invention of Man. So the nature of God is only a shallow mystery. The deep mystery is the nature of Man.

Had life and humans first come to be hundreds or even thousands of years ago, we might know most of what's important about our past. There might be very little of significance about our history that's hidden from us. Our reach might extend easily to the beginning. But instead, our species is hundreds of thousands of years old, the genus *Homo* millions of years old, primates tens of millions of years old, mammals over 200 million years old, and life about 4 billion years old.

Our written records carry us only a millionth of the way back to the origin of life. Our beginnings, the key events in our early development, are not readily accessible to us. No firsthand accounts have come down to us. They cannot be found in living memory or in the annals of our species. Our time-depth is pathetically, disturbingly shallow. The overwhelming majority of our ancestors are wholly unknown to us. They have no names, no faces, no foibles. No family anecdotes attach to them. They are unreclaimable, lost to us forever. We don't know them from Adam. If an ancestor of yours of a hundred generations ago—never mind a thousand or ten thousand—came up to you on the street with open arms, or just tapped you on the shoulder, would you return the greeting? Would you call the authorities?

We ourselves, the writers of this book, have so short a reach into our family histories that we can peer clearly only two generations back, dimly three, and almost not at all beyond that. We do not know even the names—much less the occupations, countries of origin, or personal histories—of our great-great-grandparents. Most people on Earth, we think, are similarly isolated in time. For most of us, no records have preserved the memories of our ancestors of even a few generations back.

A vast chain of beings, human and nonhuman, connects each of us with our earliest predecessors. Only the most recent links are illuminated by the feeble searchlight of living memory. All the others are plunged into varying degrees of darkness, more impenetrable the farther from us they are in time. Even those fortunate families who have managed to keep meticulous records range no more than a few dozen generations into the past. And yet a hundred thousand generations ago our ancestors were still recognizably human, and ages of geological time stretch back before them. For most of us, the searchlight progresses forward as the generations do, and as the new ones are born, information about the old ones is lost. We are cut off from our past, separated from our origins, not through some amnesia or lobotomy, but because of the brevity of our lives and the immense, unfathomed vistas of time that separate us from our coming to be.

We humans are like a newborn baby left on a doorstep, with no note explaining who it is, where it came from, what hereditary cargo of attributes and disabilities it might be carrying, or who its antecedents might be. We long to see the orphan's file.

Repeatedly, in many cultures, we invented reassuring fantasies about our parents—about how much they loved us, about how heroic and larger than life they were.[3] As orphans do, we sometimes blamed ourselves for having been abandoned. It must have been our fault. We were too sinful, perhaps, or morally incorrigible. Insecure, we clung to these stories, imposing the strictest penalties on any who dared to doubt them. It was better than nothing, better than admitting our ignorance of our own origins, better than acknowledging that we had been left naked and helpless, a foundling on a doorstep.

As the infant is said to feel it is the center of its Universe, so we were once sure, not just of our central position, but that the Universe was *made* for us. This old, comfortable conceit, this safe view of the world has been crumbling for five centuries. The more we understood of how the world is put together, the less we needed to invoke a God or gods, and the more remote in time and causality any divine intervention had to be. The cost of coming of age is giving up the security blanket. Adolescence is a roller coaster ride.

When, beginning in 1859, our very origins, it was suggested, could be understood by a natural, unmystical process—requiring no God or gods—our aching sense of isolation became nearly complete. In the words of the anthropologist Robert Redfield, the Universe began to "lose its moral character" and became "indifferent, a system uncaring of man."[4]

Moreover, without a God or gods and the attendant threat of divine punishment, will not humans be as beasts? Dostoyevsky warned that those who reject religion, however well-intentioned they may be, "will end by drenching the earth with blood."[5] Others have noted that drenching has been in progress since the dawn of civilization—and often in the name of religion.

The distasteful prospect of an indifferent Universe—or worse, a meaningless Universe—has generated fear, denial, ennui, and the sense that science is an instrument of alienation. The cold truths of our scientific age are uncongenial to many. We feel stranded and alone. We crave a purpose to give meaning to our existence. We do not want to hear that the world was not made for us. We are unimpressed with moral codes contrived by mere mortals; we want one handed down from on high. We are reluctant to acknowledge our relatives. They are strangers to us still. We feel ashamed: After imagining our Antecedent as King of the Universe, we are now asked to accept that we come from the lowest of the low—mud, and slime, and mindless beings too small to be seen with the naked eye.

Why concentrate on the past? Why upset ourselves with painful analogies between humans and beasts? Why not simply look to the future? These questions have an answer. If we do not know what we're capable of—and not just a few celebrity saints and notorious war criminals—then we do not know what to watch out for, which human propensities to encourage, and which to guard against. Then we haven't a clue about which proposed courses of human action are realistic, and which are impractical and dangerous sentimentality. The philosopher Mary Midgley writes,

> Knowing that I have a naturally bad temper does not make me lose it. On the contrary, it should help me to keep it, by forcing me to distinguish my normal peevishness from moral indignation. My freedom, therefore, does not seem to be particularly threatened by the admission, nor by any light cast on the meaning of my bad temper by comparison with animals. [6]

The study of the history of life, the evolutionary process, and the nature of the other beings who ride this planet with us has begun to cast a little light on those past links in the chain. We have not met our forgotten ancestors, but we begin to sense their presence in the dark. We recognize their shadows here and there. They were once as real as we are. We would not be here if not for them. Our natures and theirs are indissolubly linked despite the aeons that may separate us. The key to who we are is waiting in those shadows.

———

When we began this search into our origins, using the methods and findings of science, it was almost with a sense of dread. We were afraid of what we might find. We found instead not just room but reason for hope, as we begin to explain in this book.

The real orphan's file is long. We humans have uncovered bits and pieces, occasionally a few consecutive pages, nothing as elaborate as a complete chapter. Many of the words are blurred. Most have been lost. [7]

Here then is one version of some of the early pages of the orphan's file, the missing note that should have accompanied the foundling on the doorstep, something of our beginnings and the forgotten ancestors that are central to the outcome of our story. Like most family stories, it begins in the dark—so long ago and far away, in circumstances so unpromising, that no one could have guessed where it all would lead.

We are about to trace the history of life, and the path that led to us —how we got to be the way we are. It is fitting that we begin at the beginning. Or a little earlier.

Chapter 1

ON EARTH
AS IT IS
IN HEAVEN

How long the stars
Have been fading,
Lamplight dimming . . .

NANSEN
(748–834, China)[1]

For the forming of the earth they said "Earth."
It arose suddenly, just like a cloud, like a mist,
now forming, unfolding . . .

Popol Vuh: The Mayan Book of the Dawn of Life[2]

Nothing lives forever, in Heaven as it is on Earth. Even the stars grow old, decay, and die. They die, and they are born. There was once a time before the Sun and Earth existed, a time before there was day or night, long, long before there was anyone to record the Beginning for those who might come after.

Nevertheless, imagine you were a witness to that time:

An immense mass of gas and dust is swiftly collapsing under its own weight, spinning ever faster, transforming itself from a turbulent, chaotic cloud into what seems to be a distinct, orderly, thin disk. Its exact center smolders a dull, cherry red. Watch from on high, above the disk, for a hundred million years and you will see the central mass grow whiter and more brilliant, until, after a couple of abortive and incomplete attempts, it bursts into radiance, a sustained thermonuclear fire. The Sun is born. Faithfully, it will shine over the next five billion years—when the matter in the disk will have evolved into beings able to reconstruct the circumstances of its origin, and theirs.

Only the innermost provinces of the disk are illuminated. Farther out, the sunlight fails to penetrate. You plunge into the recesses of the cloud to see what wonders are unfolding. You discover a million small worlds milling about the great central fire. A few thousand sizable ones here and there, most circling near the Sun but some at great distances away, are destined to find each other, merge, and become the Earth.

This spinning disk out of which worlds are forming has fallen together from the sparse matter that punctuates a vast region of interstellar vacuum within the Milky Way galaxy. The atoms and grains that make it up are the flotsam and jetsam of galactic evolution—here, an oxygen atom generated from helium in the interior inferno of some long-dead red giant star; there, a carbon atom expelled from the atmosphere of a carbon-rich star in some quite different galactic sector; and now an iron atom freed for world-making by a mighty supernova explosion in the still more ancient past. Five billion years after the events we are describing, these very atoms may be coursing through your bloodstream.

Our story begins here in the dark, pullulating, dimly illuminated disk: the story as it actually turned out, and an enormous number of other stories that would have come to be had things gone just a little differently; the story of our world and species, but also the story of many other worlds and lifeforms destined never to be. The disk is rippling with possible futures.[3]

———

For most of their lives, stars shine by transmuting hydrogen into helium. It happens at enormous pressures and temperatures deep inside them. Stars have been aborning in the Milky Way galaxy for ten billion years or more—within great clouds of gas and dust. Almost all the placenta of gas and dust that once surrounded and nourished a star is quickly lost, either devoured by its tenant or spewed back into interstellar space. When they are a little older—but we are still talking about the childhood of the stars—a massive disk of gas and dust can be discerned, the inner lanes circling the star swiftly, the outer ones moving more stately and slowly. Similar disks are detectable around stars barely out of their adolescence, but now only as thin remnants of their former selves—mostly dust with almost no gas, every grain of dust a miniature planet orbiting the central star. In some of them, dark lanes, free of dust, can be made out. Perhaps half the young stars in the sky that are about as massive as the Sun have such disks. Still older stars have nothing of the sort, or at least nothing that we are yet able to detect. Our own Solar System to this day retains a very diffuse band of dust orbiting the Sun, called the zodiacal cloud, a wispy remake of the great disk from which the planets were born.

The story these observations are telling us is this: Stars formed in batches from huge clouds of gas and dust. A dense clump of material attracts adjacent gas and dust, grows larger and more massive, more efficiently draws matter to it, and is off on its way to stardom. When the temperatures and pressures in its interior become high enough, hydrogen atoms—the most abundant material in the Universe by far —are jammed together and thermonuclear reactions are initiated. When it happens on a large enough scale, the star turns on and the nearby darkness is dispelled. Matter is turned into light.

The collapsing cloud spins up, squashes down into a disk, and lumps of matter aggregate together—successively the size of smoke particles, sand grains, rocks, boulders, mountains, and worldlets. Then the cloud tidies itself up through the simple expedient of the largest objects gravitationally consuming the debris. The dust-free lanes are the feeding zones of young planets. As the central star begins to shine, it also sends forth great gales of hydrogen that blow grains back into the void. Perhaps some other system of worlds, fated to arise billions of years later in some distant province of the Milky Way, will put these rejected building blocks to good use.

In the disks of gas and dust that surround many nearby stars, we think we see the nurseries in which worlds, far-off and exotic, are accumulating and coalescing. All over our galaxy, vast, irregular, lumpy, pitch-black, interstellar clouds are collapsing under their own gravity, and spawning stars and planets. It happens about once a month. In the observable Universe—containing as many as a hundred billion galaxies—perhaps a hundred solar systems are forming every second. In that multitude of worlds, many will be barren and desolate. Others may be lush and fertile, on which beings exquisitely adapted to their several circumstances are growing up, coming of age, and attempting to piece together *their* beginnings. The Universe is lavish beyond imagining.

———

As the dust settles and the disk thins, you can now make out what is happening down there. Hurtling about the Sun is a vast array of worldlets, all in slightly different orbits. Patiently you watch. Ages pass. With so many bodies moving so quickly, it is only a matter of time before worlds collide. As you look more closely, you can see collisions occurring almost everywhere. The Solar System begins amid almost unimaginable violence. Sometimes the collision is fast and head-on, and a devastating, although silent, explosion leaves nothing but shards and fragments. At other times—when two worldlets are in nearly identical orbits with nearly identical speeds—the collisions are nudging, gentle; the bodies stick together, and a bigger, double worldlet emerges.

In another age or two, you notice that several much larger bodies are growing—worlds that, by luck, escaped a disintegrating collision in their early, more vulnerable days. Such bodies—each established in its own feeding zone—plow through the smaller worldlets and gobble them up. They have grown so large that their gravity has crushed out the irregularities; these bigger worlds are nearly perfect spheres. When a worldlet approaches a more massive body, although not close enough to collide, it swerves; its orbit is changed. On its new trajectory, it may impact some other body, perhaps smashing it to smithereens; or meet a fiery death as it falls into the young Sun, which is consuming the matter in its vicinity; or be gravitationally ejected into the frigid interstellar dark. Only a few are in fortunate orbits, neither eaten, nor pulverized, nor fried, nor exiled. They continue to grow.

Beyond a certain mass, the bigger worlds are attracting not just dust, but great streams of interplanetary gas as well. You watch them develop, eventually each with a vast atmosphere of hydrogen and helium gas surrounding a core of rock and metal. They become the four giant planets, Jupiter, Saturn, Uranus, and Neptune. You can see the characteristic banded cloud patterns emerge. Collisions of comets with their moons splay out elegant, patterned, iridescent, ephemeral rings. Pieces of an exploded world fall back together, generating a jumbled, odd-lot, motley new moon. As you watch, an Earth-sized body plows into Uranus, knocking the planet over on its side, so once each orbit its poles point straight at the distant Sun.

Closer in, where the disk gas has by now been cleaned away, some of the worlds are becoming Earth-like planets, another class of survivors in this game of world-annihilating gravitational roulette. The final accumulation of the terrestrial planets takes no more than 100 million years, about as long compared to the lifespan of the Solar System as the first nine months is relative to the lifetime of an average human being. A doughnut-shaped zone of millions of rocky, metallic, and organic worldlets, the asteroid belt, survives. Trillions of icy worldlets, the comets, slowly orbit the Sun in the darkness beyond the outermost planet.

The principal bodies of the Solar System have now formed. Sunlight pours through a transparent, nearly dust-free interplanetary space, warming and illuminating the worlds. They continue to course and careen about the Sun. But look more closely still and you can make out that further change is being worked.

None of these worlds, you remind yourself, has volition; none *intends* to be in a particular orbit. But those that are on well-behaved, circular orbits tend to grow and prosper, while those on giddy, wild, eccentric, or recklessly tilted orbits tend to be removed. As time goes on, the confusion and chaos of the early Solar System slowly settle down into a steadily more orderly, simple, regularly spaced, and, to your eyes, increasingly beautiful set of trajectories. Some bodies are selected to survive, others to be annihilated or exiled. This selection of worlds occurs through the operation of a few extremely simple laws of motion and gravity. Despite the good neighbor policy of the well-mannered worlds, you can occasionally make out a flagrant rogue worldlet on collision trajectory. Even a body with the most circumspect circular orbit has no warrantee against utter annihilation. To continue to survive, an Earth-like world must also continue to be lucky.

The role of something close to random chance in all this is striking. Which worldlet will be shattered or ejected, and which will safely grow to planethood, is not obvious. There are so many objects in so complicated a set of mutual interactions that it is very hard to tell—just by looking at the initial configuration of gas and dust, or even after the planets have mainly formed—what the final distribution of worlds will be. Perhaps some other, sufficiently advanced observer could figure it out and predict its future—or even set it all in motion so that, billions of years later, through some intricate and subtle sequence of processes, a desired outcome will slowly emerge. But that is not yet for humans.

You started with a chaotic, irregular cloud of gas and dust, tumbling and contracting in the interstellar night. You ended with an elegant, jewel-like solar system, brightly illuminated, the individual planets neatly spaced out one from another, everything running like clockwork. The planets are nicely separated, you realize, because those that aren't are gone.

———

It's easy to see why some of those early physicists who first penetrated the reality of the nonintersecting, coplanar orbits of the planets thought that the hand of a Creator was discernible. They were unable to conceive of any alternative hypothesis that could account for such magnificent precision and order. But in the light of modern understanding, there is no sign of divine guidance here, or at least nothing beyond physics and chemistry. Instead we see evidence of a time of remorseless and sustained violence, when vastly more worlds were destroyed than preserved. Today we understand something of how the exquisite precision that the Solar System now exhibits was extracted from the disorder of an evolving interstellar cloud by laws of Nature that we are able to grasp—motion, and gravitation, and fluid dynamics, and physical chemistry. The continued operation of a mindless selective process can convert chaos into order.[4]

Our Earth was born in such circumstances about 4.5 or 4.6 billion years ago, a little world of rock and metal, third from the Sun. But we musn't think of it as placidly emerging into sunlight from its catastrophic origins. There was no moment in which collisions of small worlds with the Earth ceased entirely. Even today objects from space run into the Earth or the Earth overtakes them. Our planet displays unmistakable impact scars from recent collisions with asteroids and comets. But the Earth has machinery that fills in or covers over these blemishes—running water, lava flows, mountain building, plate tectonics. The very ancient craters have vanished. The Moon, though, wears no makeup. When we look there, or to the Southern Highlands of Mars, or to the moons of the outer planets, we find a myriad of impact craters, piled one on top of the other, the record of catastrophes of ages past. Since we humans have returned pieces of the Moon to the Earth and determined their antiquity, it is now possible to reconstruct the chronology of cratering and glimpse the collisional drama that once sculpted the Solar System. Not just occasional small impacts, but massive, stupefying, apocalyptic collisions is the inescapable conclusion from the record preserved on the surfaces of nearby worlds.

By now, in the Sun's middle age, this part of the Solar System has been swept free of almost all the rogue worldlets. There is a handful of small asteroids that come near the Earth, but the chance that any of the bigger ones will hit our planet soon is small. A few comets visit our part of the Solar System from their distant homeland. Out there, they are occasionally jostled by a passing star or a nearby, massive interstellar cloud—and a shower of icy worldlets comes careening into the inner Solar System. These days, though, big comets hit the Earth very rarely.

Shortly, we will sharpen our focus to one world only, the Earth. We will examine the evolution of its atmosphere, surface, and interior, and the steps that led to life and animals and us. Our focus will then progressively narrow, and it will be easy to think of us as isolated from the Cosmos, a self-sufficient world minding its own business. In fact, the history and fate of our planet and the beings upon it have been profoundly, crucially influenced, through the whole history of the Earth and not just in the time of its origins, by what's out there. Our oceans, our climate, the building blocks of life, biological mutation, massive extinctions of species, the pace and timing of the evolution of life, all cannot be understood if we imagine the Earth hermetically sealed from the rest of the Universe, with only a little sunlight trickling in from the outside.

The matter that makes up our world came together in the skies. Enormous quantities of organic matter fell to Earth, or were generated by sunlight, setting the stage for the origin of life. Once begun, life mutated and adapted to a changing environment, partially driven by radiation and collisions from outside. Today, nearly all life on Earth runs off energy harvested from the nearest star. Out there and down here are not separate compartments. Indeed, every atom that is down here was once out there.[5]

Not all of our ancestors made the same sharp distinction we do between the Earth and the sky. Some recognized the connection. The grandparents of the Olympian gods and therefore the ancestors of humans were, in the myths of the ancient Greeks, *Uranus*,[6] god of the sky, and his wife *Gaia*, goddess of the Earth. Ancient Mesopotamian religions had the same idea. In dynastic Egypt the gender roles were reversed: *Nut* was goddess of the sky, and *Geb* god of Earth. The chief gods of the Konyak Nagas on the Himalayan frontier of India today are called *Gawang*, "Earth-Sky," and *Zangban*, "Sky-Earth." The Quiché Maya (of what is now Mexico and Guatemala) called the Universe *cahuleu*, literally "Sky-Earth."

That's where we live. That's where we come from. The sky and the Earth are one.

Chapter 2

SNOWFLAKES FALLEN ON THE HEARTH

There is not yet one person, one animal, bird, fish, crab, tree, rock, hollow, canyon, meadow, forest. Only the sky alone is there . . .

Popol Vuh: The Mayan Book of the Dawn of Life[1]

Before the High and Far-Off Times, O my Best Beloved, came the Time of the Very Beginnings; and that was in the days when the Eldest Magician was getting Things ready. First he got the Earth ready; then he got the Sea ready; and then he told all the Animals that they could come out and play.

RUDYARD KIPLING
"The Crab That Played with the Sea"[2]

I f you could drive an automobile straight down, in an hour or two you would find yourself deep inside the upper mantle of the Earth, far beneath the pediments of the continents, approaching an infernal region where the rock becomes a viscous liquid, mobile and red-hot. And if you could drive for an hour straight up, you would find yourself in the near-vacuum of interplanetary space.[3] Beneath you—blue, white, breathtakingly vast, and brimming over with life—would stretch the lovely planet on which our species and so many others have grown up. We inhabit a shallow zone of environmental clemency. Compared to the size of the Earth, it is thinner than the coat of shellac on a large schoolroom globe. But earlier, long ago, even this narrow habitable boundary between hell and heaven was unready to receive life.

———

The Earth accumulates in the dark. Although the primitive Sun is ablaze, there is so much gas and dust between the Earth and the Sun that at first no light gets through. The Earth is embedded in a black cocoon of interplanetary debris. There's an occasional flash of lightning by which you glimpse a ravaged, pockmarked, not quite spherical world. As it gathers up more and more matter, in units ranging from dust to worldlets, it becomes rounder, less lumpy.

A collision with a hurtling worldlet produces a shattering explosion, and excavates a great crater. Much of the impactor disintegrates into powder and atoms. There are vast numbers of such collisions. Ice is converted to steam. The planet is blanketed in vapor—which holds in the heat from the impacts. The temperature rises until the Earth's surface becomes entirely molten, a roiling world-ocean of lava, glowing by its own red heat, and surmounted by a stifling atmosphere of steam. These are the final stages of the great gathering in.

In this epoch, when the Earth is new, the most spectacular catastrophe in the history of our planet occurs: a collision with a sizeable world. It does not quite crack the Earth open, but it does blast a good fraction of it out into nearby space. The resulting ring of orbiting debris shortly falls together to become the Moon.

The day is only a few hours long. Gravitational tides raised in the Earth's oceans and interior by the Moon, and in the Moon's solid body by the Earth, gradually slow the Earth's rotation and lengthen the day. From the moment of its formation, the Moon has been drifting away from the Earth. Even now, it hovers over us, a baleful reminder that had the colliding world been much bigger, the Earth would have scattered in fragments through the inner solar system—a short-lived, unlucky world like so many others. Then humans would never have come to be. We would be just one more item on the immense list of unrealized possibilities.

———

Shortly after the Earth had formed, its molten interior was churning, great convection currents circulating, a world in a slow boil. Heavy metal was falling to its center, forming a massive molten core. Motions in the liquid iron began to generate a strong magnetic field.

The time came when the Solar System had pretty well been swept free of gas and dust and rogue worldlets. On Earth, the massive atmosphere—that had kept the heat in—dissipated. Indeed, the collisions themselves helped to drive that atmosphere into space. Convection still carried hot magma up to the surface, but the heat from the molten rock could now be radiated away to space. Slowly the Earth's surface began to cool. Some of the rock solidified and a thin, at first fragile crust formed, thickened, and hardened. Through blisters and fissures, magma and heat and gases continued to pour out of the interior.

Punctuated by spasmodic flurries of worlds falling out of the sky, the bombardment slowed. Each large impact produced a great dust cloud. There were so many impacts at first that a pall of fine particles enveloped the planet, prevented sunlight from reaching the surface, and in effect turned off the atmospheric greenhouse effect and froze the Earth. There seems to have been a period, after the magma ocean solidified but before the massive bombardment ended, when the once molten Earth became a frozen, battered planet. Who, scanning this desolate world, would have pronounced it fit for life? What wild optimist could have foreseen that peonies and eagles would one day spring from this wasteland?

The original atmosphere had been ejected into space by the relentless rain of worldlets. Now a secondary atmosphere trickled up from the interior and was retained. As the impacts declined, global dust palls became more rare. From the surface of the Earth the Sun would have seemed to be flickering, as in a time-lapse movie. So there was a time when sunlight first broke through the dust pall, when the Sun, Moon, and stars could first be noticed had there been anyone there to see them. There was a first sunrise and a first nightfall.

In sunny intervals, the surface warmed. Outgassed water vapor cooled and condensed; droplets of liquid water formed and trickled down to fill the lowlands and the impact basins. Icebergs continued to fall from the sky, vaporizing on arrival. Torrents of extraterrestrial rain helped form the primeval seas.

Organic molecules are composed of carbon and other atoms. All life on Earth is made from organic molecules. Clearly they had somehow to be synthesized *before* the origin of life *in order* for life to arise. Like water, organic molecules came both from down here and from up there. The early atmosphere was energized by ultraviolet light and the wind from the Sun, the flash and crackle of lightning and thunder, auroral electrons, intense early radioactivity, and the shock waves of objects plummeting groundward. When, in the laboratory, such energy sources are introduced into presumptive atmospheres of the primitive Earth, many of the organic building blocks of life are generated, and with astonishing ease.

Life began near the end of the heavy bombardment. This is probably no coincidence. The cratered surfaces of the Moon, Mars, and Mercury offer eloquent testimony to how massive and world-altering that battering was. Since the worldlets that have survived to our time —the comets and the asteroids—have sizeable proportions of organic matter, it readily follows that similar worldlets, also rich in organic matter but in much vaster numbers, fell on the Earth 4 billion years ago and may have contributed to the origin of life.

Some of these bodies, and their fragments, burned up entirely as they plunged into the early atmosphere. Others survived unscathed, their cargoes of organic molecules safely delivered to the Earth. Small organic particles drifted down from interplanetary space like a fine sooty snow. We do not know just how much organic matter was delivered to and how much was generated on the early Earth, the ratio of imports to domestic manufactures. But the primitive Earth seems to have been heavily dosed with the stuff of life[4]—including amino acids (the building blocks of proteins), and nucleotide bases and sugars (the building blocks of the nucleic acids).

Imagine a period hundreds of millions of years long in which the Earth is awash in the building blocks of life. Impacts are erratically altering the climate; temperatures are falling below the freezing point of water when the impact ejecta obscure the Sun, and then warming as the dust settles. There are pools and lakes undergoing wild fluctuations in conditions—now warm, bright, and bathed in solar ultraviolet light, now frozen and dark. Out of this varied and changeable landscape and this rich organic brew, life arises.

Presiding over the skies of Earth at the time of the origin of life was a huge Moon, its familiar surface features being etched by mighty collisions and oceans of lava. If tonight's Moon looks about as large as a nickel at arm's length, that ancient Moon might have seemed as big as a saucer. It must have been heartbreakingly lovely. But it was billions of years to the nearest lovers.

We know that the origin of life happened quickly, at least on the time scale by which suns evolve. The magma ocean lasted until about 4.4 billion years ago. The time of the permanent or near-permanent dust pall lasted a little longer. Giant impacts occurred intermittently for hundreds of millions of years after that. The largest ones melted the surface, boiled away the oceans, and flushed the air off into space. This earliest epoch of Earth history is, appropriately, called Hadean, hell-like. Perhaps life arose a number of times, only to be snuffed out by a collision with some wild, careening worldlet newly arrived from the depths of space. Such "impact frustration" of the origin of life seems to have continued until about 4 billion years ago. But by 3.6 billion years ago, life had exuberantly come to be.

———

The Earth is a vast graveyard, and every now and then we dig up one of our ancestors. The oldest known fossils, you might imagine, are microscopic, discovered only by painstaking scientific analysis. Some are. But some of the most ancient traces left by life on Earth are easily visible to the untrained naked eye—although the beings that made them were microscopic. Often meticulously preserved, they're called stromatolites; not unusual are examples the size of a basketball or a watermelon. A few are half the length of a football field. Stromatolites are *big*. Their age is read from the radioactive clocks in the ancient basaltic lava in which they are embedded.

They still grow and flourish today—in warm bays, lagoons, and inlets in Baja California, Western Australia, or the Bahamas. They're composed of successive layers of sediment generated by mats of bacteria. The individual cells live together. They must know how to get on with the neighbors.

We glimpse the earliest lifeforms on Earth and the first message conveyed is not of Nature red in tooth and claw, but of a Nature of cooperation and harmony. Of course, neither extreme is the whole truth; and, examining modern stromatolites more closely, we find single-celled microbes freely swimming in and around the mats. Some of them are busily devouring their fellows. Perhaps they too were there from the beginning.

Some stromatolite communities are photosynthetic; they know how to convert sunlight, water, and carbon dioxide into food. Even today, we humans are unable to build a machine that can perform this transformation with the efficiency of a photosynthetic microbe, much less a liverwort. Yet 3.6 billion years ago the stromatolitic bacteria could do it.

Exactly what happened between the time of the first seas, rich in organic molecules and future prospects, and the time of the first stromatolites is beyond our present ability to reconstruct. Stromatolite-forming microbes could hardly have been the first living things. Before there were colonial forms, there must, it seems, have been individual, free-living, one-celled organisms. And before that, something even simpler. Perhaps before the first photosynthetic organisms, there were little beings that could eat the organic matter littering the landscape: Eating food seems to be a great deal less demanding than manufacturing it. And those little beings themselves had ancestors . . . and so on, back to the earliest molecule or molecular system able to make crude copies of itself.

Why did colonial forms develop so early? Maybe it was because of the air. Oxygen, generated today by green plants, must have been in short supply before the Earth was covered by vegetation. But ozone is generated from oxygen. No oxygen, no ozone. If there's no ozone, the searing ultraviolet light (UV) from the Sun will penetrate to the ground. The intensity of UV at the surface of the Earth in those early days may have reached lethal levels for unprotected microbes, as it has on Mars today. We are concerned—and for good reason—that chlorofluorocarbons and other products of our industrial civilization will reduce the amount of ozone by a few tens of percent. The predicted biological consequences are dire. How much more serious it must have been to have no ozone shield at all.

In a world with deadly UV reaching the surface of the waters, sunblock may have been the key to survival—as it may become again. Modern stromatolite microorganisms secrete a kind of extracellular glue that helps them to stick together and also to adhere to the ocean floor. There would have been an optimum depth, not so shallow as to be fried outright by unfiltered UV, and not so deep that the visible light is too feeble for photosynthesis. There, partly shielded by seawater, it would have been advantageous for the organisms to put some opaque material between themselves and the UV. Suppose, in reproducing, the daughter cells of one-celled organisms did not separate and go their individual ways, but instead remained attached to one another, generating—after many reproductions—an irregular mass. The outer cells would take the brunt of the ultraviolet damage; the inner ones would be protected. If all the cells were spread out thinly on the surface of the sea, all would die; if they were clustered together, most of the interior cells would be sheltered from the deadly radiation. This may have been a potent early impetus for a communal way of life. Some died that others might live.*

There are no earlier fossils known, in part because there's very little of the Earth's surface surviving from much before 3.6 billion years ago. Almost all the crust from that epoch has been carried deep into

---

* Although not in consequence of some policy of conscious altruism. Any individual that goes along with the stromatolitic arrangement is much more likely to find itself safely on the inside rather than perilously on the outside. A communal policy benefits most constituent cells—not entirely risk-free, since those on the outside will be fried, but as if a cost-benefit analysis had been performed for the average cell.

our planet's interior and destroyed. In a rare 3.8-billion-year-old sediment from Greenland, there is some evidence from the kinds of carbon atoms present that life may have been widespread even then. If so, life happened sometime between about 3.8 and maybe 4.0 billion years ago. It could not have arisen much earlier. So—because of the inhospitability of the Hadean Earth, and the need for adequate time to evolve the stromatolite-building microbes—the origin of life must be confined to a comparatively narrow window in the expanse of geological time. Life seems to have arisen very quickly.

Tentatively, tortuously, the orphan is trying to figure out, to the nearest hundred million years, when the family tree took root. "How" is much harder than "when." Deadly environmental perils, a kind of huddling together for mutual protection, and the deaths—of course, neither willing nor unwilling—of vast numbers of little beings were characteristic of life almost from the beginning. Some microbes were saving their brethren. Others were eating the neighbors.

———

When life was first emerging, the Earth seems to have been mainly an ocean planet, the monotony broken, here and there, by the ramparts of large impact craters. The very beginnings of the continents date back about 4 billion years. Being made of lighter rock, then as now, they sat high on the moving, continent-sized plates. Then as now, the plates apparently were being extruded out of the Earth, carried across its surface as on a great conveyor belt, until plummeting back into the semifluid interior. Meanwhile, new plates were emerging. Vast quantities of mobile rock were slowly exchanged between the surface and the depths. A great heat engine had been established.

By about 3 billion years ago the continents were becoming larger. They were transported halfway around the Earth by the crustal plate machinery, opening one ocean and closing another. Occasionally, continents would crash into each other in exquisite slow motion, the crust would buckle and crinkle, and mountain ranges would be thrust up. Water vapor and other gases spewed out, mainly along mid-ocean ridges and volcanoes at the edges of plates.

Today we can readily detect the growth of continents, their relative motion over the Earth's surface (sometimes called continental drift), and the subsequent transport of the ocean floor down into the interior, in a style of motion called plate tectonics. The continents tend to stay afloat even when their underlying plates plunge down to destruction. Still, time takes its toll even on continents. Some old continental crust is always being carried to the depths and only bits and pieces of truly ancient continents have survived to our time—in Australia, Canada, Greenland, Swaziland, Zimbabwe.

Greenhouse gases and stratospheric fine particles, both generated by volcanoes, can, respectively, warm or cool the Earth. The changing configuration of the continents determines rainfall and monsoon patterns, and the circulation of warming and cooling ocean currents. When the continents are all aggregated together, the variety of marine environments is limited; when they are scattered over the globe, there are many more kinds of environments, especially those near shore, where a surprising number of the fundamental biological innovations seem to have been made. Thus the history of life, and many of the steps that led to us humans, were governed by great sheets and columns of circulating magma—driven by the heat from long-gone worlds that fell together to make our planet, from the sinking of liquid iron to form the Earth's core, and from the decay of radioactive atoms originally forged in the death throes of distant stars. Had these events gone a little otherwise, a different amount of heat would have been generated, a different pace or style of plate tectonics elicited, and, from the vast array of possible futures, a different course followed in the evolution of life. Not humans, but some very different species might now be the dominant form of life on Earth.

We know next to nothing about the configuration of the continents over the first 4 billion years. They may many times have been scattered over the oceans and reaggregated into a single mass. For at least 85 percent of Earth history, a map of our planet would have seemed wholly unfamiliar—as if of another world. The earliest well-substantiated reconstruction we can manage dates to as recent a time as 600 million years ago. The Northern Hemisphere then was mostly ocean; in the South, a single massive continent, plus fragments of future continents, drifted across the face of the Earth at about an inch a year —much slower than a snail's pace. Trees grow vertically faster than continents move horizontally, but if you have millions of years to play with, this is quite sufficient for continents to collide and wholly alter what's on the maps.

For hundreds of millions of years, what are now the southern continents—Antarctica, Australia, Africa, and South America—plus India, were joined in a common assemblage that geologists call Gondwana.* What was later to be North America, Europe, and Asia were adrift, sailing in pieces through the world ocean. Eventually, all this floating continental debris gathered itself together into one massive supercontinent. Whether we describe it as a landlocked planet with an immense saltwater lake, or an ocean planet with an immense island is only a matter of definition. It might have seemed a friendly world: At least, you could walk anywhere; there were no distant lands across the sea. Geologists call this supercontinent Pangaea—"all Earth." It included, but of course was considerably larger than, Gondwana.

Pangaea was formed about 270 million years ago, during the Permian Period, a trying time for Earth. Worldwide, conditions had been warming. In some places the humidity was very high and great swamps formed, later to be supplanted by vast deserts. About 255 million years ago Pangaea began to shatter—because, it is thought, of the sudden rise of a superplume of molten lava through the Earth's mantle from its deep seething core. Texas, Florida, and England were then at the equator. North and South China, in separate pieces, Indochina and Malaya together, and fragments of what would later be Siberia were all large islands. Ice ages flickered on and off every 2.5 million years, and the level of the seas correspondingly fell and rose.

Towards the end of the Permian Period, the map of the Earth seems to have been violently reworked. Whole oblasts of Siberia were inundated with lava. Pangaea rotated and drifted north, moving mainland Siberia towards its present position, near the North Pole. "Megamonsoons," torrential seasonal rains on a much larger scale than humans have ever witnessed, drenched and flooded the land. South China slowly crumpled into Asia. Many volcanoes blew their tops together,

---

* You can occasionally see, on the automobile bumper stickers of geology graduate students, the nostalgic plea, "Reunite Gondwanaland." Except in a metaphorical political sense (and it's not too likely there either) it is the most hopeless of lost causes—on any but a geological time scale. But the breakup and separation of continents can go only so far. On a round Earth, what you run away from on one side you will eventually edge into on the other. A few hundred million years from now our remote descendants, if any, may witness the reaggregation of a supercontinent. Gondwanaland will at last have been reunited.

belching sulfuric acid into the stratosphere and perhaps playing an important role in cooling the Earth.[5] The biological consequences were profound—a worldwide orgy of dying, on land and at sea, the likes of which has never been seen before or since.[6]

The breakup of Pangaea continued. By 100 million years ago South America and Africa, which even today fit together like two pieces of a jigsaw puzzle, were just barely separated by a narrow strait of ocean—receding from one another at about an inch a year. North and South America were then separate continents, with no Isthmus of Panama connecting them. India was a large island headed north away from Madagascar. Greenland and England were connected to Europe. Indonesia, Malaysia, and Japan were part of the mainland of Asia. You might have strolled from Alaska to Siberia. There were great inland seas where none exists today. This time, at a glance from orbit you would have recognized it as the Earth—but with the configuration of land and water strangely altered, as if by a careless, slapdash cartographer. This was the world of the dinosaurs.

Later, the continents drifted further apart, pulled by their underlying plates. Africa and South America continued to recede from one another, opening up the Atlantic. Australia split off from Antarctica. India collided with Asia, raising the Himalayas high. This is the world of the primates.

———

Each of us is a tiny being, permitted to ride on the outermost skin of one of the smaller planets for a few dozen trips around the local star. The great internal engine of plate tectonics is indifferent to life, as are the small changes in the Earth's orbit and tilt, the variation in the brightness of the Sun, and the impact with the Earth of small worlds on rogue orbits. These processes have no notion of what has been going on over billions of years on our planet's surface. They do not care.

The longest-lived organisms on Earth endure for about a millionth of the age of our planet. A bacterium lives for one hundred-trillionth of that time. So of course the individual organisms see nothing of the overall pattern—continents, climate, evolution. They barely set foot on the world stage and are promptly snuffed out—yesterday a drop of semen, as the Roman Emperor Marcus Aurelius wrote, tomorrow a handful of ashes.[7] If the Earth were as old as a person, a typical organism would be born, live, and die in a sliver of a second. We are fleeting, transitional creatures, snowflakes fallen on the hearth fire. That we understand even a little of our origins is one of the great triumphs of human insight and courage.

Who we are and why we are here can be glimpsed only by piecing together something of the full picture—which must encompass aeons of time, millions of species, and a multitude of worlds. In this perspective it is not surprising that we are often a mystery to ourselves, that, despite our manifest pretensions, we are so far from being masters even in our own small house.

ON IMPERMANENCE

The present life of man, O king, seems to me, in comparison of that time which is unknown to us, like to the swift flight of a sparrow through the room wherein you sit at supper in winter, with your commanders and ministers, and a good fire in the midst, whilst the storms of rain and snow prevail abroad; the sparrow, I say, flying in at one door, and immediately out at another, whilst he is within, is safe from the wintry storm; but after a short space of fair weather, he immediately vanishes out of your sight, into the dark winter from which he had emerged. So this life of man appears for a short space, but of what went before, or what is to follow, we are utterly ignorant.

THE VENERABLE BEDE *Ecclesiastical History*[8]

Chapter 3

"WHAT MAKEST THOU?"

Shall the clay say to him that fashioneth it,
What makest thou?

Isaiah 45:9

The world and everything in it was made for us, as we were made for God:

For the last few thousand years, and especially since the end of the Middle Ages, this proud, self-confident assertion was increasingly common belief, held by Emperor and slave, Pope and parish priest. The Earth was a lavishly decorated stage set, designed by an ingenious if inscrutable Director, who had managed to round up, from only He knew where, a multitudinous supporting cast of toucans and mealy bugs, eels, voles, elms, yaks, and much, much more. He placed them all before us, in their opening night costumes. They were ours to do with as we pleased: drag our burdens, pull our plows, guard our homes, produce milk for our babies, offer up their flesh for our dinner tables, and provide useful instruction—bumblebees, for example, on the virtues not just of hard work, but of hereditary monarchy. Why He thought we needed hundreds of distinct species of ticks and roaches, when one or two would have been more than sufficient, why there are more species of beetles than any other kind of being on Earth, no one could say. No matter; the composite effect of life's extravagant diversity could only be understood by postulating a Maker, not all of whose reasons we could grasp, who had created the stage, the scenery, and the subsidiary players for our benefit. For thousands of years, virtually everyone, theologian and scientist alike, found this, both emotionally and intellectually, a satisfying account.

The man who wrecked this consensus did so with the utmost reluctance. He was no ideologue bent on kicking in the door of the Establishment, no firebrand. If not for a bit of happenstance he would probably have passed his days as a well-liked Church of England parson in a nineteenth-century rural, picture-postcard village. Instead he ignited a firestorm[1] that destroyed more of the old order than any violent political upheaval ever had. Through the astonishingly powerful method of science, this gentleman who was known to find lively conversation too taxing, somehow became the revolutionary's revolutionary. For more than a century, the mere mention of his name has been sufficient to unsettle the pious and rouse the bookburners from their fitful slumbers.

———

Charles Darwin was born at Shrewsbury, England, on February 12, 1809, the fifth child of Robert Waring Darwin and Susannah Wedgwood. The Darwin and Wedgwood families were allied through the close friendship of their patriarchs, Erasmus Darwin, the noted author, physician, and inventor, and Josiah Wedgwood, who had risen from poverty to found the Wedgwood pottery dynasty. These two men shared radically progressive views, even going so far as to side with the rebellious colonies in the American Revolution. "He who allows oppression," Erasmus wrote, "shares the crime." [2]

Their club was called The Lunar Society, because it met only during the full moon when the late-night ride home would be well-lit and therefore less dangerous. Among its members were William Small, who had taught Thomas Jefferson science (at the College of William and Mary in Virginia and whom Jefferson singled out [3] as having "probably fixed the destinies" of his life); James Watt, whose steam engines powered the British Empire; the chemist Joseph Priestley, the discoverer of oxygen; and an expert on electricity named Benjamin Franklin.

The poet Samuel Taylor Coleridge thought Erasmus Darwin "the most original-minded man" he had ever known. Erasmus was also making quite a name for himself as a doctor. George III invited him to become his personal physician. (Erasmus declined the honor out of an unwillingness, he said, to leave his happy home in the countryside, but perhaps the champion of American revolutionaries had political reasons as well.) His real fame, though, stemmed from a string of hit encyclopaedic rhyming poems.

Erasmus Darwin's two-volume work, *The Botanic Garden*, comprising *The Loves of the Plants*, written in 1789, and its eagerly awaited sequel, *The Economy of Vegetation*, were runaway best-sellers. They were so successful that he decided to tackle the animal kingdom next. The result was a 2,500-page tome, this one in prose, entitled *Zoonomia: or, the Laws of Organic Life*. In it he asked this prescient question:

When we revolve in our minds, first the great changes which we see naturally produced in animals after their nativity as in the production of the butterfly from the crawling caterpillar or of the frog from the subnatant tadpole; secondly when we think over the great changes introduced into various animals by artificial cultivation as in horses or in dogs . . .; thirdly when we revolve in our minds the great similarity of structure which obtains in all the warm-blooded animals as well as quadrupeds, birds, amphibious animals as in mankind, would it be too bold to imagine that all warm-blooded animals have arisen from one living filament (archetype, primitive form)?[4]

Erasmus Darwin believed that "There are three great objects of desire, which have changed the forms of many animals by their exertions to gratify them: hunger, security and lust." Especially lust. The lilting refrain of his last effort, *The Temple of Nature: or, The Origin of Society*,[5] was "And hail THE DEITIES OF SEXUAL LOVE." The capitalization is his. Elsewhere, he observed that the stag had developed horns to fight other males for "the exclusive possession of the female." There's no question that he was on to something. But his was a kind of disordered originality, a brilliance that could not be bothered by methodical research. Science exacts a substantial entry fee in effort and tedium in exchange for its insights. Erasmus was unwilling to ante up.

His grandson Charles, who would pay those dues, read *Zoonomia* twice; once when he was eighteen and again a decade later, after he'd been around the world. He took pride in his grandfather's precocious anticipation of some of the ideas that would make Jean-Baptiste de Lamarck famous twenty years later. However, Charles "was much disappointed" by Erasmus' failure to investigate, carefully and rigorously, whether there was any truth to his inspired speculations.

Lamarck had been a soldier, a self-taught botanist, and the zoologist who had gone on to develop the precursor of the modern natural history museum. When everyone else was thinking in terms of thousands of years, he was contemplating millions. He believed that the idea of the living world walled up into separate compartments called species was an illusion; species are slowly transmogrifying, one into another, he taught, and this would be immediately apparent to us if our lives were not so brief and fleeting.

Lamarck is best known for arguing that an organism could inherit the acquired characteristics of its ancestors. In his most famous example, the giraffe strains to nibble at the leaves on the higher branches of the tree, and somehow the slightly elongated neck that attends the stretching is passed on to the next generation. Lamarck could not have been knowledgeable of the family history of many generations of giraffes, but he did have relevant data that he chose to ignore: For thousands of years, Jews and Moslems have been ritually circumcising their sons, with no break in continuity, and yet not one case is known of a Jewish or Islamic boy born without a foreskin. Queen bees and drones do no work, and have not for geological ages; yet worker bees whose parents are queens and drones (and never other workers) do not seem to be growing more indolent, generation after generation; instead, they are proverbially industrious.[6] Domestic and farm animals have their tails docked, their ears clipped, or their flanks branded for generations, but the newborn show no signs of these mutilations. Chinese women had their feet cruelly bound and deformed for centuries, but infant girls obstinately persisted in being born with normal appendages.[7] Despite such counterexamples, Charles would take seriously, for his entire life, the notion of Lamarck and his grandfather Erasmus that acquired characteristics could be inherited.

The mechanism by which discrete hereditary units, the genes, are reshuffled and passed on to the next generation, the way in which those genes are randomly altered, their molecular nature, and their wonderful ability to encode long chemical messages and replicate those messages precisely—all this was wholly unknown to Darwin. To attempt an understanding of the evolution of life when heredity was still an almost complete mystery would require either an exceptionally foolish or an exceptionally able scientist.

———

Josiah Wedgwood and Erasmus Darwin had long entertained the hope that someday their children would formalize through marriage the bonds of affection that already united their two families. Of the two, only Erasmus lived to see it happen. His son, Robert, a generous but moody physician, a great big, fat man, a silhouette out of Dickens, who alternately comforted and terrified the patients of his far-flung practice, married Susannah Wedgwood. She was widely admired for her "gentle, sympathising nature" and the active role she took in her husband's scientific interests. Susannah suffered an agonizing death from a gastrointestinal affliction out of sight but within earshot of her eight-year-old son, Charles. Writing near the end of his own life, he could recall nothing about his mother "except her death-bed, her black velvet gown, and her curiously constructed work-table."

In this autobiographical memoir, conceived as a gift for his children and grandchildren, and written "as if I were a dead man in another world looking back at my own life," Charles Darwin admitted "that in many ways I was a naughty boy . . . I was much given to inventing deliberate falsehoods, and this was always done for the sake of causing excitement." He boasted to another boy that he "could produce variously coloured polyanthuses and primroses by watering them with certain coloured fluids, which was of course a monstrous fable." Even at that tender age he had begun to speculate on the variability of plants. His life-long absorption in the natural world was under way. He became a passionate collector of the bits and pieces of Nature that form the gritty detritus in the pockets of children everywhere. He was particularly mad for beetles, but his sister convinced him that it would be immoral to take a beetle's life merely for collecting. Dutifully, he confined himself to gathering up only the recently deceased. He watched the birds and recorded his observations of their behavior. "In my simplicity," he later wrote, "I remember wondering why every gentleman did not become an ornithologist."

At the age of nine he was sent to study at Dr. Butler's day school. "Nothing could have been worse for the development of my mind," Darwin later wrote. Butler believed that school was no place for curiosity or excitement about learning. For that, Charles looked to a well-thumbed copy of *Wonders of the World,* and to the members of his family who patiently answered his many questions. As an old man he could still recall the delight he felt when an uncle had explained to him how the barometer works. His older brother, Erasmus—named after their grandfather—transformed the garden toolhouse into a chemistry lab and allowed Charles to help him with his experiments. This earned Charles the nickname "Gas" at school and an angry public rebuke from Dr. Butler.

Charles was doing so poorly at school that when it was time for Erasmus to go off to Edinburgh University, his father decided to send Charles with him. The boys were supposed to study medicine. Here, too, Charles found the lectures oppressively dull. He couldn't bear to dissect anything, and the experience of seeing a botched operation on a child, "long before the blessed days of chloroform," was to haunt him for the rest of his life. But it was in Edinburgh that he first found friends who shared his passion for science.

After two sessions at Edinburgh, Robert Darwin became resigned to the fact that Charles was not cut out for a medical career. Perhaps he would make a good clergyman? Dutiful Charles had no objections, but just the same, he thought he should check up on Church of England dogma before agreeing to commit his life to instilling it in others. "Accordingly I read with care Pearson on the *Creed*, and a few other books on divinity; and as I did not then in the least doubt the strict and literal truth of every word in the Bible, I soon persuaded myself that our Creed must be fully accepted."

Charles spent the next three years at Cambridge University, where he managed to get better grades. But still he felt a restless dissatisfaction with the curriculum. His happiest moments there were spent in pursuit of his adored beetles, now dead or alive.

> I will give a proof of my zeal: one day, on tearing off some old bark, I saw two rare beetles, and seized one in each hand; then I saw a third and new kind, which I could not bear to lose, so that I popped the one which I held in my right hand into my mouth. Alas! it ejected some intensely acrid fluid, which burnt my tongue so that I was forced to spit the beetle out, which was lost, as was the third one. [8]

It was as a beetle hunter that the first published reference to Charles Darwin was made. "No poet ever felt more delighted at seeing his first poem published than I did at seeing, in Stephen's *Illustrations of British Insects*, the magic words, 'captured by C. Darwin, Esq.' "

At Cambridge he had been persuaded to take a course in geology taught by Adam Sedgwick. Darwin told Professor Sedgwick of the curious but credible claim made to him by a laborer that a "large, worn tropical Volute shell" (the spiral-shaped shell of a warm-water mollusc) had been found embedded in an old Shrewsbury gravel pit. Sedgwick was incurious and dismissive; it must have been dumped there by someone. Darwin remembered in his *Autobiography*,

> But then, [Sedgwick added,] if [the shell was] really embedded there it would be the greatest misfortune for geology, as it would overthrow

all that we know about the superficial deposits of the Midland Counties. These gravel-beds belong in fact to the glacial period, and in after years I found in them broken arctic shells. But I was then utterly astonished at Sedgwick not being delighted at so wonderful a fact as a tropical shell being found near the surface in the middle of England. Nothing before had ever made me thoroughly realise, though I had read various scientific books, that science consists in grouping facts so that general laws or conclusions may be drawn from them.[9]

At about that time, Darwin's cousin brought him around to one of the Rev. John Steven Henslow's botany lectures. This was "a circumstance which influenced my career more than any other." A handsome man in his early thirties, Henslow had the great teacher's genius for making his subject come alive, so much so that the same students returned year after year to attend courses they had already completed. Moreover, he exhibited an exceptional sensitivity to the feelings of his students. The novice's "foolish" question was answered with respect. All were welcome to the open house he held every week, and there were regular invitations to dinner with his family. Darwin wrote, "during the latter half of my time at Cambridge I took long walks with him on most days; so that I was called by some of the dons 'the man who walks with Henslow.' " Darwin judged his knowledge "great in botany, entomology, chemistry, mineralogy, and geology." He added that Henslow was "deeply religious, and so orthodox that he told me one day he should be grieved if a single word of the Thirty-nine Articles [of the Anglican faith] were altered."

Ironically, it was Henslow who left the message "informing me that Captain FitzRoy was willing to give up part of his own cabin to any young man who would volunteer to go with him without pay as naturalist to the Voyage of the *Beagle*." Henslow wrote of "a trip to Tierra del Fuego, and home by the East Indies . . . Two years . . . I assure you I think you are the very man they are in search of."

The scene is not hard to imagine: The twenty-two-year-old races home from college breathless with excitement. He squirms in his chair while Father, an intimidating man in the best of circumstances, harangues him with a litany of past indulgences and harebrained schemes. First, doctor, then, clergyman, now, this? Afterwards, what congregation will want you? They must have first offered it to others and been turned down . . . Doubtless something is seriously wrong with the vessel . . . Or the expedition . . .

And then, after much discussion: "If you can find any man of common sense, who advises you to go, I will give my consent."[10] The chastened son regards the situation as hopeless and sends Henslow polite regrets.

The next day he rides over to the Wedgwoods' for a visit. Uncle Josiah—named after Charles' grandfather's boon companion—sees the voyage as a once-in-a-lifetime opportunity. He drops what he's doing to write Charles' father a point-by-point refutation of his objections. Later that same day, Josiah worries that a personal appearance might accomplish what a note might not. He grabs Charles and gallops over to the Darwin household to try to convince the young man's father to let him go. Robert keeps his word and agrees. Touched by his father's generosity and feeling a little guilty over past extravagances, Charles seeks to reconcile him, saying, "I should be deuced clever to spend more than my allowance whilst on board the *Beagle*."

"But they tell me you are very clever," his father answers with a smile.

Robert Darwin had given his blessing, but some obstacles still remained. Captain Robert FitzRoy was having second thoughts about sharing such close quarters for such an extended period of time. A relation of his had known the young Darwin at Cambridge. He said he wasn't a bad sort, but did FitzRoy, the high Tory, know that he'd be rooming for two years with a Whig? And then there was the pesky problem of Darwin's nose. FitzRoy was, as were many of his contemporaries, a believer in phrenology, which held that the shape of the skull was indicative of intelligence and character, or their absence. Some adherents expanded this doctrine to include noses. To FitzRoy, Darwin's nose proclaimed at a glance grave deficiencies in energy and determination. After the two men had spent a little time together, though, FitzRoy, despite his reservations, decided to take a chance on the young naturalist. Darwin wrote, "I think he was afterwards well satisfied that my nose had spoken falsely."

The *Beagle*'s earlier survey mission to South America had been such an unpleasant experience, the weather so consistently rotten, that her Captain had committed suicide before it was over. The British admiralty office in Rio de Janeiro turned to the twenty-three-year-old Robert FitzRoy to assume command. By all accounts he did brilliantly. He was at the helm when the *Beagle* resumed her survey of Tierra del

Fuego and the islands nearby. After the theft of one of the *Beagle*'s whale boats, FitzRoy kidnapped five of the local people, who were called Fuegians by the British. When he gave up hope of recovering the boat and humanely released his hostages, one of them, a little girl they called Fuegia Basket, didn't want to leave—or so the story goes. FitzRoy had been wondering about bringing some Fuegians back to England so they might learn its language, mores, and religion. Upon returning home, FitzRoy imagined, they would provide a liaison with other Fuegians and become loyal protectors of British interests at the strategic southern tip of South America. The Lords Commissioners of the Admiralty granted FitzRoy permission to bring the Fuegians to England. Although they were vaccinated, one died of smallpox. Fuegia Basket, a teenaged boy they called Jemmy Button, and a young man they called York Minster survived to study English and Christianity with a clergyman in Wandsworth, and to be presented by FitzRoy to the King and Queen.

Now it was time for the Fuegians—whose real names no one in England had bothered to learn—to go back; and for the *Beagle* to resume her survey of South America and "to determine more accurately . . . the longitude of a large number of oceanic islands as well as of the continents."[11] This assignment was expanded to include "observations of longitude right round the world." She would sail down the east coast of South America, up the west coast, cross the Pacific, and circumnavigate the planet before returning home to England. Once the *Beagle* had been re-commissioned under Captain FitzRoy's command, he took measures to insure that this new expedition would be very different from the previous one. Largely at his own expense, he had the 90-foot square-rigger completely re-fit. He resurfaced her hull, raised her deck, and festooned her bowsprit and her three tall masts with state-of-the-art lightning conductors. He tried to learn everything he could about weather and became one of the founders of modern meteorology in the process. On December 27, 1831, the *Beagle* was finally ready to sail.

On the eve of her departure, Darwin had suffered an anxiety attack and heart palpitations. There would be episodes of these symptoms, gastrointestinal distress, and profound bouts of exhaustion and depression throughout his life. Much speculation has been offered on the cause of these spells. They've been attributed to a psychosomatic reaction to the traumatic loss of his mother at so tender an age; to anxieties about the reactions his life's work might elicit from God and the public; to an unconscious tendency to hyperventilate; and, strangely, although the symptoms pre-date his marriage by many years, to the pleasure he took in his beloved wife's genius for nursing the sick. The sequence of events also makes implausible the contention that his illness was due to a South American parasite acquired during the *Beagle*'s voyage. We simply do not know. His symptoms caused this explorer to be mainly housebound for the last third of his life.

Darwin's personal library on the journey included two books, each a *bon voyage* gift. One was an English translation of Humboldt's *Travels* that Henslow had given him. Before Darwin left Cambridge he had read Humboldt's *Personal Narrative* and Herschel's *Introduction to the Study of Natural Philosophy*, which together evoked in Darwin "a burning zeal to add even the most humble contribution to the noble structure of Natural Science."[12] The other gift was from the Captain. It was Volume I of Charles Lyell's *Principles of Geology*, and FitzRoy would live to regret bitterly his choice of going-away present.

The scientific revelations of the European Enlightenment had posed disturbing challenges to the biblical account of the Earth's origin and history. There were those who tried to reconcile the new data and new insights with their faith. They held that Noah's flood was the primary agent responsible for the present configuration of the Earth's crust. A big enough flood, they thought, could transform the Earth's geology in just forty days and forty nights, consistent with an Earth only a few thousand years old. With a little spin control on a literal reading of the Book of Genesis, they felt they had managed to pull it off.

Lyell had been a lawyer for as long as he could stand it. When he was thirty years old, he abandoned the law for geology, his true passion. He wrote *Principles of Geology* to advance the "Uniformitarian" view that the Earth has been shaped by the same gradual processes that we observe today, but operating not merely over a few weeks, or a few thousand years, but ages. There were distinguished geologists who held that floods and other catastrophes might explain the Earth's landforms, but that the Noachic flood wasn't enough. It would take *many* floods, *many* catastrophes. These scientific Catastrophists were comfortable with Lyell's long time scales. But for the biblical literalists Lyell posed an awkward problem. If Lyell was right, the rocks were saying that the Bible's six days of Creation, and the age of the Earth deduced by adding up the "begats," were somehow in error. It was through this apparent hole in *Genesis* that the *Beagle* would sail into history.

Hired mainly as FitzRoy's companion and sounding board, Darwin was obliged to bear with equanimity the Captain's politically conservative, racist, and fundamentalist harangues. For most of the voyage, the two men managed to maintain a truce with regard to their philosophical and political differences. However, Darwin was simply unable to let FitzRoy's opinion on one particular issue go unchallenged:

> [A]t Bahia, in Brazil, he defended and praised slavery, which I abominated, and told me that he had just visited a great slave-owner, who had called up many of his slaves and asked them whether they wished to be free, and all answered "No." I then asked him, perhaps with a sneer, whether he thought that the answers of slaves in the presence of their master was worth anything? This made him excessively angry, and he said that as I doubted his word we could not live any longer together.[13]

Darwin fully expected to be kicked off the ship. But when the gunroom officers heard of the row, they vied with each other for the privilege of sharing their quarters with him. FitzRoy calmed down and actually apologized to Darwin, rescinding the eviction. Possibly, Darwin's evolutionary views emerged, in part, out of his exasperation with FitzRoy's inflexible conventionalism, and the necessity of the young man to suppress for five years the counterarguments that were welling up inside him.[14]

Perhaps it was the legacy of his grandfathers that enabled Darwin to detect the inconsistencies and injustices that other members of his social class would not see. At the very beginning of his book, *The Voyage of the Beagle*, he tells of a place not far from Rio de Janeiro:

> This spot is notorious from having been, for a long time, the residence of some runaway slaves, who, by cultivating a little ground near the top, contrived to eke out a subsistence. At length they were discovered, and a party of soldiers being sent, the whole were seized

with the exception of one old woman, who, sooner than again be led into slavery, dashed herself to pieces from the summit of the mountain. In a Roman matron this would have been called the noble love of freedom: in a poor negress it is mere brutal obstinacy.[15]

Darwin had been lured to South America by the prospect of discovering new birds and new beetles, but he couldn't help noticing the carnage the Europeans were inflicting. Colonial arrogance, the institution of slavery, the extirpation of countless species for the enrichment and entertainment of the invaders, the first depredations of the tropical rain forest—in short, many of the crimes and stupidities that haunt us today—troubled Darwin at a time when Europe was confident that colonialism was an unalloyed benefit for the uncivilized, that the forests were inexhaustible, and that there would always be enough egret feathers for every millinery shop until the Day of Judgment. In part because of these sensitivities, in part because Darwin always wrote as clearly and directly as he could—striving to communicate to the greatest number of people—*The Voyage of the Beagle* is still a stirring and accessible adventure story.

However, this book has watershed status because it was during the course of the expedition it recounts that Darwin began to amass the great body of evidence—not intuition, but data—that makes the case for evolution by natural selection. "At last gleams of light have come," he was later to write, "and I am almost convinced that species are not (it is like confessing a murder) immutable."

The Galapagos is an archipelago of thirteen good-sized islands and many smaller ones lying off the coast of Ecuador. If all the species on Earth were immutable, then why did the beaks of the otherwise very similar finches on islands separated by no more than fifty or sixty miles of ocean vary so dramatically? Why narrow, tiny, pointy beaks on the finches of one island and larger, parrot-like curved beaks on the finches of the next? "Seeing this gradation and diversity of structure in one, small intimately related group of birds," he later wrote in *The Voyage*, "one might really fancy that, from an original paucity of birds in this archipelago, one species had been taken and modified for different ends." (These volcanic islands, we now know, are less than 5 million years old.) And it wasn't just the finches that raised such problems, but the giant tortoises and the mockingbirds, too.

Back in England, Henslow and Sedgwick had been reading Darwin's letters aloud at meetings of scientific societies. When Darwin returned home in October 1836, he found he had acquired something of a reputation as an explorer and naturalist. His father was now well pleased with him, and all talk of a parsonage ceased. The same month he met the geologist, Lyell, for the first time. Though not without its rough spots, it was to be a lifelong friendship.

Darwin made important contributions to geology. His interpretation of coral reefs—that they mark the locations of slowly subsiding sea-mounts that had once been islands—was substantiated on the *Beagle* and corresponds to the modern understanding. In 1838 he published a paper arguing that earthquakes, volcanoes, and the thrusting up of islands are all caused by slow, intermittent, but irresistible global motions in the semi-liquid interior of the Earth. This "almost prophetic"[16] thesis, as far as it goes, is part and parcel of modern geophysics. In his 1838 Presidential Address to the Geological Society, William Whewell mentioned Darwin's name (in the context of this work) more than twice as often as any other geologist, living or dead. In geology, following Lyell, as in biology, Darwin championed the idea that profound changes are worked little by little over vast intervals of time.

In 1839, he married his cousin, Emma Wedgwood. Through ten children and more than four decades they shared a deep, loving, and almost entirely harmonious relationship. During their early married life he was writing down, but certainly not for publication, his first tentative sketch for a theory of evolution. Their rare differences were over religion. "Before I was engaged to be married," he wrote in his autobiography, "my father advised me to conceal carefully my doubts, for he said that he had known extreme misery thus caused with married persons."[17] A few weeks after their wedding, she wrote to him:

> May not the habit in scientific pursuits of believing nothing till it is proved influence your mind too much in other things which cannot be proved in the same way, and which if true are likely to be above our comprehension?

Years later, Darwin wrote at the bottom of Emma's letter,

> When I am dead, know that many times,
> I have kissed and cried over this.[18]

He tried his best to avoid the public version of this domestic tension. Our past was then a dark and shameful secret. To expose it would have been perceived by many as an affront to the prevailing religious norms and as an assault against human dignity. But to suppress it would have been to reject the data because the implications were disturbing. Darwin recognized that if he was to convince anyone he would have to support his argument with a compelling body of evidence.

In 1844, a sensational book, fundamentally pseudoscience, called *Vestiges of the Natural History of Creation* was published. Robert Chambers, the encyclopedist and amateur geologist who was its anonymous author, claimed that he had traced human ancestry all the way back to . . . frogs. Chambers' reasoning was half-baked (although no more so than Erasmus Darwin's) but its audacity attracted a great deal of attention. Nagging doubts about Creation were beginning to bubble to the surface, and Darwin felt that he should write down his own theory in as irrefutable a form as possible. He expanded a short essay, begun two years before, into a two-part work entitled "On the Variation of Organic Beings under Domestication and in the Natural State" and "On the Evidence Favourable and Opposed to the View That Species Are Naturally Formed Races Descended from Common Stock." However, he was not ready to publish. He wrote a letter to Emma that he asked be considered as a codicil to his will. In the event of his death, he wanted her to

> devote £400 to its publication and further will yourself . . . take trouble in promoting it—I wish that my sketch be given to some competent person, with this sum to induce him to take trouble in its improvement and enlargement. [19]

He felt he was on to something important, but feared—perhaps especially in view of his frequent bouts of illness—that he would not live to complete the work.

In what superficially seems an odd next move, he now put his evolutionary studies aside and for the next eight years devoted his life almost exclusively to barnacles. His great friend, the botanist Joseph Hooker, would later observe to Darwin's son, Francis, "Your father had Barnacles on the brain from Chili [Chile] onwards!" [20] It was this exhaustive project that really earned him his credentials as a naturalist. Another close friend, the anatomist and brilliant polemicist Thomas Henry Huxley, observed that Darwin

> never did a wiser thing . . . Like the rest of us, he had no proper training in biological science, and it has always struck me as a remarkable instance of his scientific insight, that he saw the necessity of giving himself such training, and of his courage, that he did not shirk the labour of obtaining it . . . It was a piece of critical self-discipline, the effect of which manifested itself in everything [he] wrote afterwards, and saved him from endless errors of detail.[21]

Darwin had not been the only scientist to get a jolt from Chambers' *Vestiges*. Alfred Russel Wallace, a surveyor who had become a naturalist, was also unimpressed with Chambers' arguments, but also intrigued by the notion that there was a knowable process at work in the evolution of life. In 1847, he traveled to the Amazon in search of factual support for this idea. A fire on the ship taking him back to England consumed every one of his specimens. Wallace persevered, setting off to the Malay Peninsula to gather a new collection. In the September 1855 issue of *Annals and Magazine of Natural History*, his paper "On the Law Which Has Regulated the Introduction of New Species" appeared.

By this time, Darwin had been wrestling with such problems for two decades. Now, it was entirely possible that his claims of priority to the solution of life's greatest mystery would be snatched away. If science were in the business of conferring sainthood, the conduct of Darwin and Wallace towards one another would have earned it for them both. Darwin wrote a letter of hearty congratulation to Wallace in which he mentioned how long he'd been working on the same problem.

Darwin's friends Huxley and Hooker prodded him to quit stalling and write the paper that would make an ironclad case for evolution. He complied and was nearing its completion in 1858, while Wallace, now in Indonesia and sick with malaria, tossed and turned, grappling with the question "Why do some die and some live?"[22] Emerging from his stupor, he understood natural selection. He wrote "On the Tendency of Varieties to Depart Indefinitely from the Original Type" and promptly mailed it to Darwin, asking him to use his judgment about what should be done with it. Darwin was distressed to see how very close Wallace's work was to his own writings of 1839 and 1842. In 1844 he had combined them into an essay, but it remained unpublished. Darwin turned to his friends for guidance on how to deal ethically with this dilemma. Hooker and Lyell came up with a wise solution: Present both the Wallace paper and a version of Darwin's unpublished 1844 essay at the next meeting of the Linnaean Society and publish them together in the Society's *Proceedings.* [23] Thereafter, Wallace always spoke of evolution as being Darwin's theory and Darwin always credited Wallace with its independent discovery. Darwin now applied himself to the task of writing the book that would cause so much trouble.

On November 24, 1859, *The Origin of Species* was published. The first edition of 1,250 copies was snapped up by the booksellers. Darwin had been careful to make only one reference to humans in the whole book: "Light will be thrown on the origin of man and his history." [24] Anything more from his pen on this delicate matter would have to wait another twelve years, for the publication of *The Descent of Man.* His restraint fooled no one. Given its formidable armamentarium of data, there could be no reconciling *The Origin* with a literal rendition of Genesis.

Chapter 4

A GOSPEL
OF DIRT

I detest all systems that depreciate human nature. If it be a delusion that there is something in the constitution of man that is venerable and worthy of its author, let me live and die in that delusion, rather than have my eyes opened to see my species in a humiliating and disgusting light. Every good man feels his indignation rise against those who disparage his *kindred* or his *country*; why should it not rise against those who disparage his *kind?*

THOMAS REID
letter of 1775[1]

When I view all beings not as special creations, but as the lineal descendants of some few beings which lived long before the first bed of the Cambrian [geological] system was deposited, they seem to me to become ennobled.

CHARLES DARWIN
The Origin of Species, Chapter XV[2]

"M ankind has conducted an experiment of gigantic proportions," Darwin wrote in *The Origin of Species*. He was struck
by the success of "husbandry," as it is tellingly called, in generating
new varieties of animals and plants useful for humans. Nature provides the varieties and we select who shall reproduce, which traits we
want preferentially to propagate into future generations. By transfer-
ring pollen from flower to flower with a camel's hair brush, or by
letting the stallion in with the mare, humans take it upon themselves
to determine who shall mate with whom. Indigestible crops, weakling
horses, scrawny turkeys, sheep with knotty coats, and cows that are
grudging with their milk are discouraged from reproducing. Generation after generation, by cumulative selection, humans impress their
interests on the heredity of the plants and animals whose breeding
they control. But Nature, too, selects those plants and animals which
by its lights happen to be more favorably adapted than their fellows;
such fortunate beings preferentially reproduce, leave more offspring
and, as time goes on, supplant the competition. Artificial selection
helps us to understand how natural selection works.

The ability of the environment to nurture and sustain large populations—the so-called carrying capacity—is of course finite. As the
number of organisms increases, not all will be able to survive. There
will be a stringent competition for scarce resources. Slight differences
in ability, imperceptible to a casual observer, may spell life or death to
the organism. Natural selection is a great sieve, straining out the vast
majority and permitting only a tiny vanguard to pass its heredity on to
the next generation. Natural selection is far more ruthless than the
most callous and resolute animal breeder in determining the genetic
makeup of future generations. And instead of the measly few thousand years since the domestication of animals began in earnest, natural selection has been working for billions.

Consider the diverse specializations that, through artificial selection, we've generated in dogs—greyhounds and borzois for speed, to outrun the wolves; collies for herding sheep; beagles, pointers, and setters for hunting; Labrador retrievers for helping fishermen gather their nets; guide dogs for the blind; bloodhounds for tracking criminals; terriers for worrying prey out of burrows; mastiffs for guard duty; and the original Pekinese (of which only a dwarf remnant remains) for war. We did all that, in only a few thousand years, by meddling with the sex lives of dogs. We evolved cauliflower, rutabaga, broccoli, brussels sprouts, and the now common and luxuriant cabbage from the sorry wild cabbage (these vegetables, like the different breeds of dogs, remain interfertile). Now think of a much more rigorous, much more stringent selection operating on all of Nature over an expanse of time a million times longer—and established not by the conscious meddling of dog or plant breeders with some idea of what kind of dog or plant they're aiming for, but by a blind, purposeless, and changing environment. If artificial selection represents an experiment of gigantic proportions, what must be the dimensions of the experiment that natural selection has performed? Isn't it plausible that all the elegantly adaptive diversity of life on Earth could thereby be sifted and extracted? Indeed, it is the only known process that adapts organisms to their environments.[3]

Here are the passages from Darwin's *Origin of Species* in which he first develops the point and counterpoint of artificial and natural selection:

> One of the most remarkable features in our domesticated races is that we see in them adaptation, not indeed to the animal's or plant's own good, but to man's use or fancy. Some variations useful to him have probably arisen suddenly, or by one step . . . But when we compare the dray-horse and race-horse, the dromedary and camel, the various breeds of sheep fitted either for cultivated land or mountain pasture, with the wool of one breed good for one purpose, and that of another breed for another purpose; when we compare the many breeds of dogs, each good for man in different ways; when we compare the game-cock, so pertinacious in battle, with other breeds so little quarrelsome, with "everlasting layers" [of eggs] which never desire to sit, and with the bantam so small and elegant; when we compare the host of agricultural, culinary, orchard, and flower-garden races of plants, most useful to man at different seasons and for different purposes, or so beautiful in his eyes, we must, I think, look further than to mere

variability. We cannot suppose that all the breeds were suddenly produced as perfect and as useful as we now see them; indeed, in many
cases, we know that this has not been their history. The key is man's
power of accumulative selection: nature gives successive variations;
man adds them up in certain directions useful to him. In this sense
he may be said to have made for himself useful breeds.

. . . [H]ardly any one is so careless as to breed from his worst animals . . .

If there exist savages so barbarous as never to think of the inherited
character of the offspring of their domestic animals, yet any one
animal particularly useful to them, for any special purpose, would be
carefully preserved during famines and other accidents, to which
savages are so liable, and such choice animals would thus generally
leave more offspring than the inferior ones; so that in this case there
would be a kind of unconscious selection going on . . .

Man . . . can never act by selection, excepting on variations which
are first given to him in some slight degree by nature . . .

This preservation [in Nature] of favourable individual differences
and variations, and the destruction of those which are injurious, I
have called Natural Selection, or the Survival of the Fittest. Variations neither useful nor injurious would not be affected by natural
selection . . .

When we see leaf-eating insects green, and bark-feeders mottled-
grey; the alpine ptarmigan white in winter, the red-grouse the colour
of heather, we must believe that these tints are of service to these
birds and insects in preserving them from danger . . .

If it profit a plant to have its seeds more and more widely dissemi-
nated by the wind, I can see no greater difficulty in this being effected
through natural selection, than in the cotton-planter increasing and
improving by selection the down in the pods on his cotton-trees . . .

There is no reason why the principles which have acted so efficiently under domestication should not have acted under nature. In
the survival of favoured individuals and races, during the constantly-
recurrent Struggle for Existence, we see a powerful and ever-acting
form of Selection. The struggle for existence inevitably follows from

the high geometrical ratio of increase which is common to all organic beings. This high rate of increase is proved by calculation,—by the rapid increase of many animals and plants during a succession of peculiar seasons, and when naturalised in new countries. More individuals are born than can possibly survive. A grain in the balance may determine which individuals shall live and which shall die,—which variety or species shall increase in number, and which shall decrease, or finally become extinct . . . The slightest advantage in certain individuals, at any age or during any season, over those with which they come into competition, or better adaptation in however slight a degree to the surrounding physical conditions, will, in the long run, turn the balance.[4]

In his 1858 paper in the Linnaean Society *Proceedings*, he asks us to imagine a being who could continue selecting, with unfailing attention, for a single desired characteristic over "millions of generations." Natural selection implies—in effect, although not literally—that such a being exists. "We have almost unlimited time" for evolution, he wrote.

Darwin then went on to propose that, over such immense periods of time, continuing natural selection may generate such a divergence of an organism from its parental stock as to constitute a new species. Giraffes develop long necks because those whose necks are—by some spontaneous genetic variation—a little longer are able to browse on the topmost foliage, flourish when others are ill-fed, and leave more offspring than their shorter-necked fellows. He pictured a vast family tree, symbolic of the varied forms of life, slowly growing, branching, and anastomosing, organisms evolving to produce all the "exquisite adaptations" of the natural world.

There is "grandeur," he thought, in the fact that "from so simple a beginning, endless forms most beautiful and most wonderful have been, and are being evolved."

Analogy would lead me one step farther, namely, to the belief that all animals and plants are descended from some one prototype. But analogy may be a deceitful guide. Nevertheless all living things have much in common, in their chemical composition, their cellular structure, their laws of growth, and their liability to injurious influences. . . . [O]n the principle of natural selection with divergence of

character, it does not seem incredible that, from such low and intermediate form, both animals and plants may have been developed;
and, if we admit this, we must likewise admit that all the organic
beings which have ever lived on this earth may be descended from
some one primordial form.

And how did such a primordial form arise? In 1871, Darwin wistfully
imagined, in a letter to his friend Joseph Hooker, "But if (and oh! what
a big if!) we could conceive in some warm little pond, with all sorts of
ammonia and phosphoric salts, light, heat, electricity, &c., present,
that a proteine compound was chemically formed, ready to undergo
still more complex changes . . ."[5]

If such a thing were possible, why isn't it happening today? Darwin
immediately foresaw one reason: "At the present day, such matter
would be instantly devoured or absorbed, which would not have been
the case before living creatures were formed." In addition, we now
know that the absence of the oxygen molecule in the atmosphere of
the primitive Earth made the formation and survival of organic molecules then much more likely. (And vastly more organic molecules
were falling from the sky than do so today in our tidied-up and regu-
larized Solar System.) That warm little pond—or something like it—
laboratory experiments show, could have quickly produced the amino
acids. Amino acids, energized a little, readily join up to make something like "a proteine compound." In related experiments, simple nucleic acids are made. Darwin's guess, as far as it went, is today pretty
well confirmed. The building blocks of life were abundant on the early
Earth, although we certainly cannot yet say we fully understand the
origin of life. But we humans, starting with Darwin, have only just
begun to look into the matter.

————

The publication of *The Origin of Species* met, as might have been
expected, with a passionate response, both pro and con, including a
stormy meeting of the British Association for the Advancement of
Science shortly after publication. The larger debate can perhaps best
be glimpsed by disinterring the literary reviews of the day. These mag-
azines, generally published monthly, covered the widest range of top-
ics—fiction and nonfiction, prose and poetry, politics, philosophy,

religion, and science. Reviews of twenty printed pages were not uncommon. Almost all articles were unsigned, although many were written by the leading figures in their fields. Comparable publications in
the English language seem sparse today, although *The Times* of London's *Literary Supplement* and *The New York Review of Books* perhaps
come closest.

The Westminster Review of January 1860 recognized that Darwin's
book might be of historic significance:

> If the principle of Modification by Natural Selection should be ad-
> mitted to anything like the extent to which Mr. Darwin would carry
> it . . . a grand and almost untrodden field of inquiry will be opened
> . . . Our classifications will come to be, as far as they can be so made,
> genealogies; and will then truly give what may be called the plan of
> creation.[6]

The Edinburgh Review of April 1860 (in an unsigned critique by the
anatomist Richard Owen) took a less charitable view:

> The considerations involved in the attempt to disclose the origin of
> the worm are inadequate to the requirements of the higher problem
> of the origin of man . . . To him, indeed, who may deem himself
> devoid of soul and as the brute that perisheth, any speculation, point-
> ing, with the smallest feasibility, to an intelligible notion of the way
> of coming in of a lower organised species, may be sufficient, and he
> need concern himself no further about his own relations to a Creator
> . . . Mr. Darwin offers us . . . intellectual husks . . . endorsed by his
> firm belief in their nutritive sufficiency.[7]

The reviewer praises scientists "who trouble the intellectual world
little with their beliefs, but enrich it greatly with their proofs," and
contrasts them to Darwin, who is said to have no more than "a discur-
sive and superficial knowledge of nature."

Professor Owen is much impressed by the work of Cuvier on the
mummified ibises, cats, and crocodiles "preserved in the tombs of
Egypt," which prove "that no change in their specific characters has
taken place during the thousands of years . . . which had elapsed . . .
since the individuals of those species were the subjects of the mum-
mifier's skill." Cuvier's data, it is said, were of "far higher value" than

the "speculations" of Darwin. But the mummified animals of ancient Egypt walked the Earth only a split second ago on the geological time scale—not nearly long enough ago to show major evolutionary change, which characteristically requires millions of years. Owen's review ripples with florid scorn: "Prosaic minds," it says, "are apt to bore one by asking for our proofs, and one feels almost provoked, when seduced to the brink of such a draught of forbidden knowledge as the [evolutionists] offer, to have the Circean cup dashed away" by more knowledgeable experts of a different opinion.

Other commentators raised more substantial objections: No example of a beneficial mutation or hereditary change is known, it was said; Darwin must invoke enormous intervals of time before the epoch of the dinosaurs, and yet no sign of life could be found in the earlier geological record; transitional forms between one species and the other were said to be wholly lacking in the geological record. In fact Darwin stressed the almost total ignorance in his time of the nature of hereditary transmission and mutation, and he himself pointed to the sparseness of the geological record as a problem for the theory (although he also said he would produce the transitional fossils when his opponents showed him all the intermediate forms between wild dogs and greyhounds, say, or bulldogs). Since then, not only have the laws of inheritance by genes and chromosomes (which are made entirely of nucleic acids) been carefully worked out, but their detailed molecular structure is known; we even understand how a mutation can be caused by the substitution of a single atom for another. The geological record has been extended not only to before the time of the dinosaurs, but we now have spotty glimpses of life through the preceding 3.5 billion years. Despite his exhaustive studies of artificial selection, Darwin did not know of a single case history of natural selection in the wild; today we know of hundreds.[8] The fossil evidence remains sparse, though: A few more transitional forms are now known—*Archeopteryx*, for example, a halfway house between reptile and bird—but still not nearly enough to show even the majority of the important evolutionary pathways. But the most powerful evidence for evolution comes, as we will see, from a science whose very existence was unknown in Darwin's time—molecular biology.

A critique in *The North American Review* for April 1860 attempts to refute Darwin by a kind of unselfconscious sophism: The very long periods of geological time required for evolution are declared "virtually infinite." Darwin himself used similarly loose mathematical language. Then the review goes on to assert that "the difference between such a conception and that of the strictly infinite, if any, is not appreciable." Infinity, however, belongs not to science but to metaphysics, so the reviewer concludes that the theory of evolution is not scientific but metaphysical—"resting altogether upon the idea of 'the infinite,' which the human mind can neither put aside nor comprehend."[9] This last point would seem to apply, especially, to the reviewer. In fact, any two numbers, no matter how large or small, are equally distant from infinity, and 4.5 billion years is a respectably finite period of time. Infinity does not enter the evolutionary perspective. The speciousness of this argument (and other critiques) gives us a sense of how anxious people were to reject Darwin's ideas. (His later suggestion that all living things including humans were *still* evolving, and that in the far future our descendants would not be human, was dismissed even by sympathetic reviewers as going too far.)

In *The London Quarterly Review* of July 1860, in an article called "Darwin's Origin of Species," Darwin is anonymously taken to task by his adversary Samuel Wilberforce, the Anglican Bishop of Oxford—among many other things, for "wantonness of conjecture" and "extravagant liberty of speculation." His "mode of dealing with nature" is condemned as

> utterly dishonourable to all natural science, as reducing it from its present lofty level as one of the noblest trainers of man's intellect and instructors of his mind, to being a mere idle play of the fancy, without the basis of fact or the discipline of observation.

He is accused of circumventing "the obstinacy of fact" by waving a magic wand and saying, " 'Throw in a few hundreds of millions of years more or less, and why should not all these changes be possible . . . ?' "

The terrible implication is drawn that Darwin's unexpressed supposition was that "man" might be only "an improved ape." (Wilberforce on this point was not far from the mark; this is close to what Darwin thought.) That natural selection might apply to humans is denounced as "absolutely incompatible" with "the Word of God." Moreover, "man's derived supremacy over the earth; man's power of articulate speech; man's gift of reason; man's free-will and responsibility; man's fall and man's redemption; the Incarnation of the Eternal Son; the indwelling of the Eternal Spirit, all are equally and utterly irreconcilable with the degrading notion of the brute origin of him who was created in the image of God, and redeemed by the Eternal Son." The idea of evolution tends "inevitably to banish from the mind most of the peculiar attributes of the Almighty." Darwin's insights are compared to "the frenzied inspiration of the inhaler of mephitic gas." His views are contrasted by Bishop Wilberforce with those of "a far greater philosopher," Professor Owen, whom he quotes, a little tangentially, as advising teenagers:

> Oh! you who possess it in all the supple vigour of lusty youth, think well what it is that He has committed to your keeping. Waste not its energies; cull them not by sloth; spoil them not by pleasures! The supreme work of creation has been accomplished that you might possess a body—the sole erect—of all animal bodies the most free— and for what? for the service of the soul . . . Defile it not.[10]

The North British Review of May 1860, no less hostile, begins its critique: "If notoriety be any proof of successful authorship, Mr. Darwin has had his reward." Darwin is compared with writers who "seem ever distrustful of views of nature which, even remotely, tend to set them or their readers in direct relation with a personal God." As in many of the negative reviews, this one acknowledges Darwin's reputation as an accomplished naturalist and praises his felicity of style. He is, though, a "charlatan" and guilty of "unbelief in the governing Creator." The book's "seeming depth is only darkness." He is accused of setting a throne "somewhere, above Olympus, and the goddess of the author's devotion is seated on it." This goddess is Natural Selection. "The 'chance' of heathenism has developed into a higher form . . . Mr. Darwin's work," *The North British Review* concludes, "is in direct antagonism to all the findings of a natural theology, formed on legitimate inductions in the study of the works of God; and it does open violence to everything which the Creator Himself has told us in the Scriptures of truth." The publication of *The Origin of Species* is said to have been a "mistake." "Its author would have done well to science, and to his own fame, had he, being determined to write it, put it away among his papers, marked, 'A Contribution to Scientific Speculation in 1720' "—that being the reviewer's estimate of how retrogressive and passé Darwin's argument was.[11]

The process of natural selection, extracting order out of chaos as if by magic, was counterintuitive and disturbing to many, and Darwin was repeatedly accused of something not far short of idolatry. He answered the charge in these words:

> It has been said that I speak of natural selection as an active power or Deity; but who objects to an author speaking of the attraction of gravity as ruling the movements of the planets? Every one knows what is meant and is implied by such metaphorical expressions; and they are almost necessary for brevity. So again it is difficult to avoid personifying the word Nature; but I mean by Nature, only the aggregate action and product of many natural laws, and by laws the sequence of events as ascertained by us. With a little familiarity such superficial objections will be forgotten . . .
>
> As man can produce, and certainly has produced, a great result by his methodical and unconscious means of selection, what may not natural selection effect? Man can act only on external and visible characters: Nature, if I may be allowed to personify the natural preservation or survival of the fittest, cares nothing for appearances, except in so far as they are useful to any being. She can act on every internal organ, on every shade of constitutional difference, on the whole machinery of life. Man selects only for his own good: Nature only for that of the being which she tends . . .
>
> It may metaphorically be said that natural selection is daily and hourly scrutinising, throughout the world, the slightest variations; rejecting those that are bad, preserving and adding up all that are good; silently and insensibly working . . . We see nothing of these slow changes in progress, until the hand of time has marked the lapse of ages, and then so imperfect is our view into long-past geological ages, that we see only that the forms of life are now different from what they formerly were.

Darwin was criticized by some for being a teleologist—for believing that Nature was working with some long-term end in view—and, conversely, by others for constructing a Nature in which random, purposeless variation is key. ("The law of higgledy-piggledy," the astronomer John Herschel dismissively called it.) People had real difficulty grasping the concept of natural selection. His motives, sincerity, honesty, and ability were all questioned. Many who criticized him did not understand his argument or the cumulative power of the data he invoked in its support. Many—including some of the most distinguished scientists of the day, among them, painfully, Adam Sedgwick, his old geology professor—rejected Darwin's insight, not because the evidence was against it, but because of where it led: seemingly, to a world in which humans were degraded, souls denied, God and morality scorned, and monkeys, worms, and primeval ooze elevated; "a system uncaring of man." Thomas Carlyle called it "a Gospel of dirt."

None of these moral and theological criticisms is compelling, Darwin, Huxley, and others labored to show: In astronomy, we no longer believe that an angel pushes each planet around the Sun; the inverse square law of gravitation and Newton's laws of motion suffice. But no one considers this a demonstration of the nonexistence of God, and Newton himself—except for a private reservation about the notion of the Trinity—was close to the conventional Christianity of his day. We are free to posit, if we wish, that God is responsible for the laws of Nature, and that the divine will is worked through secondary causes. In biology those causes would have to include mutation and natural selection. (Many people would find it unsatisfying, though, to worship the law of gravity.)

As the debate proceeded over the years, natural selection seemed less strange and less threatening. Increasing numbers of scientists, literary figures, and even clergymen were won over. But by no means all. In July 1871, *The London Quarterly Review*—which eleven years earlier had published Bishop Wilberforce's anonymous diatribe—remained unreconstructed, wholly missing Darwin's point. "Why should natural selection favor the preservation of useful varieties only? Such action cannot be referred to blind force; it can belong to mind alone." Not only are evolution and natural selection rejected, but so is the newly discovered law of the conservation of energy,[12] one of the foundations of modern physics.

Some of the underlying emotional reasons for rejecting natural selection were later vividly expressed by the playwright George Bernard Shaw:

[T]he Darwinian process may be described as a chapter of accidents. As such, it seems simple, because you do not at first realize all that it involves. But when its whole significance dawns on you, your heart sinks into a heap of sand within you. There is a hideous fatalism about it, a ghastly and damnable reduction of beauty and intelligence, of strength and purpose, of honor and aspiration, to such casually picturesque changes as an avalanche may make in landscape, or a railway accident in a human figure. To call this Natural Selection is a blasphemy, possible to many for whom Nature is nothing but a casual aggregation of inert and dead matter, but eternally impossible to the spirits and souls of the righteous . . . If this sort of selection could turn an antelope into a giraffe, it could conceivably turn a pond full of amoebas into the French Academy.[13]

Fine words. But what if undreamed-of powers lie hidden in "inert and dead matter," given 4 billion years of preserving what works? Such objections address (and far from compellingly) only the philosophical and social implications of natural selection, and not the evidence for it.

Naive Darwinists, including many capitalists, have self-servingly argued that oppression of the weak and the poor is a justified application of natural selection to human affairs. Naive biblical literalists, including some high officials charged with safeguarding the environment, have self-servingly argued that the destruction of non-human life is justified because the world will shortly end anyway, or because of the injunction in Genesis that we have "dominion . . . over every living thing."[14] But neither evolution nor the sacred books of various religions are invalidated because dangerous conclusions have been mistakenly drawn from them.

By the 1870s and 1880s, the evidence amassed by Darwin was changing many minds. Reviews were acknowledging "the certainty of the action of natural selection," and even the possibility that humans evolved from some lower animal.[15] However, some of the conclusions of Darwin's 1871 book, *The Descent of Man*, stuck in the craws of even the most sympathetic reviewers. The debate, we find, had moved into a new arena:

We deny [animals] . . . the power of reflecting of their own existences, or of inquiring into the nature of objects and their causes. We deny that they know that they know, or know themselves in knowing. In other words, we deny them *reason*.

We return to this new level of debate later, and here note only how quickly many of the theological reservations about evolution had dissipated as Darwin's argument became better understood. "Nothing is more remarkable," he wrote in his *Autobiography*, "than the spread of scepticism or rationalism during the latter half of my life."[16]

———

Of innumerable modern examples of natural selection in the real world, we select one—of interest because it involves humans and because it is the outcome of an experiment, although one performed inadvertently and under tragic circumstances. Malaria is endemic among nearly half the people of the world (just before World War II, the number was two thirds of all humans). It is a serious illness associated, in the absence of appropriate medicine or natural immunity, with high mortality. Even today several million people die from it each year. When the plasmodium parasite causing malaria is injected (usually by mosquito bite) into the bloodstream, it eventually invades the red blood cells that carry oxygen from the lungs to every cell of the body. The red blood cells are rendered sticky, adhere to the walls of very small blood vessels, and are prevented from being circulated to the spleen—which destroys plasmodium parasites. This is good for the parasites and bad for the humans.

People in malarial zones of tropical Africa, as elsewhere, have an adaptation to malaria: the sickle-cell trait. Under the microscope some of the red blood cells do look a little bit like sickles or croissants. But in someone with the sickle-cell trait, the altered red blood cells are surrounded by needle-like microscopic filaments that work, it is suggested, a little like a porcupine's quills. The parasites are impaled or otherwise damaged, and the red blood cells—protected from the parasites' sticky proteins—are then carried to the "untender mercies" of the spleen. With the parasites dead, many of the red blood cells return to their normal state, "unruffled" by the experience.[17] However, when the genes for this trait are inherited from both parents, serious anemia, obstruction of the small blood vessels, and other infirmities often result. The trade-off, it is natural to think, is that it's better for a part of the population to be seriously anemic than for most of the population to be dead of malaria.

In the seventeenth century slave traders from Holland arrived in the Gold Coast of West Africa (present-day Ghana). They bought or captured slaves in large numbers and transported them to two Dutch colonies—Curaçao in the Caribbean and Surinam in South America. There is no malaria in Curaçao, so the sickle-cell trait conferred anemia but no compensating advantage to the slaves brought there. But malaria is endemic in Surinam, and the sickle-cell trait was often the difference between life and death.

If now, some three centuries later, we examine the descendants of these slaves, we find that those in Curaçao show hardly any incidence of the trait, while it remains prevalent in Surinam. In Curaçao the sickle-cell trait was "selected against"; in Surinam, as in West Africa, it was "selected for." We see natural selection operating on very short time scales, even for such slowly reproducing beings as humans.[18] As always, there is a range of hereditary predispositions in a given population; the environment elicits some but not others. Evolution is the product of a hand-in-hand interplay between heredity and environment.

————

At the end of his life, Darwin called himself a theist, a believer in a First Cause. He had doubts, though:

> [C]an the mind of man, which has, as I fully believe, been developed from a mind as low as that possessed by the lowest animal, be trusted when it draws such grand conclusions?[19]

Evolution in no way *implies* atheism, although it is *consistent* with atheism. But evolution is clearly inconsistent with the literal truth of certain revered books. If we believe the Bible was written by people, and not dictated word-for-word to a flawless stenographer by the Creator of the Universe, or if we believe God might on occasion resort to metaphor for clarity, then evolution should pose no theological problem. But whether it poses a problem or not, the evidence for evolution —*that* it has happened, apart from the debate on whether uniformitarian natural selection fully explains *how* it happened—is overwhelming.

The Darwinian perspective is central to all of modern biology, from investigations of the molecular structure of DNA to studies of the behavior of apes and men.[20] It connects us with our long-forgotten ancestors and our swarm of relatives, the millions of other species with whom we share the Earth. But the price exacted has been high, and there are still—especially in the United States—those who refuse to pay, and for very human and fathomable reasons. Evolution suggests that if God exists, God is fond of secondary causes and factotum processes: getting the Universe going, establishing the laws of Nature, and then retiring from the scene. A hands-on Executive seems to be absent; power has been delegated. Evolution suggests that God will not intervene, whether beseeched or not, to save us from ourselves. Evolution suggests we're on our own—that if there is a God, that God must be very far away. This is enough to explain much of the emotional anguish and alienation that evolution has worked. We long to believe that there's someone at the helm.

———

Darwin's transcendantly democratic insight that all humans are descended from the same non-human ancestors, that we are all members of one family, is inevitably distorted when viewed with the impaired vision of a civilization permeated by racism. White supremacists seized on the notion that people with high abundances of melanin in their skin must be closer to our primate relatives than bleached people. Opponents of bigotry, perhaps fearing that there might be a grain of truth in this nonsense, were just as happy not to dwell on our relatedness to the apes. But both points of view are located on the same continuum: the selective application of the primate connection to the veldt and the ghetto, but never, ever, perish the thought, to the boardroom or the military academy or, God forbid, to the Senate chamber or the House of Lords, to Buckingham Palace or Pennsylvania Avenue. This is where the racism comes in, not in the inescapable recognition that, for better or worse, we humans are just a small twig on the vast and many-branched tree of life.

Natural selection has been misused by capitalists and communists, whites and blacks, Nazis and many others to grind this or that self-serving ideological axe. It's not surprising that feminists feared that a Darwinian perspective would provide yet another cudgel for male scientists to hit women over the head with—about alleged inferiorities in mathematics or statecraft. But for all we know, such a perspective might reveal, that the raging hormonal imbalances that propel men to violence make them less than optimal for leadership of a modern state. If we believe sexism to be a prejudicial error, that fact will emerge from scientific examination, and we should favor its rigorous scrutiny by the methods of science.

Much of the recent controversy over the application of Darwinian ideas to human behavior has been motivated by the fear of such misuse by racists, sexists, and other bigots—as indeed happened with ghoulish and tragic consequences in World War II. However, the cure for a misuse of science is not censorship, but clearer explanation, more vigorous debate, and making science accessible to everyone. If some of our proclivities are inborn, as surely must be the case, it hardly follows that we cannot learn to modify, mitigate, enhance, or redirect the resulting behavior.

———

Vice-Admiral FitzRoy had been the British Board of Trade's weatherman for more than a decade when his 1865 long-range forecast proved to be wildly, calamitously wrong. The proud, choleric FitzRoy took a terrible beating in the newspapers. When he could no longer bear the ridicule, he slit his throat, an early martyr to the predictive failures of meteorology. Although FitzRoy had spoken publicly against Darwin in the "creationism" controversy and despite the fact that the two men had not been face-to-face in eight years, Darwin took the news of FitzRoy's suicide badly. What images from the youthful adventure they shared must have come to Darwin's mind? "What a melancholy career he has run," he observed to Hooker, "with all his splendid qualities."[21]

On melancholia, too, Darwin was something of an expert. These years he was depressed, exhausted, and sick most of the time. Throughout this miserable period he was consistently productive and his relationships with Emma, the survivors among their ten children, and a great number of friends seemed none the worse for it. If anything, the letters they exchanged and their written recollections testify to an openness, an emphasis on the importance of feelings, a respect for children, a harmonious family life. His daughter remembered him saying that he hoped none of his children would ever believe something just because it was he who told it to them. "He kept up his delightful, affectionate manner towards us all his life," his son Francis wrote. "I sometimes wonder that he could, with such an undemonstrative race as we are; but I hope he knew how much we delighted in his loving words and manner . . . He allowed his grown-up children to laugh with and at him, and was generally speaking on terms of perfect equality with us."[22]

There were many who comforted themselves with the thought that in his last moments Darwin would renounce his evolutionary heresies and repent. There are still people today who piously believe that's just what happened. Instead, Darwin faced death calmly and apparently without regret, saying on his deathbed "I am not the least afraid to die."[23]

The family wished to bury him on their estate at Down, but twenty Members of Parliament, with the support of the Anglican Church, appealed to them to allow him to be interred at Westminster Abbey, a few feet away from Isaac Newton. You've got to hand it to the Church of England. It was an act of consummate grace. For you, they seemed to be saying, who have done the most to raise doubts about the truth of what we say, we reserve the highest honor—a respect for the correction of error that is, incidentally, characteristic of science when it is faithful to its ideals.

HUXLEY AND THE GREAT DEBATE

Thomas Henry Huxley was born to a large, struggling, dysfunctional family in the England of 1825, where class was destiny for almost everyone. His formal education consisted of two years of elementary school. But he had an insatiable hunger for knowledge and legendary self-discipline. At age seventeen, on an impulse, Huxley entered an open competition given by a local college, and was awarded the Silver Medal of the Pharmaceutical Society and a scholarship to study medicine at Charing Cross Hospital. Forty years later he was President of the Royal Society, then the foremost scientific organization in the world. He made fundamental contributions to comparative anatomy and many other fields, and was, along the way, inventor of the words "protoplasm" and "agnostic." Through his whole life he was committed to teaching science to the public. (More than one member of the upper classes was known to don shabby clothes in order to gain admittance to his lectures for working people.) He taught that a fair scientific examination of the facts demolished European claims of racial superiority.[24] At the end of the American Civil War, he wrote that while the slaves might now be free, half of the human species—women—had yet to be emancipated.[*]

One of Huxley's interests had been the idea that all animals, including us, were "automata," carbon-based robots, whose "states of consciousness . . . are immediately caused by molecular changes of the brain-substance."[25] Darwin closed his last letter to him with these words: "Once again, accept my cordial thanks, my dear old friend. I wish to God there were more automata in the world like you."[26]

"If I am to be remembered at all," Huxley confided late in life,

---

[*] "[G]irls have been educated either to be drudges or toys, beneath men; or a sort of angels above him . . . The possibility . . . that women are meant . . . to be men's comrades, their fellows, and their equals, so far as Nature puts no bar on that equality, does not seem to have entered into the minds of those who have had the conduct of the education of girls." The first step to a better world, he said, was "Emancipate girls." Their hair "will not curl less gracefully outside the head by reason of there being brains within."[29]

"I would rather it should be as 'a man who did his best to help the people' than by any other title."[27] What he is actually best remembered for is delivering the punch line in the decisive debate that gained acceptance for Darwin's ideas.

———

The Huxley/Wilberforce debate is the grand climactic scene in the 1930s Hollywood movie version that might be imagined of Darwin's life:

A small item on the front page of The Daily Oxonian: *"Annual Meeting of British Association for the Advancement of Science to Be Held Tomorrow." The dateline reads June 29, 1860. Front page begins to spin like a roulette wheel.*

Dissolve to reveal that we are following the highly imaginative, although slightly shady Robert Chambers (played by Joseph Cotten) as he makes his way down an Oxford street. He is jostled by another man and just as he turns in annoyance, he realizes that it is none other than the pugnacious Thomas Henry Huxley (Spencer Tracy), whose conviction with regard to the truth of his friend Darwin's controversial theory is so fierce it will one day earn him the nickname "Darwin's Bulldog."

Rascal that he is, Chambers can't resist asking Huxley if he'll be attending Draper's reading at the British Association meeting. The title is to be "The Intellectual Development of Europe with Reference to the Views of Mr. Darwin." Huxley claims he's too busy.

Knowingly, Chambers allows that " 'Soapy Sam' Wilberforce is sure to be there."

Huxley, growing more defensive, insists that it would be a waste of time.

Chambers says slyly, "Deserting the cause, Huxley?"

Piqued, Huxley makes his excuses and walks off.

The following day. The doors to the great hall are thrown open. The place is packed but only one voice is heard. We pan in for a tight close-up of the Bishop of Oxford, Samuel Wilberforce (George Arliss). Fingers in lapels, he turns pointedly to Huxley (who is of course there, despite his protestations of scheduling conflicts) and with arch courtesy begs to know "whether it is through your grandfather or your grandmother that you claim your descent from a monkey?" *Grasping the smarmy nuance of "grand*mother," *the crowd utters low "ooh's" and turns its attention to Huxley.*

Still seated, Huxley turns to the man next to him and, almost winking, murmurs, "The Lord hath delivered him into mine hands." Rising and looking Wilberforce squarely in the eye, he says: "I would rather be the offspring of two apes than be a man and afraid to face the truth."

The crowd has never seen a bishop insulted to his face before. Stunned reaction. Ladies faint. Men shake their fists. Chambers in the crowd, positively gleeful. But wait. There's someone else standing up. Why, it's Vice-Admiral Robert FitzRoy (Ronald Reagan), back in England after his term as Governor of New Zealand. "I was arguing with Charles Darwin and his crazy ideas thirty years ago on the Beagle." And then, brandishing his Bible: "This and this alone is the source of all truth." More clamor.

Now it's Hooker's turn (Henry Fonda). Sincerely, "I knew this theory fifteen years ago. I was then entirely opposed to it; I argued against it again and again; but since then I have devoted myself unremittingly to natural history; in its pursuit I have traveled around the world. Facts in this science which before were inexplicable to me became one by one explained by this theory, and conviction has been thus gradually forced upon an unwilling convert."

The camera pulls out of the great hall. Dissolve to a close-up of a finch perched on the branch of a tree. A bearded man (Ronald Colman), kindly, dressed in rural gentleman's hat and cape, but with a muffler despite the June weather, is staring lovingly up at the bird. He hardly seems to hear the voice of his wife (Billie Burke), high-pitched, affectionate, calling from the great house, off-camera: "Charles . . . CHARLES . . . Trevor is here with news from that meeting at Oxford." He casts one appreciative look back at the finch before finally walking off to the house . . .[28]

Chapter 5

LIFE IS JUST A THREE-LETTER WORD

Who first drives life to begin its journey?

The *Kena Upanishad*
(8th to 7th centuries B.C., India)[1]

Who's aware of mutability?
Not even Buddhas.

DAITETSU
(1333–1408, Japan)[2]

I n a shaft of sunlight, even when the air is still, you can sometimes see a tribe of dust motes dancing. They move in zigzag paths as if animated, motivated, propelled by some small but earnest purpose. Some of the followers of Pythagoras, the ancient Greek philosopher, thought that each mote had its own immaterial soul that told it what to do, just as they thought that each human has a soul that gives us direction and tells us what to do.[3] Indeed, the Latin word for soul is *anima*—it is something similar in many modern languages—from which come such English words as "animate" and "animal."

In fact, those motes of dust make no decisions, have no volition. They are instead the passive agents of invisible forces. They're so tiny that they're battered about by the random motion of molecules of air, which have a slight preponderance of collisions first on one side of the mote and then on the other, propelling them, with what looks to us as some mix of intention and indecision, through the air. Heavier objects —threads, say, or feathers—cannot much be jostled by molecular collisions; if not wafted by a current of air, they simply fall.

The Pythagoreans deceived themselves. They did not understand how matter works on the level of the very small, and so—from a specious and oversimple argument—they deduced a ghostly spirit that pulls the strings. When we look around us at the living world, we see a profusion of plants and animals, all seemingly designed for specific ends and single-mindedly devoted to their own and their offspring's survival—intricate adaptations, an exquisite match of form to function. It is natural to assume that some immaterial force, something like the soul of a dust mote, but far grander, is responsible for the beauty, elegance, and variety of life on Earth, and that each organism is propelled by its own, appropriately configured, spirit. Many cultures all over the world have drawn just such a conclusion. But might we here, as did the ancient Pythagoreans, be overlooking what actually goes on in the world of the very small?

We can believe in animal or human souls without holding to evolution, and vice versa. But if we examined life more closely, might we be able to understand at least a little of how it works *and* how it came to be, purely in terms of its constituent atoms? Is something "immaterial" present? If so, is it in every beast and vegetable, or just in humans? Or is life no more than a subtle consequence of physics and chemistry?

———

One educated look at how the molecule is shaped and you can figure out what it's for. Even at the molecular level, function follows form. Before us is a detailed blueprint of breathtaking precision for building complex molecular machines. The molecule is very long and composed of two intertwined strands. Running the length of each strand is a sequence made of four smaller molecular building blocks, the nucleotides—which humans conventionally represent by the letters A, C, G, and T. (Each nucleotide molecule actually looks like a ring, or two connected rings, made of atoms.) On and on the sequence goes, for billions of letters. A short segment of it might read something like this:

ATGAAGTCGATCCTAGATGGCCTTGCAGACACCACCTTCC
    GTACCATCACCACAGACCTCCT . . .

Along the opposite strand there's an identical sequence, except that wherever nucleotide A was in the first strand, it's T in the second; and instead of G it's always C. And vice versa. Like this:

TACTTCAGCTAGGATCTACCGGAACGTCTGTGGTGGAAGG
    CATGGTAGTGGTGTCTGGAGGA . . .

This is a code, a long sequence of words written out in an alphabet of only four letters. As in ancient human writing, there are no spaces between the words. Inside this molecule there are, written in a special language of life, detailed instructions—or rather, two copies of the same detailed instructions, because the information in one strand can surely be reconstructed from the information in the other, once you understand the simple substitution cipher. The message is redundant, bespeaking care, conservatism; it conveys a sense that whatever it is saying must be preserved, treasured, passed intact to future generations.

Almost every issue of leading scientific journals such as *Science* or *Nature* contains the newly uncovered ACGT sequence of some part of the genetic instructions of some lifeform or other. We're slowly beginning to read the genetic libraries. The library of our own hereditary information, the human genome, is also becoming increasingly revealed, but there's a lot to read: Every cell of your body has a full set of instructions about how to manufacture you, encoded in a very compressed format—it takes only a picogram (a trillionth of a gram) of this molecule to specify everything you've inherited from your ancestors, back to the first beings of the primeval sea. Yet, there are almost as many nucleotide building blocks, or "letters," in the microminiaturized genetic information in any of your cells as there are people on Earth.

All words in the genetic code are three letters long. So, if we insert the implicit spaces between the words, the beginning of the first message above looks like this:

ATG AAG TCG ATC CTA GAT GGC CTT GCA GAC ACC ACC
    TTC CGT ACC . . .

Since there are only four kinds of nucleotides (A, C, G, and T), there are at most only 4 x 4 x 4 = 64 possible words in this language. But if the order in which the words are put together is central to the meaning of the message, you can say a great deal with only a few dozen different words. With messages that are a billion carefully selected words long, what might be possible? You must take care in reading the message, though: With no spaces between the words, if you start reading at the wrong place, the meaning will surely change and a lucid message might be reduced to gibberish. This is one reason the giant molecule has special code words meaning "START READING HERE" and "STOP READING HERE."

As you watch the molecule closely you observe that the two strands occasionally unwind and unzip. Each copies the other, using available A, C, G, and T raw materials—like the metal type stored in an old-fashioned printer's box. Now, instead of one pair, there are two pairs of identical messages. As well as utilizing a language and embodying a complex, redundantly encoded text, this molecule is a printing press.

But what's the use of a message if nobody reads it? Through copying links and relays, the sequences of As, Cs, Gs, and Ts are revealed to be the job orders and blueprints for the construction of particular molecular machine tools. Some sequences are orders to itself—arranging for the giant molecule to twist and kink so it can then issue a particular set of instructions. Other sequences ensure that the instructions will be followed to the letter. Many three-letter words specify a particular amino acid (or a punctuation mark, like the one that signifies "START") out there in the surrounding cell, and the sequence of words encoded determines the sequence of amino acids that will make up the protein machine tools that control the life of the cell. Once such a protein is manufactured, it usually twists and folds itself into a three-dimensional shape spring-loaded for action. Sometimes another protein bends it into shape. These machine tools, at a pace determined both by the long double-stranded molecule and by the outside world, then proceed on their own to strip other molecules down, to build new ones up, to help communicate molecular or electrical messages to other cells.

This is a description of some of the humdrum, everyday action in each of the ten trillion or so cells of your body, and those of nearly every other plant, animal, and microbe on Earth. The tiny machine tools perform stupefying feats of molecular transformation. They are submicroscopic and made of organic molecules, rather than macroscopic and made of silicates or steel, but at the molecular level life was tool-using and tool-making from the start.

The long self-replicating double-stranded molecule with the complex message is a sequence of genes, a little like beads on a string.[4] Chemically, it is a nucleic acid (here, the kind abbreviated DNA, which stands for deoxyribonucleic acid). The two strands, wrapped around each other, comprise the famous DNA double helix. The nucleotide bases in DNA are called adenine, cytosine, guanine, and thymine, which is where the abbreviations A, C, G, and T come from. Their names date back to long before their key role in heredity was understood. Guanine, for example, is named unpretentiously after guano, the bird droppings from which it was first isolated. It is a double ring molecule made of five carbon atoms, five hydrogens, five nitrogens, and one oxygen. There's something like a billion guanines (and roughly equal numbers of As, Cs, and Ts) in the genes of any one of your cells.

Except for some oddball microbes, the genetic information of every organism on Earth is contained in DNA—a molecular engineer of formidable, even awesome talents. One (very long) sequence of As, Cs, Gs, and Ts contains all the information for making a person; another such sequence, nearly identical, for a chimpanzee; others, not so different, for a wolf or a mouse. In turn, the sequences for nightingales, sidewinders, toads, carp, scallops, forsythia, club mosses, seaweed, and bacteria are still more different—although even they collectively hold many sequences of As, Cs, Gs, and Ts in common. A typical gene, controlling or contributing to one specific hereditary trait, might be a few thousand nucleotides long. Some genes may comprise more than a million As, Cs, Gs, and Ts. Their sequences specify the chemical instructions for, say, manufacturing the organic pigments that make eyes brown or green; or extracting energy out of food; or finding the opposite sex.

How this complex information got *into* our cells, and how arrangements were made for its precise replication and the obedient implementation of its instructions, is tantamount to asking how life evolved. Nucleic acids were unknown when *The Origin of Species* was first published, and the messages they contain were not to be deciphered for another century. They constitute the demonstration and definitive record of evolution that Darwin sought. Scattered in the ACGT sequences of the diverse lifeforms of our planet is an incomplete history of the evolution of life—not the blood, bones, brains, and the other manufactured products of the genetic factories, but the actual production records, the master instructions themselves, slowly varying at different rates in different beings in different epochs.

Because evolution is conservative and reluctant to tamper with instructions that work, the DNA code incorporates documents—job orders and blueprints—dating back to remote biological antiquity. Many passages have faded. In some places there are palimpsests, where remains of ancient messages can be seen peeking out from under newer ones. Here and there a sequence can be found that is transposed from a different part of the message, taking on a different shade of meaning in its new surrounds; words, paragraphs, pages, whole volumes have been moved and reshuffled. Contexts have changed. The common sequences have been inherited from remote times. The more distinct the corresponding sequences are in two different organisms, the more distantly related they must be.

These are not only the surviving annals of the history of life, but also handbooks of the mechanisms of evolutionary change. The field of molecular evolution—only a few decades old—permits us to decode the record at the heart of life on Earth. Pedigrees are written in these sequences, carrying us back not a few generations, but most of the way to the origin of life. Molecular biologists have learned to read them and to calibrate the profound kinship of all life on Earth.[5] The recesses of the nucleic acids are thick with ancestral shadows.

We can now almost follow the itinerary of the naturalist Loren Eiseley:

> Go down the dark stairway out of which the race has ascended. Find yourself at last on the bottommost steps of time, slipping, sliding, and wallowing by scale and fin down into the muck and ooze out of which you arose. Pass by grunts and voiceless hissings below the last tree ferns. Eyeless and earless, float in the primal waters, sense sunlight you cannot see and stretch absorbing tentacles toward vague tastes that float in water.[6]

———

A particular sequence of As, Cs, Gs, and Ts is in charge of making fibrinogen, central to the clotting of human blood. Lampreys look something like eels (although they are far more distant relations of ours than eels are); blood circulates in their veins too; and their genes also contain instructions for the manufacture of the protein fibrinogen. Lampreys and people had their last common ancestor about 450 million years ago. Nevertheless, most of the instructions for making human fibrinogen and for making lamprey fibrinogen are identical. Life doesn't much fix what isn't broken. Some of the differences that do exist are in charge of making parts of the molecular machine tools that hardly matter—something like the handles on two drill presses being made of different materials with different brand names, while the guts of the two are identical.

Or here, to take another example, are three versions of the same message,[7] taken from the same part of the DNA of a moth, a fruit fly, and a crustacean:

Moth:
GTC GGG CGC GGT CAG TAC TTG GAT GGG TGA CCA CCT
   GGG AAC ACC GCG TGC CGT TGG . . .

Fruit fly:
GTC GGG CGC GGT TAG TAC TTA GAT GGG GGA CCG CTT
   GGG AAC ACC GCG TGT TGT TGG . . .

Crustacean:
GTC GGG CCC GGT CAG TAC TTG GAT GGG TGA CCG CCT
   GGG AAC ACC GGG TGC TGT TGG . . .

Compare these sequences and recall how different a moth is from a lobster. But these are not the job orders for mandibles or feet—which could hardly be closely similar in moths and lobsters. These DNA sequences specify the construction of the molecular jigs on which newly forming molecules are laid out under the ministrations of the molecular machine tools. Down at this level, it's not absurd that moths and lobsters might have closer affinities than moths and fruit flies. The comparison of moth and lobster suggests how slow to change, how conservative the genetic instructions can be. It's a long time ago that the last common ancestor of moths and lobsters scudded across the floor of the primeval abyss.

We know what every one of those three-letter ACGT words means —not just which amino acids they code for, but also the grammatical and lexigraphical conventions employed by life on Earth. We have learned to read the instructions for making ourselves—and everybody else on Earth. Take another look at "START" and "STOP." In organisms other than bacteria, there's a particular set of nucleotides that determine when DNA should start making molecular machine tools, which machine tool instructions should be transcribed, and how fast the transcription should go. Such regulatory sequences are called "promoters" and "enhancers." The particular sequence TATA, for example, occurs just before the place where transcription is to occur.

Other promoters are CAAT and GGGCGG. Still other sequences tell the cell where to stop transcribing.[8]

You can see that the substitution of one nucleotide for another might have only minor consequences—you could, for example, substitute one structural amino acid for another (in the "handle" of the machine tool) and in no way change what the resulting protein does. But it could also have a catastrophic effect: A single nucleotide substitution might convert the instructions for making a particular amino acid into the signal to stop the transcription; then, only a fragment of the molecular machine in question will be manufactured, and the cell might be in trouble. Organisms with such altered instructions will probably leave fewer offspring.

The subtlety and nuance of the genetic language is stunning. Sometimes there seem to be overlapping messages using the same letters in the same sequence, but with different functional import depending on how it's read: two texts for the price of one. Nothing this clever occurs in any human language. It's as if a long passage in English had two completely different meanings,[9] something like

ROMAN CEMENT TOGETHER NOWHERE . . .

and

ROMANCEMENT TO GET HER NOW HERE . . .

but much better—on and on for pages, perfectly lucid and grammatical in both modes, and, we think, beyond the skill of any human writer. The reader is invited to try.

In "higher" organisms, many long sequences seem to be nonfunctional genetic nonsense. They lie after a "STOP" and before the next "START" and generally remain ignored, forlorn, untranscribed. Maybe some of these sequences are garbled remnants of instructions that, long ago, in our distant ancestors, were important or even keys to survival, but that today are obsolete and useless.* Being useless,

---

* The silent "gh" in such English words as thought and height, or the silent "k" in knife or knight, were likewise once sounded out, but today are little more than a vestige of the evolution of language. Something similar is true for the circumflex and cedilla

these sequences evolve quickly: Mutations in them do no harm and are not selected against. Maybe a few of them are still useful, but elicited only under extraordinary circumstances. In humans some 97% of the ACGT sequence is apparently good for nothing. It's the remaining 3% that, as far as genetics goes, makes us who we are.

Startling similarities among the *functional* sequences of As, Cs, Gs, and Ts are seen throughout the biological world, similarities that could not have come about unless—beneath the apparent diversity of life on Earth—there was an underlying and fundamental unity. That unity exists, it seems clear, because every living thing on Earth is descended from the same ancestor 4 billion years ago; because we are all kin.

But how could machines of such elegance, subtlety, and complexity ever arise? The key to the answer is that these molecules are able to evolve. When one strand is making a copy of the other, sometimes a mistake occurs and the wrong nucleotide—an A, say, instead of a G —will be inserted into the newly assembled sequence. Some of them are honest replication errors—good as it is, the machinery isn't perfect. Some are induced by a cosmic ray or another kind of radiation, or by chemicals in the environment. A rise in temperature might slightly increase the rate at which molecules fall to pieces, and this could lead to mistakes. It even happens that the nucleic acid generates a substance that alters itself—perhaps thousands or millions of nucleotides away.

Uncorrected mistakes in the message are propagated down to future generations. They "breed true." These changes in the sequence of As, Cs, Gs, and Ts, including alterations of a single nucleotide, are called mutations. They introduce a fundamental and irreducible randomness into the history and nature of life. Some mutations may neither help nor hinder, occurring, for example, in long, repetitive sequences —containing redundant information—or in what we've called the handles of the molecular machine tools, or in untranscribed se-

---

which are in the course of being phased out in French, and for recent simplifications of Chinese and Japanese. The nonfunctional genetic sequences, however, are not just a few letters here and there, but reams of obsolete and/or garbled information—something like a confused account in ancient Assyrian on how to manufacture chariot axles, set in more recently generated nonsense information.

quences between STOP and START. Many other mutations are deleterious. If you're crafting superb machine tools and, while you're not looking, someone introduces a few random changes into the computer instructions for manufacture, there isn't much chance that the resulting machines, built according to the new, garbled instructions, will work better than the earlier model. Enough random changes in a complex set of instructions will cause serious harm.

But a few of the random changes, by luck, prove advantageous. For example, the sickle-cell trait we mentioned in the last chapter is caused by the mutation of a single nucleotide in the DNA, generating a difference of a single amino acid in the hemoglobin molecules that nucleotide helps code for; this in turn changes the shape of the red blood cell and interferes with its ability to carry oxygen, but at the same time it eventually kills the plasmodium parasites those cells contain. A lone mutation, one particular T turning into an A, is all it takes.

And, of course, not just the hemoglobin in red blood cells, but every part of the body, every aspect of life, is instructed by a particular DNA sequence. Every sequence is vulnerable to mutation. Some of these mutations cause changes more far-reaching than the sickle-cell trait, some less. Most are harmful, a few are helpful, and even the helpful ones may—like the sickle-cell mutation—represent a tradeoff, a compromise.

This is a principal means by which life evolves—exploiting imperfections in copying despite the cost. It is not how we would do it. It does not seem to be how a Deity intent on special creation would do it. The mutations have no plan, no direction behind them; their randomness seems chilling; progress, if any, is agonizingly slow. The process sacrifices all those beings who are now less fit to perform their life tasks because of the new mutation—crickets who no longer hop high, birds with malformed wings, dolphins gasping for breath, great elms succumbing to blight. Why not more efficient, more compassionate mutations? Why must resistance to malaria carry a penalty in anemia? We want to urge evolution to get to where it's going and stop the endless cruelties. But life doesn't *know* where it's going. It has no long-term plan. There's no end in mind. There's no mind to keep an end in mind. The process is the opposite of teleology. Life is profligate, blind, at this level unconcerned with notions of justice. It can afford to waste multitudes.

——

The evolutionary process could not have gone very far, though, if the mutation rate had been too high. In any given environment, there must be a delicate balance—simultaneously avoiding mutation rates so high that instructions for essential molecular machine tools are quickly garbled, and mutation rates so low that the organism is unable to retool when changes in the external environment require it to adapt or die.

There is a vast molecular industry that repairs or replaces damaged or mutated DNA. In a typical DNA molecule, hundreds of nucleotides are inspected every second and many nucleotide substitutions or errors corrected. The corrections are then themselves proofread, so that there is only about one error in every billion nucleotides copied. This is a standard of quality control and product reliability rarely reached in, say, publishing or automobile manufacture or microelectronics. (It is unheard of that a book this size, containing around a million letters would have no typographical errors; a 1% failure rate is common in automobile transmissions manufactured in America; advanced military weapons systems are typically down for repair some 10% of the time.) The proofreading and correction machinery devotes itself to DNA segments that are actively involved in controlling the chemistry of the cell, and mainly ignores nonfunctioning, largely untranscribed, or "nonsense" sequences.

The unrepaired mutations steadily accumulating in these normally silent regions of the DNA may lead (among other causes) to cancer and other illnesses, should the "STOP" be ignored, the sequence turned on, and the instructions carried out. Long-lived organisms such as humans devote considerable attention to repairing the silent regions; short-lived organisms such as mice do not, and often die filled with tumors.[10] Longevity and DNA repair are connected.

Consider an early one-celled organism floating near the surface of the primeval sea—and thereby flooded with solar ultraviolet radiation. A small segment of its nucleotide sequence reads, let's say,

. . . TACTTCAGCTAG . . .

When ultraviolet light strikes DNA, it often binds two adjacent T nucleotides together by a second route, preventing DNA from exer-

cising its coding function and getting in the way of its ability to reproduce itself:

$$\ldots \text{TAC}\overset{\frown}{\text{T}}\text{TCAGCTAG} \ldots$$

The molecule literally gets tied up in knots. In many organisms enzy-
matic repair crews are called in to correct the damage. There are three
or four different kinds of crews, each specialized for repairing a different kind of damage. They snip out the offending segment and its
adjacent nucleotides (C$\overset{\frown}{\text{T}}$TC, say) and replace it with an unimpaired
sequence (CTTC). Protecting the genetic information and making
sure it can reproduce itself with high fidelity is a matter of the highest
priority. Otherwise, useful sequences, tried-and-true instructions, essential for the adaptation of organism to environment, may be quickly
lost by random mutation. Proofreading and repair enzymes correct
damage to the DNA from many causes, not just UV light. They probably evolved very early, at a time before ozone, when solar ultraviolet
radiation was a major hazard to life on Earth. Early on, the rescue
squads themselves must have undergone fierce competitive evolution.
Today, up to a certain level of irradiation and exposure to chemical
poisons, they work extremely well.

Advantageous mutations occur so rarely that sometimes—especially in a time of swift change—it may be helpful to arrange for an
increased mutation rate. Mutator genes in such circumstances can
themselves be selected for—that is, those varieties with active mutator
genes serve up a wider menu of organisms for selection to draw upon,
and serve them up faster. Mutator genes are nothing mysterious; some
of them, for example, are just the genes ordinarily in charge of proofreading or repair. If they fail in their error-correcting role, the mutation rate, of course, goes up. Some mutator genes encode for the
enzyme DNA polymerase, which we will meet again later; it's in
charge of duplicating DNA with high fidelity. If that gene goes bad,
the mutation rate may rise quickly. Some mutator genes turn As into
Gs; others, Cs into Ts, or vice versa. Some delete parts of the ACGT
sequence. Others accomplish a frame shift, so the genetic code is
read, three nucleotides at a time, as usual, but from a starting point
offset by one nucleotide—which can change the meaning of everything.[11]

This is a marvel of self-reflexive talent. Even very simple microorganisms have it. When conditions are stable, the precision of reproduction is stressed; when there's an external crisis that needs attending to, an array of new genetic varieties is generated. It might look as if the microbes are conscious of their predicament, but they haven't the foggiest notion of what's going on. Those with appropriate genes preferentially survive. Active mutators in placid and stable times tend to die off. They are selected against. Reluctant mutators in quickly changing times are also selected against. Natural selection elicits, evokes, draws forth a complex set of molecular responses that may superficially look like foresight, intelligence, a master Molecular Biologist tinkering with the genes; but in fact all that is happening is mutation and reproduction, interacting with a changing external environment.

———

Since favorable mutations are served up so slowly, major evolutionary change will ordinarily require vast expanses of time. There are, as it turns out, ages available. Processes that are impossible in a hundred generations may be inevitable in a hundred million. "The mind cannot grasp the full meaning of the term of a million or a hundred million years," Darwin wrote in 1844, "and cannot consequently add up and perceive the full effects of small successive variations accumulated during almost infinitely many generations." [12]

The time scale problem was formidable when Darwin wrote. Lord Kelvin, the greatest physicist of the late Victorian age, authoritatively announced that the Sun—and therefore life on Earth—could be no more than about a hundred million (later downgraded to thirty million) years old. The fact that he provided a quantitative argument, plus his enormous prestige, intimidated many geologists and biologists, Darwin included. Is it more probable, Kelvin asked,[13] that straightforward physics was in error, or that Darwin was wrong? There was in fact no error in Kelvin's physics, but his starting assumptions were mistaken. He had assumed that the Sun shines because of meteorites and other debris falling into it. There was not the faintest hint in the physics of Kelvin's time of thermonuclear reactions; even the existence of the atomic nucleus was unknown. As late as the first decade of the twentieth century it was believed that the Earth was only 100 million years old, instead of 4.5 billion, and that the mammals had supplanted the dinosaurs only 3 million years ago, instead of 65 million.

On the basis of these misconceptions, Darwin's critics argued—properly—that even if evolution worked in principle, there might not be enough time for it to do its stuff in practice.* On an Earth created less than ten thousand years ago, it was absurd to imagine that species flowed one into another, that the slow accumulation of mutations could explain the varied forms of life on Earth. It made sense, not merely as an expression of faith, but as legitimate science, to conclude that each species must have been separately created by the same Maker who had only a moment before created the Universe.

The breakup of rocks by the waves, the transport of rock powder by the winds, lava flowing down the sides of a volcano—if the Earth is only a few thousand years old, such processes cannot have much reworked the face of our planet. But the most casual look at the landforms of Earth reveals a profound reworking. So if you imagined from biblical chronology that the world was formed around the year 4000 B.C., it made sense to be a catastrophist—and believe that immense cataclysms, unknown in our time, have occurred in earlier history. The Noachic flood, as we've mentioned, was a popular example. If, though, the Earth is 4.5 billion years old, the cumulative impact of small, nearly imperceptible changes over the course of ages could wholly alter our planet's surface.

Once the time scale for the terrestrial drama had been extended to billions of years, much that had once seemed impossible could now be readily explained as the concatenation of apparently inconsequential events—the footfalls of mites, the settling of dust, the splatter of raindrops. If, in a year, wind and water rub a tenth of a millimeter off the top of a mountain, then the highest mountain on Earth can be flattened in ten million years. Catastrophism gave way to uniformitar-

---

* Before the method of radioactive dating was invented, the physicists simply had no way to get the timescales right. Darwin's son George became a leading expert on tides and gravity—in part to refute the claim that the history of the Moon proved the Earth to be too young for much biological evolution. Several different radioactive clocks found within samples from the Earth, the Moon, and the asteroids; the abundance of impact craters on nearby worlds; and our understanding of the evolution of the Sun all independently and definitively point to an Earth about 4.5 billion years old.

ianism, championed by Lyell in geology and by Darwin in biology. The accumulation of vast numbers of random mutations was now inevitable, unavoidable. Great cataclysms were discredited and special creation became, both in geology and biology, a redundant and unnecessary hypothesis.

Many advocates of uniformitarianism denied that quick and violent biological change had ever occurred. T. H. Huxley, for example, wrote, "There has been no grand catastrophe—no destroyer has swept away the forms of life of one period, and replaced them by a totally new creation: but one species has vanished and another has taken its place; creatures of one type of structure have diminished, those of another have increased, as time has passed on."[14] In the light of modern evidence, he was right in general, right for most of the history of the Earth. But he went too far; clearly it is possible to acknowledge the importance of slow, cumulative, background change without denying the possibility of occasional global cataclysms.

In recent years it has become increasingly evident that catastrophes *have* swept over the Earth, generating vast alterations both in landforms and in life. Major worldwide discontinuities in the record in the rocks are readily explained by such catastrophes; and abrupt transitions in the forms of life on Earth, occurring in the same epoch, are naturally understood as mass extinctions, times of great dyings. (Of these, the late Permian is the most extreme example, and the late Cretaceous—when the dinosaurs were all snuffed out—the best-known). Previous ecologies are then supplanted wholesale by new teams of organisms. The fossil record shows that long periods of very slow evolutionary change are often interrupted by rarer, episodic intervals of quick change, the "punctuated equilibrium" of Niles Eldredge and Stephen J. Gould.[15] We live on a planet in which both catastrophes and uniform change have played their roles. In the purported distinction between all-at-once and slow-and-steady, as in much else, the truth embraces seemingly antithetical extremes.

The case for special creation has not been strengthened by this new balance. Catastrophism is an awkward business for biblical literalists: It suggests imperfections in either the design or the execution of the Divine Plan. Mass extinctions permit the survivors to evolve quickly, occupying ecological niches formerly closed to them by the competition. The painstaking selection of mutations continues, catastrophes or no catastrophes. But the wiping out of whole species, genera, families and orders of life, the randomness of mutation, the infelicities in the molecular machinery of life, and the slow evolutionary fiddling displayed in the fossil record—of trilobites, say, or crocodiles—all reveal a tentativeness, a hesitancy, an indecision that hardly seems consistent with the *modus operandi* of an omnipotent, omniscient, "hands-on" Creator.

————

Why are many cave fish, moles, and other animals that live in perpetual darkness blind, or nearly so? At first the question seems ill-conceived, since no adaptive reward would attend the evolution of eyes in the dark. But some of these animals *do* have eyes, only they're beneath the skin and don't work. Others have no eyes at all, although anatomically it's clear that their ancestors did. The answer seems to be that they all evolved from sighted creatures that entered a new and promising habitat—a cave, say, lacking competitors and predators. There, over many generations, no penalty is paid for the loss of eyesight. So what if you're blind, as long as you live in pitch darkness? Mutations for blindness, which must be occurring all the time (there being many possible malfunctions in the genetic instructions for vision—in eye, retina, optic nerve, and brain), are not selected against. A one-eyed man has no advantage in the kingdom of darkness.

Similarly, whales have small, internal, and wholly useless pelvises and leg bones, and snakes have four vestigial internal feet. (In the mambas of Southern Africa a single claw from each rudimentary limb breaks through the scaly skin to plain view.) If you swim or slither and never walk anymore, mutations for the withering away of feet do you no harm. They are not selected against. They might even be selected for (feet can be in the way when you're pouring down a narrow hole). Or if you're a bird that finds itself on an island devoid of predators, no penalty is levied for the steady atrophy, generation after generation, of wings (until European sailors arrive and club you all to death).

Mutations are occurring all the time for the loss of all sorts of functions. If there's no disadvantage attached to these mutations, they can establish themselves in the population. Some will even be helpful —shedding formerly useful machinery, say, that is no longer worth the effort of maintaining. There must also be enormous numbers of mutations for biochemical incompetence and other major dysfunctions which result in beings that never survive their embryonic stages. They die before they're born. They're rejected by natural selection before the biologist can examine them. Relentless, draconian winnowing is occurring all around us. Selection is a school of hard knocks.

Evolution is just trial and error—but with the successes encouraged and proliferated, the failures ruthlessly extirpated, and prodigious vistas of time available for the process to work itself out. If you reproduce, mutate, and reproduce your mutations, you *must* evolve. You have no choice in the matter. You get to keep playing the game of life only if you keep winning; that is, if you keep leaving descendants (or close relatives). One break in the train of generations, and you and your particular, idiosyncratic DNA sequences are condemned without hope of reprieve.

———

The English-language edition of this book is printed in letters that trace back to western Asia, and in a language primarily derived from Central Europe. But this is solely a matter of historical accident. The alphabet might not have been invented in the ancient Near East if there had not been a thriving mercantile culture there, if there had been no need for systematic records of commercial transactions. Spanish is spoken in Argentina, Portuguese in Angola, French in Quebec, English in Australia, Chinese in Singapore, a form of Urdu in Fiji, a form of Dutch in South Africa, and Russian in the Kuriles only because of a contingent sequence of historical events, some quite unlikely. Had they run a different course, other languages might be spoken in these places today. The Spanish, French, and Portuguese languages in turn depend on the fact that the Romans had imperial ambitions; English would be very different if Saxons and Normans had not been bent on overseas conquest; and so on. Language depends on history.

That a planet the size of the Earth is a sphere and not a cube, that a star the size of the Sun mainly emits visible light, that water is a solid *and* a liquid *and* a gas on any world at the surface temperature and pressure of the Earth—these facts are all readily understood from a few simple principles of physics. They are not contingent truths. They do not depend on a particular sequence of events that could just as well have gone some other way. Physical reality has a permanence and stability, an obsessive regularity to it, while historical reality tends to be fickle and fluid, less predictable, less rigidly determined by those laws of Nature we know. Something like accident or chance seems to play a major role in issuing marching orders to the flow of historical events.

Biology is much more like language and history than it is like physics and chemistry. Why we have five fingers on each hand, why the cross-section of the tail of a human sperm cell looks so much like that of a one-celled Euglena, why our brains are layered like an onion, involve strong components of historical accident. Now you might say that where the subject is simple, as in physics, we can figure out the underlying laws and apply them everywhere in the Universe; but where the subject is difficult, as in language, history, and biology, governing laws of Nature may well exist, but our intelligence may be too feeble to recognize their presence—especially if what is being studied is complex and chaotic, exquisitely sensitive to remote and inaccessible initial conditions. And so we invent formulations about "contingent reality" to disguise our ignorance. There may well be some truth to this point of view, but it is nothing like the whole truth, because history and biology *remember* in a way that physics does not. Humans share a culture, recall and act on what they've been taught. Life reproduces the adaptations of previous generations, and retains functioning DNA sequences that reach billions of years back into the past. We understand enough about biology and history to recognize a powerful stochastic component, the accidents preserved by high-fidelity reproduction.

———

DNA polymerase is an enzyme. Its job is to assist a DNA strand in copying itself. It itself is a protein, configured out of amino acids and manufactured on the instructions of the DNA. So here's DNA controlling its own replication. DNA polymerase is now on sale at your local biochemical supply house. There's a laboratory technique, polymerase chain reaction, which unzips a DNA molecule by changing its temperature; the polymerase then helps each strand to reproduce. Each of the copies is in turn unzipped and replicates itself.[16] At every step in this repetitive process, the number of DNA molecules doubles.

In forty steps there are a trillion copies of the original molecule. Of course, any mutation happening along the way is also reproduced. So polymerase chain reactions can be used to simulate evolution in the test tube.* Something similar can be done for other nucleic acids:

In the test tube before you is another kind of nucleic acid—this one single-stranded. It's called RNA (ribonucleic acid). It's not a double helix and does not have to be unzipped to make a copy of itself. The strand of nucleotides may loop around to join itself, tail in mouth, a molecular circle. Or it may have hairpin or other shapes. In this experiment it's sitting mixed with its fellow RNA molecules in water. There are other molecules added to help it along, including nucleotide building blocks for making more RNA. The RNA is coddled, jollied, handled with kid gloves. It's extremely finicky and will do its magic only under very specific conditions. But magic it does. In the test tube not only does it make identical copies of itself, but it also moonlights as a marriage broker for other molecules. Indeed, it performs even more intimate services, providing a kind of platform or marital bed for oddly shaped molecules to join together, to fit into one another. It's a jig for molecular engineering. The process is called catalysis.

This RNA molecule is a self-replicating catalyst. To control the chemistry of the cell, DNA has to oversee the construction of factotums—a different class of molecules, proteins, which are the catalytic machine tools we've been discussing above. DNA makes proteins because it can't catalyze on its own. Certain kinds of RNA, though, can themselves serve as catalytic machine tools.[17] Making a catalyst or *being* a catalyst gives you the biggest return for the smallest investment: Catalysts can control the production of millions of other molecules. If you make a catalyst, or if you *are* a catalyst—the right kind of catalyst—you have a long lever arm on your destiny.

---

* The technique is also being used to take tiny quantities of DNA from the remains of ancient organisms—bacteria from the gut of a preserved mastodon, for example— and make enough copies so they can be studied. It has even been proposed that preserved somewhere in amber may be the remains of a bloodsucking insect that bit a dinosaur, from which we may one day learn about dinosaur biochemistry or even—this point is keenly debated—reconstruct, and in a way resuscitate, dinosaurs extinct for 100 million years. In the best of circumstances, this does not seem to be a prospect for the near future.

Now in these laboratory experiments, which are being carried out in our time, imagine many generations of RNA molecules more or less identically replicating in the test tube. Mutations inevitably occur, and much more often than in DNA. Most of the mutated RNA sequences will leave no, or fewer, copies, again because random changes in the instructions are rarely helpful. But occasionally a molecule comes into existence that aids its own replication. Such a newly mutated RNA might replicate faster than its fellows or with greater fidelity. If we were uncaring about the fates of individual RNA molecules—and while they may arouse feelings of wonder, they seldom elicit sympathy—and wished only for the advancement of the RNA clan, this is just the kind of experiment we would perform. Most lines would perish. A few would be better adapted and leave many copies. These molecules will slowly evolve. A self-replicating, catalytic RNA molecule may have been the first living thing in the ancient oceans about 4 billion years ago, its close relative DNA being a later evolutionary refinement.

In an experiment with synthetic organic molecules that are *not* nucleic acids, two closely related species of molecules are found to make copies of themselves out of molecular building blocks provided by the experimenter. These two kinds of molecules both cooperate and compete: They may aid each other's replication, but they are also after the same limited pool of building blocks. When ordinary visible light is made to shine on this submicroscopic drama, one of the molecules is observed to mutate: It changes into a somewhat different molecule that breeds true—it makes identical copies of itself, and not its pre-mutation ancestor. This new variety, it turns out, is much more adept at replicating itself than the other two hereditary lines. The mutant line rapidly out-competes the others, whose numbers precipitously fall.[18] We have here, in the test tube, replication, mutation, replication of mutations, adaptation, and—we do not think it is too much to say—evolution. These are not the molecules that make us up. They are probably not the molecules involved in the origin of life. There may well be many other molecules which reproduce and mutate better. But what prevents us from calling this molecular system alive?

Nature has been performing similar experiments, and building on its successes, for 4 billion years.

Once even crude replication becomes possible, an engine of enormous powers has been let loose into the world. For example, consider that primitive organic-rich ocean of the Earth. Suppose we were to drop a single organism (or a single self-replicating molecule) into it, considerably smaller than a contemporary bacterium. This tiny being divides in two, as do its offspring. In the absence of any predators and with inexhaustible food supplies, their numbers would increase exponentially. The being and its descendants would take only about one hundred generations to eat up all the organic molecules on Earth. A contemporary bacterium under ideal conditions can reproduce once every fifteen minutes. Suppose that on the early Earth the first organism could reproduce only once a year. Then in only a century or so, all the free organic matter in the whole ocean would have been used up.

Of course, long before that, natural selection would be brought to bear. The genre of selection might be competition with others of your kind—for example, for foodstuffs in an ocean with dwindling stocks of preformed molecular building blocks. Or it might be predation—if you don't look out, some other being will mug you, strip you down, pull you to pieces, and use your molecular parts for its own ghastly purpose.

Major evolutionary advance might take considerably more than one hundred generations. But the devastating power of exponential replication becomes clear: When the numbers are small, organisms may only infrequently come into competition; but after exponential replication, enormous populations are produced, stringent competition occurs, and a ruthless selection comes into play. A high population density generates circumstances and elicits responses different from the more friendly and cheerful lifestyles that pertain when the world is sparsely populated.

The external environment is continuously changing—in part because of the enormous population growth when conditions are favorable, in part because of the evolution of other organisms, and in part because of the ticking geological and astronomical clockwork. So there's never such a thing as a permanent or final or optimum adaptation of a lifeform to "the" environment. Except in the most protected and static surrounds, there must be an endless chain of adaptations. However it feels on the inside, it might very well be described from the outside as a struggle for existence and a competition between adults to ensure the success of their offspring.

You can see that the process tends to be adventitious, opportunistic —not foresighted, not with any future end in view. The evolving molecules do not plan ahead. They simply produce a steady stream of varieties, and sometimes one of the varieties turns out to be a slightly improved model. No one—not the organism, not the environment, not the planet, not "Nature"—is mulling the matter over.

This evolutionary shortsightedness can lead to difficulties. It might, for example, cast aside an adaptation that is perfectly suited for the next environmental crisis a thousand years from now (about which, of course, no one has a glimmering). But you have to get from here to there. One crisis at a time is life's motto.

ON IMPERMANENCE

If we lived forever, if the dews of Adashino never vanished, if the crematory smoke on Toribeyama never faded, men would hardly feel the pity of things. The beauty of life is in its impermanence. Man lives the longest of all living things . . . and even one year lived peacefully seems very long. Yet for such as love the world, a thousand years would fade like the dream of one night.

KENKO YOSHIDA, *Essays in Idleness* (1330–1332)[19]

Chapter 6

US AND THEM

Let there be no strife, I pray thee, between me
and thee . . . for we be brethren.

Genesis 13:8

There are no compacts between lions and men.

H O M E R, *The Iliad*[1]

Whether there were many instances of the origin of life on Earth or only one is a deep and perhaps impenetrable mystery. For all we know, there may have been millions of dead ends and false starts, unmourned ancient genealogies snuffed out as new ones arose. But it seems very clear that there's only one hereditary line leading to all life *now* on Earth. Every organism is a relative, a distant cousin, of every other. This is manifest when we compare how all the organisms on Earth do business, how they're built, what they're made of, what genetic language they speak, and especially how similar their blueprints and molecular job orders are. All life is kin.

In our imagination, let's cast our eyes back to the earliest organisms. They could not have been so purebred and pampered a line of self-replicating molecules as contemporary DNA or RNA—superbly efficient in the replication and proofreading of their messages, but reproducing only under the meticulously controlled conditions upon which modern organisms insist. The first living things must have been rough-and-ready, slow, careless, inefficient—just barely good enough to make crude copies of themselves. Good enough to get started.

At some point, probably extremely early on, organisms had to be more than a single molecule, no matter how talented that molecule might be. For very precise instructions to be followed to the letter, for reproduction to occur with high fidelity, other molecules were needed —to scour building blocks from the adjacent waters and bend them to your purpose; or, like DNA polymerase, to be midwife in the replication process; or to proofread a newly minted set of genetic instructions. But it did you no good if such accessory molecules kept drifting out to sea. What you needed was a kind of trap to keep useful molecules captive. If only you could surround yourself with a membrane that, like a one-way valve, lets in the molecules you need and doesn't let them out . . . There are molecules that do that—that, for example, are attracted to water on one side of them, but are repelled, absolutely revolted by water on the other. They're common in Nature. They tend to make little spheres. And they're the basis of cell membranes today.

The earliest cells, although able simultaneously to multiply and divide, could not possibly have been conscious in anything like the sense that humans are. Still, they had certain behavioral repertoires. They knew how to copy themselves, of course; how to convert molecules from the outside, different from them, into molecules on the inside that *were* them. They were preoccupied with improvements in the precision of replication and the efficiency of metabolism. Some could even distinguish sunlight from darkness.

Breaking down molecules taken in from the outside, that is, digesting food, can be done safely only in a step-by-step fashion, each step controlled by a given enzyme, and each enzyme controlled by its own ACGT sequence, or gene. The genes then must work together in exquisite harmony; otherwise none of them will propagate into the future. In digesting a molecule of sugar, for example, the meticulously choreographed action of dozens of enzymes is required, each picking up where the last one left off, each enzyme manufactured by a particular gene. The defection of a single gene from the common enterprise can be fatal to all of them. An enzyme chain is only as strong as its weakest link. On this level, genes are single-mindedly dedicated to the general welfare of their tribe.

Early enzymes had to be discriminating; they had to take care not to decompose the very similar molecules that constituted the lifeform they were part of. If you digest yourself—the sugars that are part of your DNA, say—you don't leave many descendants. If you don't digest others—convenient repositories of organic raw materials and finished molecular goods—you may not leave many descendants either. Cells of 3.5 billion years ago must have possessed some knowledge of the difference between "me" and "you." And "you" was more expendable than "me." A dog-eat-dog or, at least, a microbe-eat-microbe world. But wait . . .

A time came—perhaps 2 or 3 billion years ago—when one being could incorporate another whole. One would nuzzle up to the other, the cell walls or membranes would pucker, and the littler fellow would find itself inside the bigger. Attempts at digestion, with varying success, doubtless ensued. Suppose you are a largish one-celled organism in the primitive oceans who in this way gobbles up some photosynthetic bacteria, tiny specialists who know how to use sunlight, carbon dioxide, and water to manufacture sugars and other carbohydrates. You'll leave more descendants if you're better than your competitors in acquiring sugar (a key building block needed to replicate your genetic instructions and to power all you do).

But suppose also that these ingested bacteria—the latest, sturdy, rustproof models—do not succumb to your digestive enzymes. For all they know, they've found their way into a molecular Garden of Eden. You protect them from many of their enemies; because you're transparent, sunlight shines into you for them; and there's plenty of water and carbon dioxide around. So inside you, the bacteria continue to do their photosynthetic magic. Some sugars leak out of them, for which you are grateful. Some of them die and their interior molecules spill out, available for your use. Others of them flourish and multiply. When the time comes for you to reproduce, some of them wind up inside your offspring. Not yet *de jure* (because nothing of this arrangement is yet encoded in the nucleic acids), but certainly *de facto*, an accommodation has been reached between your descendants and theirs. [2]

It's a good deal for both parties. They open up a little fast-food concession stand inside your body, at hardly any cost to you. You provide a stable and protected environment for them (so long as you take care not to digest your guests). After many generations have passed, you've evolved into quite a different kind of being, with little green photosynthetic power plants inside of you reproducing when you do, clearly part of you, but also clearly different. You've become a partnership. This seems to have happened a half dozen times or more in the history of life, each instance leading to a different major group of plants. [3]

Today every green plant contains such inclusions, called chloroplasts. They are still rather like their free-living one-celled bacterial ancestors. Nearly every bit of green in the natural world is due to chloroplasts. They are the photosynthetic engines of life. We humans pride ourselves on being the dominant lifeform on this planet, but these tiny beings—unobtrusive, the perfect guests—are in a sense running the show. Without them, almost all life on Earth would die.

They've made many concessions to their hosts. They've achieved a working mutual assistance pact of long duration, called symbiosis. Each partner relies on the other. Still, the chloroplasts are recognizably a latecomer to the cell. The clearest sign of their separate origin is the difference between their nucleic acids and the plant cell's own nucleic acids, although long ago they had a common ancestor. The signature of their separate, early evolution before joining forces is plain. The original chloroplast seems to have come from a photosynthetic bacterium very much like those living in stromatolite communities today. [4]

———

You look at these little one-celled beings under the microscope and you're struck by their apparent self-assurance. They seem to know with such certainty what they're about. They swim toward the light or attack prey or struggle to escape from predators. Because they're transparent, you can see their internal parts, the DNA-driven protoplasmic clockwork, making them go. Their ability to transmute the food they come across into the molecules they need—for energy, for parts, for reproduction—is downright alchemical. The plants among them convert air and water and sunlight into themselves not haphazardly, but according to specific recipes, the mere writing out of which would fill many volumes on organic chemistry and molecular biology. Each of them is only one cell; no organs, no brains, no snappy conversation, no poetry, no higher spiritual values—and yet they can do, without any apparent conscious awareness, far more along these chemical lines than can our vaunted technology.

And there's something else they can do that we can't. They can live forever. Or nearly so. These asexual, one-celled organisms reproduce by fission—not nuclear, but biological fission. A little furrow, an indentation, appears and ripples down the middle of the organism. The internal parts are divided more or less evenhandedly, and suddenly we have before us not one organism but two. It has split in half. We now see two smaller beings, each nearly identical to its single parent and genetically the same, identical twins. Quickly, each grows to adult size. Later, the process continues. Except for the odd mutation, remote descendants are perfect facsimiles of their ancestors. In a real sense, the ancestors never died. At no point along the way are there corpses of aged parents. If there are no accidents, no drop of poison released by other microbes, no extremes of temperature, no running out of food, no encounters with a big, bad amoeba, then they continue to live on, the natural slow falling to pieces of their organic body parts mitigated or reversed by their frequent reproduction.

These ubiquitous, invisible, and most humble organisms are immortal—at least by human standards. There are enough natural vicissitudes that they cannot go for too long without encountering one disaster or another. But at least some of them live for more lifetimes than the most extravagant and credulous disciple of reincarnation or "multiple life regression" ever imagined. The current official record is held by a laboratory stock of the one-celled organism called paramecium, familiar to high-school biology students. Eleven thousand successive generations of paramecia have been carefully nurtured in the test tube, with no senescence or aging apparent. [5] (In humans, eleven thousand generations would take us all the way back to the dawn of our species.) Except for the slow buildup of mutations, the paramecia at the end of this train of generations were genetically identical to those at the beginning. In a way, the longing for immortality, so characteristic of Western civilization, is a longing for the ultimate regression into the past—to our single-celled ancestors in the seething primeval ocean.

———

We have, so far in this saga, not come within a billion years of our epoch. But even in so remote a time, many of the major themes and variations of present life on Earth had been clearly articulated. Some of the fossils of that time are indistinguishable in form from some contemporary organisms, the stromatolites being the most famous example. Others are wildly different. There has surely been a growing biochemical sophistication over the aeons, in enzyme chemistry, fidelity of DNA replication, and many other matters that must be indetectable in mere fossils; still, it seems astonishing that any organism should be unchanged—even if just in its gross anatomy—over 3.5 billion years. We can recognize again a stolid conservatism in living things. And yet quick and fundamental change sometimes happens. The picture that emerges is of a rich menu of candidate adaptations offered up by mutation for consideration by natural selection. But only under sentence of death (or what in the evolutionary perspective is the same thing, the threat of no descendants) are these mutational propositions seriously taken up and tried out. Except for cosmetic touches, new kinds of life are ordinarily discouraged. Change is grudging.

You can see the same classes of molecules used over and over again for completely different purposes. Today, for example, the same complex organic molecule is used, with minor variations, as the green pigment that sips sunlight in plants; as the red pigment that carries oxygen through the bloodstreams of animals; as the agent that makes shrimps and flamingos pink; and in a widely used enzyme that helps wheedle energy safely out of sugar. The energy is banked, against future need, in molecules nearly identical to the nucleotides A, C, G, and T of the genetic code. While these are molecules of breathtaking versatility, their repeated use and recycling reveals parsimony as a way of life.

It's as if, for every million dyed-in-the-wool conservative organisms, there's one radical who's out to change things (although usually very small things); and for every one of the radicals, only one in a million actually knows what it's talking about—providing a significantly better survival plan than the one currently fashionable. And yet the evolution of life is determined by these revolutionaries.

Given enough food, microorganisms reproduce so quickly that they can evolve in the time between putting them on a shelf for storage and retrieving them for further examination. The speed with which bacteria "acquire" resistance to antibiotics cautions restraint about prescribing them too frequently. The antibiotic does not usually induce adaptive mutations; instead, it acts as a fierce agent of selection, killing off all bacteria except a favored few that, by chance, are immune to the medicine—a strain that earlier, for other reasons, might not have competed successfully with its fellows. The fact that bacteria quickly evolve resistance to antibiotics (or insects to DDT) reflects the enormous diversity of forms and biochemistries always churning subsurface in the microbial world. There is a continuing war of measure and countermeasure, raging between host and parasite—in this case, between the pharmaceutical companies, generating new antibiotics, and the microbes, generating new resistant strains to replace their more vulnerable ancestors.

———

Well-developed even by 3.5 billion years ago, we've argued, was the distinction between the inside and the outside, me and you, us and them, a rudimentary consciousness of self. If you're in the habit of eating organic molecules dissolved in the primeval oceans, you're also used to eating the molecules that make up other beings; after all, they're the same molecules. But then you'd better take care you don't eat yourself. You may not have pity or compassion for other organisms. That's probably not how a microbe views the world. But you must make some fine distinctions. You may lack sentimental feelings for your chloroplasts, but if you digest them, you're in trouble. If the distinction is too difficult for you to make—if you can't figure out the difference between "me" and "you," if you can't control your digestive enzymes—then you'll leave fewer offspring, or none. There's not yet any thinking through. There may be no feelings of any sort. Nevertheless, organisms are beginning to *behave* as if they had wants, needs, preferences, emotions, drives, instincts.

If you're living in a group, it will help neither them nor you if you set about eating your fellows. You may be a ruthless, implacable predator, but you must also be a pushover for your relatives and neighbors. So all of you may suffuse your outer membranes with a chemical that serves for species recognition. When you taste this molecule emanating from another microbe, you become very affable. "Friend," the chemical says. "Sister." Other chemicals carry different information. Some bacteria routinely produce their own chemical warfare agents, antibiotics that are harmless to themselves and others of their own strain, but deadly to bacteria of different strains, foreigners. A delicate balance has evolved between hostility to the outside group and cooperation with the inside group. Them and us. The first intimations of xenophobia and ethnocentrism evolved early.

Big carnivores enjoy their work. (One-celled carnivores may also.) They don't hunt because they have an academic knowledge of nutrition: They hunt, it seems, because hunting is a delight; because stalking, chasing, maiming, killing, dismembering, and eating are the pleasures of life; because the urge to do so is irresistible. Fat cats and lazy dogs, stuffed with hors d'oeuvres, their gustatory needs provided for, nevertheless sometimes heed an ancient call, and the urban pet owner finds a dead mouse or pigeon proudly laid at her feet. The machinery is hardwired; the computer is preprogrammed. An appropriate stimulus can set it off. Its hunting proclivities finding no other outlet, the dog fetches a stick or a Frisbee, and the cat swats at a cobweb or pounces on a ball of wool.

Even so formidable and elegant an example of hardwiring as a cat hunting a rat, though, depends a great deal on past experience. In a set of classic experiments, the psychologist Z. Y. Kuo[6] showed that almost all kittens who witness their mother killing and eating a rodent eventually do so themselves. However, when kittens are *raised* in the same cage with a rat, never seeing any other rat, and never seeing a cat kill a rat, then they almost never kill rats themselves. When kittens have a rat for a littermate and also witness their mothers killing rats *outside* the cage, about half of them learn to kill—but they tend to kill only the kinds of rat they had seen their mothers kill, and not the kinds that they grew up with. Finally, when kittens are given an electric shock each time they see a rat, they soon learn not to kill rats—indeed, to run in terror from them.

So even such basic hardwiring as the predation program in cats is malleable. Of course humans are not cats. But we might be tempted nevertheless to guess that childhood experience, education and culture can do much to mitigate even deep inborn proclivities.

Starting with the early microbes, the behavioral machinery for hunting and escaping, and for altering these inclinations according to experience, were developing. Predators slowly evolved into larger, faster, and smarter models, with new options (for example, feinting). Potential prey likewise evolved larger, faster, and smarter models with other options (for example, "playing dead")—because those who didn't were more often eaten. Many strategies were devised; the successful ones were retained: protective camouflage, body armor, ink or sprayed noxious liquids to cover an escape, poisonous stings, and exploiting niches where there were as yet no predators—a shallow hole in the ocean floor, perhaps, or a sanctuary in a seashell, or a homestead on an untenanted island or continent. Another strategy was simply to produce so many progeny that at least some survive. Again, no potential prey plans such adaptations; it's just that after a while the only prey left are the ones who act as if they *had* planned it all out. No matter how fine your intentions, how benign and contemplative your inclinations, if you're potential prey you're forced by natural selection into adopting countermeasures.

By around 600 million years ago, many multicellular animals started walling themselves in, surrounding their soft bodies with shells and carapaces, learning to do small-scale civil engineering, building defenses out of silicate and carbonate rock. Lifestyles of clams, oysters, crabs, lobsters, and many other armored animals, some now extinct, developed then. Since, with rare exceptions, soft parts of dead animals decompose quickly and hard parts or their imprints survive longer— sometimes even long enough to be noticed by paleontologists hundreds of millions of years in the future—the evolution of body armor made these distant creatures knowable to their remote collateral relatives.

The warfare between predator and prey extends to the plant kingdom as well. Plants load themselves with poisons to discourage animals from eating them. The animals evolve detoxification chemistry and special organs—the liver, most prominently—to keep pace with the plants. What we like about coffee, for example, are the toxins that have evolved to deter insects and small mammals from consuming coffee beans.[7] But we have sophisticated livers.

Of course, predators need not be bigger than their prey. Disease microbes can be formidable predators—not only attacking and eventually killing the organisms that bear them, but also taking over their hosts, changing their behavior to spread the disease microorganisms to other hosts. One of the most striking examples is the rabies virus. On being injected into the bloodstream of a placid, people-loving dog, they head straight for the limbic system of the dog's brain, where the control buttons for rage reside. There, they set about converting the poor animal into a marauding, snarling, vicious predator that now bites the hand that feeds it. Rabid animals are afraid of no one. At the same time, other rabies viruses are dispatched to inactivate the nerves for swallowing, to put the saliva-manufacturing machinery into overdrive, and to invade the saliva in huge numbers. The dog is furious, although it has no idea why. A pawn of the viruses within it, it's helpless to resist the impulse to attack. If the attack is successful, the viruses in the dog's saliva enter the bloodstream of the victim through the lesion or laceration, and then set about taking over this new host. The process continues.

The rabies virus is a brilliant scenarist. It knows its victims, and how to pull their strings. It circumvents their defenses—infiltrating, outflanking, accomplishing a coup d'état within beings so much larger, you might have thought them invulnerable.*

---

* Humans are newly evolved. Our availability on a global scale as hosts for parasites is very recent. In the absence of medical countermeasures, we might expect, sometime

In influenza or the common cold, it's not an incidental adjunct of the infection that we cough and sneeze, but rather central to the proliferation of the virus responsible, and under its control. Some other examples of microbes pulling the strings:

> A toxin produced by the cholera bacterium interferes with reabsorption of liquid from the bowel, thus resulting in profuse diarrhea that spreads the infection . . . Tobacco mosaic virus causes its host to enlarge cell membrane pores so that the virus can pass through to uninfected cells . . . A lancet fluke is effectively transmitted from ants to sheep because it induces an infected ant to climb to the top of a blade of grass and grab on, never to let go. A fluke causes snail hosts to crawl to exposed sections of beach where they are easy prey for the gulls that are the next host in the life cycle.[8]

Over many generations of life-and-death interaction between predator and prey, a kind of permanent arms race is established. For every offensive advance there is a defensive counter, and vice versa. Measure and countermeasure. Rarely does anyone become safer.

Some prey grow up together, swarm together, school together, herd together, flock together. There's safety in numbers. The strongest can be brought in to intimidate or defend against a large predator. The attacker can be mobbed by the entire group of prey. Lookouts can be posted. Danger calls can be agreed upon and coordinated, escape strategies chosen. If the prey are quick, they can dart before the predator, outrace and confuse it, or draw it away from especially vulnerable members of the group. But there is also a selective advantage for cooperation among the *predators*—for example, one group flushing prey toward another that lies in ambush. For prey and predator alike, community life may be more rewarding than solitude.

To play the escalating evolutionary game of predator and prey, complex behavioral repertoires are eventually needed. Each must detect the other at a distance, and a high premium is established on supplanting local senses such as touch and taste by more long-range senses such as smell, sight, hearing and echo-location. A talent for

---

in the future, the evolution of new kinds of microorganisms that pull our strings more artfully than any rabies virus could ever do.

remembering the past develops in the heads of small animals. Some simple cases of contingency planning, imagining what your response might be to a variety of circumstances ("I'll do Z if it does A; I'll do Y if it does B") may already have been in the genes; but expanding that talent into more complex branched contingency trees, new logic for future needs, greatly aids survival. Indeed, to find and eat anyone— even organisms that take *no* evasive action—requires, especially when the supply is sparse, a predator to know a great deal.

Basing all your behavior on a pre-programmed set of instructions written in the ACGT language places no undue demands—as long as the environment is the one you were evolved for. But no pre-programmed set of instructions, no matter how elaborate, no matter how successful in the past, can guarantee continuing survival in the face of rapid environmental change. Evolution through natural selection involves only the most remote, generalized, almost metaphorical kind of learning from experience. Something else is needed. When you hunt food; when mobility is high and organisms can roam among very different environments; when social relations with your own kind as well as predator/prey interactions become intricate; when you're required to process enormous amounts of information about the external world—at such times, especially, it pays to have a brain. With a brain you can remember past experiences and relate them to your present predicament. You can recognize the bully who picks on you and the weakling you can pick on, the warm burrow or protected rock crevice to which you have safely fled before. Opportunistic scenarios for gathering food, or hunting, or escaping may occur to you at a critical moment. Neural circuitry develops for data processing, pattern recognition, and contingency planning. There are premonitions of forethought.

The style of evolution of brains—and much else—is not usually a matter of steady progression. Instead, the fossil record speaks of short periods of rapid and radical evolution, punctuating immense periods of time in which the sizes of brains hardly change at all. This seems true from the evolution of the earliest mammals to the evolution of our own species.[9] It's as if there's a rare concatenation of events— perhaps changes in the DNA sequence and the external environment together—that provides an adaptive opportunity. The new environmental niches are quickly filled, and for a long time subsequent evolution is devoted to consolidating the gains. Major advances in neural architecture—in the brain's ability to process data, to combine information from different senses, to improve its model of the nature of the outside world, and to think things through—may be very expensive. For many animals these are such broad-gauge talents, requiring so many separate evolutionary steps, that the major benefits may come only in the far future, while evolution is transfixed by the here-and-now. Nevertheless, even tiny advances in thinking are adaptive. Spurts in brain size have happened sufficiently often in the history of life for us to conclude, from this fact alone, that brains are useful to have around.

Feeling, in mammals at least, is mainly controlled by lower, more ancient parts of the brain, and thinking by the higher, more recently evolved outer layers.[10] A rudimentary ability to think was superimposed on the pre-existing, genetically programmed behavioral repertoires—each of which probably corresponded to some interior state, perceived as an emotion. So when unexpectedly it is confronted with a predator, before anything like a thought wells up, the potential prey experiences an internal state that alerts it to its danger. That anxious, even panicky state comprises a familiar complex of sensations, including, for humans, sweaty palms, increased heartbeat and muscle tension, shortened breath, hairs standing on end, a queasiness in the belly, an urgent need to urinate and defecate, and a strong impulse either for combat or retreat.* Since in many mammals fear is produced by the same adrenaline-like molecule, it may feel pretty much the same in all of them. That's at least a reasonable first guess. The more adrenaline in the bloodstream, up to a certain limit, the more fear the animal feels. It's a telling fact that you can artificially be made to feel this precise set of sensations just by being injected with some adrenaline—as sometimes happens at the dentist's (to shorten the clotting time of your blood, another useful adaptation when you're confronting a predator. Of course you may also be generating some of your own adrenaline at the dentist's.) Fear *has* to have an emotion tone about it. It *has* to be unpleasant.

---

* It's not hard to see how the components of this "fight-or-flight" response are all adaptive—evolved to get you through the crisis. That feeling of cold and emptiness at the pit of your stomach, for example, results from a reallocation of blood from digestion to the muscles.

If the predator's eye/retina/brain combination is geared especially to detect motion, the prey often have, in their repertoire of defenses, the tactic of standing frozen, stock-still, for long periods of time. It's not that squirrels, say, or deer understand the physiology of their enemies' visual systems; but a beautiful resonance between the strategies of predator and prey has been established by natural selection. The prey animal may run; play dead; exaggerate its size; erect its hairs and shout; produce foul-smelling or acrid excretions; threaten to counterattack; or try a variety of other, useful survival strategies—all without conscious thought. Only then may it notice an escape route or otherwise bring into play whatever mental agility it possesses. There are two nearly simultaneous responses: one, the ancient, all-purpose, tried-and-true, but limited and unsubtle hereditary repertoire; and the other, the brand-new, generally untried intellectual apparatus— which can, however, devise wholly unprecedented solutions to urgent current problems. But large brains are new. When "the heart" counsels one action and "the head" another, most organisms opt for heart. The ones with the biggest brains more often opt for head. In either case, there are no guarantees.

———

Obliged to accommodate to every twist and turn in the environment they depend upon, living things evolve to keep up. By painstaking, small steps, through the passage of immense vistas of geological time, via the deaths of innumerable slightly maladapted organisms, uncomplaining and unlamented, life—in its interior chemistry, external form, and menu of available behavior—became increasingly complex and capable. These changes, of course, are reflected in (indeed, caused by) a corresponding elaboration and sophistication of the messages written in the ACGT code, down there at the level of the gene. When some splendid new invention comes along—bony cartilage as body armor, say, or the ability to breathe oxygen—the genetic messages responsible proliferate across the biological landscape as the generations pass. At first no one has these particular sequences of genetic instructions. Later, large numbers of beings all over the Earth live by them.

It's not hard to imagine that what's *really* going on is an evolution of genetic instructions, battles between the genetic instructions of competing organisms, genetic instructions calling the shots—with the plants and animals little more, or maybe nothing more, than automata. The genes arrange for their own continuance. As always, the "arranging" is done with no forethought; it's merely that those beautifully coordinated genetic instructions that, by chance, give superior orders to the living thing they inhabit make more living things motivated by the same instructions.

Think again of the changes in our behavior caused by the incursion of a rabies or an influenza virus (made of nucleic acids wearing a coat of protein). Surely much more profound control over us is exercised by our *own* nucleic acids. When you strip away the fur and feathers, the physiological and behavioral particularities, life is revealed to be the preferential replication of some ACGT messages rather than other, competing messages; a conflict of genetic recipes; a war of words.

In this perspective,[11] it's the genetic instructions that are being selected and that are evolving. Or you might with nearly equal justice say it's the individual organisms, under the tight control of the genetic instructions, that are being selected and that are evolving. There is no room here for group selection—the natural and attractive idea that species are in competition with one another, and that individual organisms work together to preserve their species as citizens work together to preserve their nation. Acts of apparent altruism are instead attributed chiefly to kin selection. The mother bird slowly flutters from the fox, one wing bent as if broken, in order to lead the predator away from her brood. She may lose her life, but multiple copies of very similar genetic instructions will survive in the DNA of her chicks. A cost-benefit analysis has been made. The genes dictate to the outer world of flesh and blood with wholly selfish motives, and real altruism —self-sacrifice for a non-relative—is deemed a sentimental illusion.[12]

This, or something quite like it, has become the prevailing wisdom in the field of animal (and plant) behavior. It has considerable explanatory power. At the human level it helps to explain such varied matters as nepotism and the fact that foster children are much more likely (in America, for example, about a hundred times more likely[13]) to be fatally abused than children living with their natural parents.

The cooperation of the cells in stromatolites and other colonial organisms can be understood as selfish at the level of the gene, since they're all close relatives. Cooperation of the chloroplast and the cell with which it forms a symbiotic attachment—is this selfish? The cell that eats its chloroplasts is at a competitive disadvantage. It refrains from eating them not because it has even a glimmer of altruistic feeling for the chloroplasts, but because it's dead without them. It forgoes the pleasures of a chloroplast meal for a substantial future benefit. It exercises restraint on short-term, selfish behavior. It practices impulse control. Selfishness still prevails, but we are made aware of the distinction between short- and long-term selfishness.

For most social animals, and for obvious reasons, the animals you grow up with tend to be close relatives. So if you cooperate, if you show what superficially might seem like altruism, it's naturally directed toward close kin and can therefore be explained as kin selection. An organism might forego its own replication, for example, and devote its life to improving the chances of the survival and reproduction of close relatives—those with very similar DNA sequences. If all that counts is which sequences will be widely represented in the life of the future, those species with a flair for altruism might do well. They can help ensure that much of their genetic *information* is passed on, even if none of their atoms wind up in the bodies of the next generation.[14]

The geneticist R. A. Fisher described heroism as a predisposition inclining its bearer toward "an increased probability of entering an occupation not easily to be reconciled with family life." Nevertheless, Fisher argued, heroism—in humans or in other animals—might carry a selective advantage by preserving the very similar genetic sequences of close relatives, enabling such sequences to be passed on to future generations. This is one of the first clear articulations of kin selection. Parents sacrificing themselves for a child can be understood on similar grounds. The hero or the devoted parent will be doing simply what feels "right," without attempting any calculus of benefit versus risk to the gene pool. But the reason it feels "right," Fisher proposed, is that extended families characterized by conscientious parenting and heroes aplenty will tend to do very well.*

Animals may be willing to make real sacrifices for close relatives, but not for more distant kin. Think of it this way: Imagine sleeping soundly at night, knowing that your children are starving, homeless, or gravely ill. For almost all of us, it would be unthinkable. But forty

---

* True, of course, only for sexual organisms. Asexual beings, reproducing by splitting in two, cannot enhance the fitness of their descendants through a spirit of self-sacrifice.

thousand children die each day of easily preventable hunger, neglect, or disease. Institutions such as the United Nations Children's Fund are in place that could save these children—with innoculations against illness, with a few cents a day worth of salts and sugar. But the money is unavailable. Other needs are deemed more pressing. The children continue to die while we sleep well. They are far away. They are not ours. Now tell us you don't believe in the reality of kin selection.

Still, if you find yourself among others of your own species who are not your near kin, surely it is to your advantage to cooperate against a common enemy. You can draw upon behavior evolved for kin selection in order that a group of animals not closely related can cohere and survive.* And if altruism is one of your talents, you might find yourself practicing it even on animals of another species. Dogs are known to risk their lives to save humans—surely no close relatives. Nor does the hope of future reward explain their behavior.

How are we to understand well-attested cases of dolphins saving drowning humans by repeatedly nuzzling them up to the surface and pushing them toward shore? Is the dolphin unable to distinguish the thrashing human from an infant dolphin in trouble? This is highly unlikely; dolphins are discerning observers. What about cases of abandoned or strayed human infants being raised by wolf mothers that have lost their pups, or birds of a different species brooding cuckoo eggs? Why do drivers swerve to avoid hitting a dog on the road, although they thereby put their own children in the back seat at increased risk? What about youngsters dashing back into the burning house to rescue the cat? Such cases of courage and care directed to other species may derive from a misdirected kin selection, but they do happen and they do save lives. Shouldn't we then expect to find altruistic behavior much more frequently directed toward other members of the *same* species, even if they're not close relatives?

Consider two groups, one composed of unrelentingly selfish individualists, the other of solid citizens who are occasionally willing to sac-

---

* Humans do this routinely. Large multi-ethnic states are revealingly called "fatherland" or "motherland." Leaders encourage patriotic fervor—the word "patriotic" comes from the Greek for father. Especially in monarchies, it was easy to pretend that the nation was a family. The distant and powerful king was like many fathers. Everyone understood the metaphor.

rifice themselves for (even distantly related) others. Against a common enemy, can we not imagine circumstances in which the latter group fares better than the former? Obvious disadvantages also accrue to a community of strict altruists constantly throwing their lives away in order to benefit total strangers. Such a group would not last long—if only because any tendency toward selfishness would quickly spread.

What if there's a critical size for the group to work? When membership is below some rough threshold, certain functions of the group begin to fail. For example, the bigger the group, the better huddling together for warmth works,[15] or mobbing a predator;[16] and below a certain size, group benefits become increasingly unavailable. It's not hard to imagine wholly selfish genes that cause defections from community service—a refusal to mob a predator, say, because it might be dangerous. If these genes proliferate, the point will be reached where almost nobody has the gumption to mob, and the danger posed to everyone by predators has increased. Thus, for longer-term reasons that are selfish at the level of the genetic instructions, short-term altruism may be adaptive, and might be selected for—even if the members of the group are not near relatives. In closely knit communities, individual selection and what looks very much like group selection are both elicited.

Many examples thought to demonstrate group selection have, with an almost maddening ingenuity, been explained at least equally well by a new school of biologists and game theorists. Some explanations seem quite plausible, but not all. For example, when a predator threatens a group of Thomson gazelles, one or two may leap in conspicuous high arcs near the predator. This is called stotting. The group selectionist view is straightforward: The individual calls attention to itself and risks being eaten in order to save the group. (But suppose stotting were never invented; could the predator eat more than one Thomson gazelle anyway? Compared to other species of gazelles ignorant of stotting, *are* fewer eaten thanks to stotting?) The prevailing individual selectionist view is that the stotter is advertising its own gymnastic abilities and reminding the predator that less athletic gazelles are easier to eat. It stots for crassly selfish reasons.[17] (But then why don't *most* Thomson gazelles stot when stalked? Why doesn't such selfishness spread through the herd? Does the predator in fact turn its attention from the stotter to a less conspicuous gazelle?)

Like the classic optical illusions—is it a candelabra, or two faces in profile?—the same data can be understood from two quite different perspectives (although neither may be fully satisfying). Each may have its own validity and utility.[18] Individual selection and group selection must ordinarily go together (or, in scientific speech, be highly correlated); otherwise evolution would never occur. We might argue that individual selection must have some primacy, because you can have individuals without a group, but not vice versa. However there are many animals, primates among them, where the individual cannot survive without the group.

Strict selfishness and strict altruism are, it seems to us, the maladaptive ends of a continuum; the optimum intermediate position varies with circumstance, and selection inhibits the extremes. And if it's too difficult for the genes to figure out on their own what the optimum mix is for each novel circumstance, might it not be advantageous for them to delegate authority? For this again, brains are needed.

Consider kin selection once more. Never mind the nagging question about how well birds, say, can distinguish uncles from cousins; especially in small groups, it doesn't much matter—everyone's a pretty close relative, and kin selection works in a statistical sense, even if you occasionally put yourself on the line for some unrelated neighbor. It makes sense, in terms of the preservation of multiple copies of closely related genetic instructions, to accept a 40% chance of dying to save the life of a sibling (who has 50% of the same genes you have); or a 20% chance to save an uncle or a niece or a grandchild (who share 25% of your genes); or a 10% chance of dying to save the life of a first cousin (who has 12.5% of exactly the same genes that you do). Well, then, what about giving up the means of affording another child in order to preserve the families of many second cousins? What about donating ten percent of your income so a gaggle of third cousins have enough to eat? Might it pay to abstain from a few luxuries so fourth cousins can be educated? What about writing a letter of recommendation for an undistinguished fifth cousin?

Kin selection is also a continuum, and in its arcane calculus *some* sacrifice must be worthwhile to aid the most far-flung and distant members of your family. But since we are all related, some sacrifice must be justified to save anyone on Earth—and not only those of our own species. Even on its own terms, kin selection extends far beyond close relatives.

Typically, any two members of a small community of primates in the wild have 10 to 15% of their genes in common [19] (and about 99.9% of their ACGT sequences in common, it requiring only a single nucleotide difference to make one gene, composed of thousands of nucleotides, different from another). So any random member of the group is pretty likely to be your parent or child or sibling, uncle, aunt, nephew, niece, or first or second cousin. Even if you can't distinguish one from the other, it makes good evolutionary sense to make real sacrifices for them—and to accept something like a 10% chance of dying in order to save the life of any one of them.

In the annals of primate ethics, there are some accounts that have the ring of parable. Consider, for example, the macaques. Also known as rhesus monkeys, they live in tightly knit cousins' clubs. [20] Since the macaque you save is statistically likely to share many of your genes (assuming you're another macaque), you're justified in taking risks to save it, and a fine discrimination of shades of consanguinity is unnecessary. In a laboratory setting, [21] macaques were fed if they were willing to pull a chain and electrically shock an unrelated macaque whose agony was in plain view through a one-way mirror. Otherwise, they starved. After learning the ropes, the monkeys frequently refused to pull the chain; in one experiment only 13% would do so—87% preferred to go hungry. One macaque went without food for nearly two weeks rather than hurt its fellow. Macaques who had themselves been shocked in previous experiments were even less willing to pull the chain. The relative social status or gender of the macaques had little bearing on their reluctance to hurt others.

If asked to choose between the human experimenters offering the macaques this Faustian bargain and the macaques themselves—suffering from real hunger rather than causing pain to others—our own moral sympathies do not lie with the scientists. But their experiments permit us to glimpse in non-humans a saintly willingness to make sacrifices in order to save others—even those who are not close kin. By conventional human standards, these macaques—who have never gone to Sunday school, never heard of the Ten Commandments, never squirmed through a single junior high school civics lesson— seem exemplary in their moral grounding and their courageous resistance to evil. Among the macaques, at least in this case, heroism is the norm. If the circumstances were reversed, and captive humans were offered the same deal by macaque scientists, would we do as well?[22] In human history there are a precious few whose memory we revere because they knowingly sacrificed themselves for others. For each of them, there are multitudes who did nothing.

———

T. H. Huxley remarked that the most important conclusion he had gleaned from his anatomical studies was the interrelatedness of all life on Earth. The discoveries made since his time—that all life on Earth uses nucleic acids and proteins, that the DNA messages are all written in the same language and all transcribed into the same language, that so many genetic sequences in very different beings are held in common—deepen and broaden the power of this insight. No matter where we think we are on that continuum between altruism and selfishness, with every layer of the mystery we strip away, our circle of kinship widens.

Not from some uncritical sentimentalism, but out of tough-minded scientific scrutiny, we find the deepest affinities between ourselves and the other forms of life on Earth. But compared to the differences between any of us and any other animal, all humans, no matter how ethnically diverse, are essentially identical. Kin selection is a fact of life, and is very strong in animals that live in small groups. Altruism is very close to love. Somewhere in these realities, an ethic may be lurking.

ON IMPERMANENCE

Insignificant
mortals, who are as leaves are, and now flourish and grow warm
with life, and feed on what the ground gives, but then
fade away and are dead.

HOMER, *The Iliad*[23]

Chapter 7

WHEN FIRE WAS NEW

Not I, but the world says it:
All is one.

H E R A C L I T U S[1]

The oxygen in the air is generated by green plants. They vent it into the atmosphere and we animals greedily breathe it in. So do many microbes and the plants themselves. We, in turn, exhale carbon dioxide into the atmosphere, which the green plants eagerly inhale. In a profound but largely unremarked intimacy, the plants and animals live off each other's bodily wastes. The atmosphere of the Earth connects these processes, and establishes the great symbiosis between plants and animals. There are many other cycles that bind organism to organism and that are mediated by the air—cycles in nitrogen, for example, or sulfur. The atmosphere brings beings all over the world into contact; it establishes another kind of biological unity to the planet.

The Earth started out with an atmosphere essentially free of the oxygen molecule. As bacteria and other one-celled organisms arose, 3.5 billion years ago or earlier, some harvested sunlight, breaking water molecules apart in the first stage of photosynthesis. The oxygen, a waste gas, was simply released into the air—like emptying a sewer into the ocean. Resolutely independent, liberated from reliance on nonbiological sources of organic matter, the photosynthetic organisms proliferated. By the time there got to be enormous numbers of them, the air was full of oxygen.

Now oxygen is a peculiar molecule. We breathe it, depend on it, die without it, and so naturally have a good opinion of it. In respiratory distress, we want more oxygen, purer oxygen. As modern words ("inspire," literally, breathe in; "aspire," breathe toward; "conspire," breathe with; "perspire," breathe through; "transpire," breathe across; "respire," breathe again; and "expire," breathe out) and Latin proverbs (such as *Dum spiro, spero,* while I breathe, I hope) remind us, we associate many aspects of our nature with breathing. The word "spirit" —in all its incarnations ("spiritual," "spirited," alcoholic "spirits," "spirits" of ammonia, and so forth)—also derives from the same Latin word for breath. Our fixation with breathing comes ultimately from considerations of energy efficiency: The oxygen we respire makes us about ten times more efficient in extracting energy from food than, say, yeast are; they know only how to ferment—breaking sugar down to some intermediate product such as ethyl alcohol rather than all the way back to carbon dioxide and water.*

But as a blazing log or a burning coal reminds us, oxygen is dangerous. Given a little encouragement, it can vandalize the intricate, painstakingly evolved structure of organic matter, leaving little more than some ash and a puff of vapor. In an oxygen atmosphere, even if you don't apply heat, oxidation, as it's called, slowly corrodes and disintegrates organic matter. Even much sturdier materials such as copper or iron tarnish and rust away in oxygen. Oxygen is a poison for organic molecules and doubtless was poisonous to the beings of the ancient Earth. Its introduction into the atmosphere triggered a major crisis in the history of life, the oxygen holocaust. The idea of organisms that gasp and choke to death after being exposed to a whiff of oxygen seems counterintuitive and bizarre, like the Wicked Witch of the West in *The Wizard of Oz* melting away to nothing when a little water falls on her. It's the ultimate version of the adage "One man's meat is another man's poison." †

Either you adapted to the oxygen, or you hid from it, or you died. Many died. Some reconciled themselves to live underground, or in marine muds, or in other environments where the deadly oxygen could not reach. Today all of the most primitive organisms—that is, the ones least related by genetic sequence to the rest of us—are mi-

---

* A biochemical imperfection exploited by the beer, wine, and liquor industries, which profitably manufacture this addictive and dangerous drug, C_2H_5OH (where C stands for a carbon atom, O for oxygen, and H for hydrogen). Millions of people worldwide die from imbibing it each year. Or, looked at another way, distillers have been exploited by the fermenting bacteria and yeast, who have gotten us to arrange for their growth and reproduction on a worldwide, industrial scale—because we love to drink ourselves senseless on microbial wastes. If they could speak, perhaps they would boast about how cleverly they've domesticated the humans. Yeasts also colonize dark, moist, oxygen-poor parts of the human body, another way in which we serve them.

† Another example was given by the ancient Greek philosopher Heraclitus: "The sea," he said, "is most pure and most polluted water: for fish, drinkable and life-preserving; for men, undrinkable and death-dealing."[2]

croscopic and anaerobic; they prefer to live, or are forced to live, where the oxygen isn't. Most organisms on Earth these days deal well with oxygen. They have elaborate mechanisms to repair the chemical damage done by oxygen, as—gingerly, held at molecular arm's length it is used to oxidize food, extract energy, and drive the organism at high efficiency.

Human cells, and many others, deal with oxygen through a special, largely self-contained molecular factory called a mitochondrion, which is in charge of dealing with this poison gas. The energy extracted by oxidizing food is stored in special molecules and safely shipped to workstations throughout the cell. Mitochondria have their own kind of DNA—circles, or daisy chains, of As, Cs, Gs, and Ts, rather than double helices, instructions different at a glance from those that run the cell proper. But they're enough like the DNA of the chloroplasts to make it clear that mitochondria also were once free-living bacteria-like organisms. The central role of cooperation and symbiosis in the early evolution of life is again evident.

Luckily for us, biochemical solutions were found to the oxygen crisis. If not, perhaps the only life on Earth today other than photosynthetic plants would be slithering in ooze and sucking at thermal vents in the abyssal depths. We have risen to the challenge and surmounted it—but only at enormous cost in the deaths of our ancestors and collateral relatives. These events show that there is no inherent foresight or wisdom in life that prevents it from making, in the short term at least, catastrophic mistakes. They also demonstrate that, long before civilization, life was producing toxic wastes on a massive scale, and for that miscalculation paying stiff penalties.

Through some such biochemical oversight, had things gone a little differently, perhaps all life on Earth would have been extinguished. Or perhaps some devastating asteroidal or cometary impact would have killed off all those tentative, fumbling microbes. Then, as we've said, organic molecules—both those synthesized on Earth and those falling from the skies—might have led to a new origin of life and an alternative evolutionary future. But the day comes when the gases leaking out of volcanos and fumaroles are no longer hydrogen-rich, no longer easy to make organic molecules from. Part of the reason is the oxygen atmosphere itself, which oxidizes these gases. Also, there gets to be a time when extraterrestrial organic molecules arrive so infrequently that they are an insufficient source of the stuff of life. Both these conditions seem to have been satisfied by around 2 or 3 billion years ago. Thereafter, if every living thing were to be wiped out, no new life could arise. The Earth would remain a desolate waste-land of a world into the remote future—until the Sun dies.

———

Back then, around 2 billion years ago or a little before, the oxygen in the Earth's atmosphere—steadily increasing, to be sure, over preced-ing ages of geological time—began quickly to approach its present abundance. (In today's air, one in every five molecules is O_2.)

The first eukaryotic cell evolved a little earlier. *Our* cells are eukary-otes, which in Greek means, roughly, "good nuclei," or "true nuclei." As usual, we chauvinistic humans admire it because we have it. But they've been very successful. Bacteria and viruses are not eukaryotes, but flowers, trees, worms, fish, ants, dogs, and people are; all the algae, fungi, and protozoa, all the animals, all the vertebrates, all the mam-mals, all the primates. One of the key distinctions of the eukaryotic cell is that the governing machinery, the DNA, is encapsulated and set apart in a cell nucleus. As in a medieval castle, two sets of walls protect it from the outside world. Special proteins bond and contort the DNA, enveloping and embracing it, so a double helix that un-coiled would be about a meter long is compressed into a submicro-scopic chamber at the heart of the cell. Perhaps the nucleus evolved —in the oxygen-rich vicinities of photosynthetic organisms—in part to protect DNA from oxygen while the mitochondria were busily ex-ploiting it.

Each long DNA double helix is called a chromosome. Humans have 23 pairs of chromosomes. The total number of As, Cs, Gs, and Ts is about 4 billion pairs of letters in our double-stranded hereditary in-structions. The information content is roughly that of a thousand different books with the size and fineness of print of the one you're reading at this moment. While the variation from species to species is large, similar numbers apply to many other "higher" organisms.

Those same proteins that surround the DNA (themselves manufac-tured, of course, on instructions from the DNA) are responsible for switching genes on and off, in part by uncovering and covering the DNA. At appointed times, the exposed ACGT information of the DNA makes copies of certain sequences and dispatches them as messages out of the nucleus into the rest of the cell; in response to the commands in these telegrams, new molecular machine tools, the enzymes, are manufactured. They in turn control all the metabolism of the cell and all its interactions with the outside world. As with the children's game called "Telephone" in America and "Grandmother's Whispers" in Britain—in which a message is whispered successively by each player into the ear of the next—the longer the sequence of relays, the more likely it is that the communication will be garbled.

It's a little like a kingdom with the distant DNA, isolated and guarded in the nucleus, as the monarch. The chloroplasts and mitochondria play the role of proudly independent dukedoms whose continuing cooperation is essential to the well-being of the realm.* Everybody else, every other molecule or complex of molecules working for the cell, has as its sole obligation punctilious obedience to orders. Great care must be taken that no message is mislaid or misunderstood. Occasionally, decisions are delegated to other molecules by the DNA, but generally every machine in the cellular toolshop is on a short tether.

However, even to the rank-and-file molecular workers in the cell, the monarch often seems half-witted and his decrees garbled and meaningless. As we've mentioned, most DNA of humans and other eukaryotes is genetic nonsense which the START and STOP instructions—like prudent assistants to a mad president—duly ignore. Immense reams of nonsense are in effect thoughtfully preceded by the notice "DRIVEL AHEAD. PLEASE IGNORE," and followed by the message "END OF DRIVEL." Sometimes the DNA goes into a stuttering frenzy in which the same ravings are repeated over and over. In the kangaroo rat of the American Southwest, for example, the sequence AAG is repeated 2.4 billion times, one after the other; TTAGGG, 2.2 billion times; and ACACAGCGGG, 1.2 billion times. Fully half of all the genetic instructions in the kangaroo rat are these three stutters.[4] Whether repetition plays another role—maybe some

---

* The genetic code of the mitochondrion is just a little different from that of the nucleus—as if it had evolved so that the nuclear DNA could not tell the mitochondria what to do, a token of independence. For example, AGA means STOP for mitochondrial nucleic acids, whereas for the nucleic acids that hail from the nucleus of a cell, it codes for a particular amino aid, arginine.[3] The mitochondria simply ignore instructions from the capital, which to them are mainly gibberish with occasional lucid passages; they follow the commands of their own feudal leader, the mitochondrial DNA.

internecine struggle for control by different gene complexes inside the DNA—is unknown. But superposed on precision replication and repair, and the meticulous preservation of DNA sequences from ages past, there is an element in the life of the eukaryotic cell that seems a little like farce.[5]

Some 2 billion years ago, several different hereditary lines of bacteria seem to have begun stuttering—making full copies of parts of their hereditary instructions over and over again; this redundant information then gradually specialized, and, excruciatingly slowly, nonsense evolved into sense.[6] Similar repetitions arose early in the eukaryotes. Over long periods of time, these redundant, repetitive sequences undergo their own mutations, and sooner or later there will be, by chance, rare short passages among them that begin to make sense, that are useful and adaptive. The process is much easier than the classic imaginary experiment of the monkeys poking at typewriter keys long enough that eventually the complete works of William Shakespeare emerge. Here, even the introduction of a very short new sequence—representing only a punctuation mark, say—may be able to increase the survival chances of the organism in a changing environment. And here, unlike the monkeys at their typewriters, the sieve of natural selection is working. Those sequences that are slightly more adaptive (to continue the metaphor, we might say those sequences that correspond even slightly to Shakespearean prose—"TO BE OR," immersed in gibberish, would be a start) are preferentially replicated. Out of randomly changing nonsense, the accidental bits of sense are preserved and copied in large numbers. Eventually, a great deal of sense emerges. The secret is remembering what works. Just such a drawing forth of meaning from random sequences of nucleotides must have happened in the very earliest nucleic acids, around the time of the origin of life.

An illuminating computer experiment analogous to the evolution of a short DNA sequence was performed by the biologist Richard Dawkins. He starts with a random sequence of twenty-eight English-language letters (spaces are counted as letters):

WDLTMNLT DTJBKWIRZREZLMQCO P.

His computer then repeatedly copies this wholly nonsensical message. However, at each iteration there is a certain probability of a mutation, of a random change in one of the letters. Selection is also simulated, because the computer is programmed to retain any mutations that move the sequence of letters even slightly toward a pre-selected goal, a particular, quite different sequence of twenty-eight letters. (Of course natural selection does not have some final ACGT sequence in mind, but—in preferentially replicating sequences that improve, even by a little, the fitness of the organism—it comes down to the same thing.) Dawkins's arbitrarily chosen twenty-eight-letter sequence, toward which his selection was aiming, was

METHINKS IT IS LIKE A WEASEL.

(Hamlet, feigning madness, is teasing Polonius.)

In the first generation, one mutation in the random sequence occurs, changing the "K" (in DTJBKW . . .) to an "S." Not much help yet. By the tenth generation, it reads

MDLDMNLS ITJISWHRZREZ MECS P,

and by the twentieth,

MELDINLS IT ISWPRKE Z WECSEL.

After thirty generations, we are at

METHINGS IT ISWLIKE B WECSEL,

and by forty-one generations, we're there.

"There is a big difference," Dawkins concludes, "between cumulative selection (in which each improvement, however slight, is used as a basis for future building), and single-step selection (in which each new 'try' is a fresh one). If evolutionary progress had had to rely on single-step selection, it would never have got anywhere."[7]

Randomly varying the letters is an inefficient way to write a book, you might be thinking. But not if there are an enormous number of copies, each changing slightly generation upon generation, the new instructions constantly tested against the demands of the outside world. If human beings were devising the volumes of instruction contained in the DNA of the given species, we would, we might offhand imagine, just sit down and write the thing out, front to back, and tell the species what to do. But in practice we are wholly unable to do this, as is DNA. We stress again, the DNA hasn't the foggiest notion a priori about which sequences are adaptive and which are not. The evolutionary process is not omnicompetent, far-seeing, crisis-avoiding, top-down. It is instead trial-and-error, short-term, crisis-mitigating, bottom-up. No DNA molecule is wise enough to know what the consequences will be if one segment of a message is changed into another. The only way to be sure is to try it out, keep what works, and run with it.

The more you know how to do, the more advanced you are—and, you might think, the better your chances for survival. But the DNA instructions for making a human being comprise some 4 billion nucleotide pairs, while those for a common one-celled amoeba contain 300 billion nucleotide pairs. There is little evidence that amoebae are almost a hundred times more "advanced" than humans, although the proponents of only one side of this question have been heard from to date. Again, some, maybe even most, of the genetic instructions must be redundancies, stutters, untranscribable nonsense. Again we glimpse deep imperfections at the heart of life.

Sometimes another organism inconspicuously slips through the defenses of the eukaryotic cell and steals into the heavily guarded inner sanctum, the nucleus. It attaches itself to the monarch, perhaps to the end of a time-tested and highly reliable DNA sequence. Now messages of a very different sort are dispatched out of the nucleus, messages that order the manufacture of a different nucleic acid, that of the infiltrator. The cell has been subverted.

Besides mutation, there are other ways (including infection and sex, to which we turn shortly) whereby new hereditary sequences arise. The net result is that a huge number of natural experiments are performed in every generation to test the laws, doctrine, and dogma encoded in the DNA. Each eukaryotic cell is such an experiment. Competition among DNA sequences is fierce; those whose commands work even slightly better become fashionable, and everyone has to have one.

The earliest known eukaryotic plankton floating on the surface of the oceans date to about 1.8 billion years ago; the earliest eukaryotes with a sex life to 1.1 billion years ago; the great burst of eukaryote evolution (that would lead to algae, fungi, land plants and animals, among others) to about the same epoch; the earliest protozoa to about 850 million years ago; and the origin of the major animal groups and the colonization of the land to about 550 million years ago.[8] Many of these epochal events may be tied to the increasing atmospheric oxygen. Since the oxygen is generated by plants, we see life forcing its own evolution on a massive scale. Of course, we can't be sure of the dates; next week paleontologists may discover examples still more ancient. The sophistication of life has increased greatly over the last 2 billion years, and the eukaryotes have done extremely well—as we have only to look around us to verify.

But the eukaryotic kind of life, very different from the rough-and-ready first organisms, is exquisitely dependent on the near-perfect functioning of an elaborate molecular bureaucracy, whose responsibilities include covering up the fits of incompetence in the DNA. Some DNA sequences are too fundamental to the central processes of life to be able safely to change. Those key instructions simply stay fixed, precisely replicated, generation after generation, for aeons. Any significant alteration is simply too costly in the short term, whatever its ostensible virtues may be in the long, and the carriers of such change are wiped out by selection. The DNA of eukaryotic cells reveals segments that clearly and specifically come from the bacteria and archaebacteria of long ago. The DNA inside us is a chimera, long ACGT sequences having been adopted wholesale from quite different and extremely ancient beings, and faithfully copied for billions of years. Some of us—much of us—is *old*.

———

Eventually, there got to be many beings whose cells had specialized functions, just as, for example, the chloroplasts or mitochondria within a given cell have specialized functions. Some cells were in charge of, say, disabling and removing poisons; others were the conduits of electrical impulses, part of a slowly evolving neural apparatus in charge of locomotion, breathing, feelings, and—much later— thoughts. Cells with quite diverse functions interacted harmoniously. Still larger beings evolved separate internal organ systems, and again survival depended on the cooperation of very different constituent parts. Your brain, heart, liver, kidneys, pituitary, and sexual organs generally work together well. They are not in competition. They make a whole that is much more than the sum of the parts.

Our ancestors and collateral relatives were restricted to the seas until about 500 million years ago, when the first amphibian crawled out onto the land. A significant ozone layer may not have developed until about then. These two facts are probably related. Earlier, deadly ultraviolet light from the Sun reached the surface of the land, frying any intrepid pioneer attempting to homestead there.* Ozone, as we've mentioned, is produced from the oxygen in the upper air by the Sun's radiation. So that reckless oxygen pollution of the ancient atmosphere, generated by the green plants, seems to have had another accidental and this time salutary consequence: It made the land habitable. Who would have figured?

Hundreds of millions of years later, a rich biology filled almost every nook and cranny of the land. The moving continental plates now carried with them cargoes of plants and animals and microbes. When new continental crust appeared, it was quickly colonized by life. When old continental crust was carried down into the Earth's interior, we might be worried that its living cargo would be carried down with it. But the conveyor belt of plate tectonics moves only an inch a year. Life is quicker. Ancient fossils, though, can't jump off the conveyer belt. They are destroyed by plate tectonics. The precious records and remains of our ancestors are swept down into the semi-fluid mantle and cremated. We are left with the odd remnants that by accident escaped.

Before there was enough oxygen, or anything combustible, fire was impossible, an unrealized potential, latent in matter (just as the release of nuclear energy was unrealized during the tenure of humans on Earth until 1942–1945). There must, therefore, have been an age of the first flame, a time when fire was new. Perhaps it was a dead fern, ignited by a flash of lightning. Since plants colonized the land long before animals, there was no one to notice: Smoke rises; suddenly, a tongue of red flickers upward. Perhaps a little thicket of vegetation has caught fire. The flame isn't a gas, or a liquid, or a solid.

---

* Seawater itself is opaque to ultraviolet light beyond a certain depth, and the early oceans were very likely covered by a slick of ultraviolet-absorbing organic molecules. The seas were safe.

It's some other, some fourth state of matter that physicists call plasma. Never before had Earth been touched by fire.

Long before humans made use of fire, plants did. When the population density is high and plants of different species are closely packed together, they fight—for access to nutrients and underground water, but especially for sunlight. Some plants have invented hardy, fire-resistant seeds, along with stems and leaves that readily burst into flames. Lightning strikes, an intense fire burns out of control, the seeds of the favored plant survive, and the competition—seeds and all —has been burned to a crisp. Many species of pines are the beneficiaries of this evolutionary strategy. Green plants make oxygen, oxygen permits fire, and fire is then used by some green plants to attack and kill their neighbors. There is hardly any aspect of the environment that has not been used, one way or another, in the struggle for existence.

A flame looks unearthly, but in this neck of the Cosmos it's unique to Earth. Of all the planets, moons, asteroids, and comets in our Solar System, there is fire only on Earth—because there are large amounts of oxygen gas, O_2, only on Earth. Fire was, much later, to have profound consequences for life and intelligence. One thing leads to another.

———

The human pedigree wends its tortuous way back to the beginning of life 4 billion years ago. Every being on Earth is our relative, since we all come from that same point of origin. And yet, precisely because of evolution, no lifeform on Earth today is an ancestor of ours. Other beings did not stop evolving because a pathway that would someday lead to humans had just been generated. No one knew what branch in the evolutionary tree was going where, and no one before humans could even raise the question. The beings from whom our ancestral line deviated continued to evolve, inside and out, or became extinct. Almost all became extinct. We know from the fossil record something of who our predecessors were, but we cannot bring them into the laboratory for interrogation. They are no more.

Luckily, though, there are organisms alive today that are similar— in some cases, very similar—to our ancestors. The beings that left stromatolite fossils probably performed photosynthesis and in other respects behaved as contemporary stromatolitic bacteria do. We learn about them by examining their surviving close relatives. But we cannot be absolutely sure. For example, ancient organisms were not necessarily and in all respects simpler than modern ones. Viruses and parasites, in general, show signs of having evolved by loss of function from some more self-sufficient forebear.

Many features in the biological landscape arrived late. Sex, for example, doesn't seem to have evolved until three quarters of the history of life till now had passed. Animals big enough for us to see—had we been there—animals made of many different kinds of cells, also do not seem to have emerged until almost three quarters of the way between the origin of life and our time. Except for microbes, there were no beings on the land until something like 90%, and no creatures with big brains for their body sizes until about 99% of the history of life thus far was over.

Enormous gaps yawn through the fossil record, although less so now than in Darwin's time. (If there were more paleontologists in the world, we'd doubtless be a little further along.) From the comparatively low rate of discovery of new fossils, we know that huge numbers of ancient organisms have not been preserved. There's something poignant about all those species—some ancestral to humans, on some sturdy trunk of our family tree, most not—about whom we know nothing, not a single example of them having survived, even in fossil form, to our own time.

Even when the incompleteness of the fossil record is taken into account, the diversity or "taxonomic richness" of life on Earth is found to have been steadily increasing, especially in the last 100 million years. [9] Diversity seems to have peaked just as humans were really getting going, and has since declined markedly—in part because of the recent ice ages, but in larger part because of the depredations of humans, both intentional and inadvertent. We are destroying the diversity of beings and habitats out of which we emerged. Something like a hundred species become extinct each day. Their last remnants die out. They leave no descendants. They are gone. Unique messages, painstakingly preserved and refined over eons, messages that a vast succession of beings gave up their lives to pass on to the distant future are lost forever.

———

More than a million species of animals are now known on Earth, and perhaps 400,000 species of eukaryotic plants. There are at least thousands of known species of other organisms, non-eukaryotes, including bacteria. Doubtless we have missed many, probably most. Some estimates of the number of species range beyond 10 million; if so, we have even glancing acquaintance with less than 10% of the species on Earth. Many are becoming extinct before we even know of their existence. Most of the billions of species of life that have ever lived are extinct. Extinction is the norm. Survival is the triumphant exception.

We've sketched the changes on the Earth's surface at the end of the Permian Period, some 245 million years ago; they resulted in the most devastating biological catastrophe so far displayed in the fossil record. Perhaps as many as 95% of all the species then living on Earth became extinct.* Many kinds of filter-feeding animals attached to the ocean floor, beings that had for hundreds of millions of years characterized life on Earth, disappeared. Ninety-eight percent of the families of crinoids became extinct. We don't hear much about crinoids these days; sea lilies are their surviving remnant. Wholesale extinctions also occurred among the amphibians and reptiles that had settled the land. On the other hand, sponges and bivalves (like clams) did comparatively well in the late Permian extinction—one consequence of which is that they are still plentiful on Earth today.

Following mass extinctions it typically takes 10 million years or more for the variety and abundance of life on Earth to recover—and then, of course, there are different organisms around, perhaps better adapted to the new environment, perhaps with better long-term prospects, or perhaps not. In the millions of years following the end of the Permian Period, volcanism subsided and the Earth warmed. This killed off many land plants and animals that had been adapted to the late Permian cold. Out of this set of cascading climatic consequences, conifers and ginkgoes emerged. The first mammals evolved from reptiles in the new ecologies established after the Permian extinctions.

Of all the species of animals alive at the end of the Permian, only about twenty-five of them, it is estimated, have left any descendants at all; ten of which account for 98% of the contemporary families of

---

* Ninety-five percent seems awfully close to 100%, and it's disquieting to be reminded that the great rumbling, internal tectonic engine can inadvertently kill off so many of us up here because of some hiccups down there.

vertebrates, which comprise about forty thousand species.[10] The rate of evolutionary change is full of fits and starts, blind alleys and sweeping change—the latter driven often by the first filling of a previously untenanted ecological niche. New species appear quickly and then persist for millions of years. In only the last 2% or 3% of the history of life on Earth, the extravagant diversification of the placental mammals has produced

> shrews, whales, rabbits and mice, anteaters, sloths, armadillos, horses, pigs and antelopes, elephants, sea cows, wolves, bears, tigers, seals, bats, monkeys, apes, and men[11]

For the vast bulk of Earth history, until just recently, not one of these beings had existed. They were present only potentially.

Think of the genetic instructions of a given being, perhaps a billion ACGT nucleotide pairs long. Randomly change a few nucleotides. Perhaps these will be in structural or inactive sequences and the organism is in no way altered. But if you change a meaningful DNA sequence, you change the organism. Most such changes, as we keep saying, are maladaptive; except in rare instances, the bigger the change, the more maladaptive it is. For all of mutation, gene recombination, and natural selection put together, the continuing experiment of evolution on Earth has brought into being only a minute fraction of the range of possible organisms whose manufacturing instructions could be specified by the genetic code. The vast bulk of those beings, of course, would be not merely maladapted, not just freaks, but wholly inviable. They could not be born alive. Nevertheless, the total number of possible functioning, living beings is still vastly greater than the total number of beings who have ever been. Some of those unrealized possibilities must be, by any standard we wish to adopt, better adapted and more capable than any Earthling who has ever lived.

——

Sixty-five million years ago most of the species on Earth were snuffed out—probably because of a massive cometary or asteroidal collision. Among those killed off were all the dinosaurs, which had for nearly 200 million years—from before the breakup of Gondwanaland—been the dominant species, the ubiquitous masters of life on Earth. This extinction event removed the chief predators of a small, fearful, cowering nocturnal order of animals called the mammals. If not for that collision—a late step in the tidying up of interplanetary space of the remaining worlds on eccentric orbits—we humans and our primate ancestors would never have come to be. And yet, if that comet had been on a slightly different trajectory, it might have missed the Earth entirely. Perhaps, in its many relays around the Sun, its ices would all have melted and its rocky and organic contents slowly spewed as fine powder into interplanetary space. Then all it would have provided for life on Earth would have been a periodic shower of meteors, perhaps admired by some newly-evolved, curious, large-brained reptile.

On the scale of the Solar System, the extinction of the dinosaurs and the rise of the mammals seem to have been a very near thing. The causality corridor, figuratively speaking, was only inches wide. Had the comet been traveling a little slower or faster or headed in a slightly different direction, no collision would then have occurred. If other comets that in our real history missed the Earth had been on slightly different trajectories, they would have hit the Earth and killed off life in some different epoch. The cosmic collision roulette, the extinction lottery, reaches into our own time.

At the depth in the fossil record above which there are no more dinosaurs, there is, worldwide, a telltale thin layer of the element iridium, which is abundant in space but not on the Earth's surface. There also are tiny grains bearing the signs of a collosal impact. This evidence tells us of a high-speed collision of a small world with the Earth which distributed fine particles worldwide. The remains of the impact crater may have been discovered in the Gulf of Mexico near the Yucatan Peninsula. But something else is found in this layer as well: soot. Planet-wide, the time of this great impact was also the time of a global fire. The debris from the impact explosion, spewed out into the high atmosphere and falling back through the air all over the Earth—a continuous meteor shower filling the sky—illuminated the ground far more brightly than the noonday Sun. Land plants everywhere on Earth burst into flames, all at once. Most of them were consumed. There is an odd causal nexus connecting oxygen, plants, giant impacts, and world-immolating fire.

There are many ways in which such an impact could have extinguished long-established and, if we may call them that, self-confident forms of life. After the initial burst of light and heat, a thick pall of impact dust enveloped the Earth for a year or more. Perhaps even more important than the world fire, the lowered temperatures, and a planet-wide acid rain was the absence for a year or two of enough light for photosynthesis. The primary photosynthesizing organisms in the oceans (then as now covering most of the Earth) are little one-celled plants called phytoplankton. They are especially vulnerable to lowered light levels because they lack major food reserves. Once the lights get turned out their chloroplasts can no longer generate carbohydrates from sunlight, and they die. But these little plants are the principal diet of one-celled animals that are eaten by larger, shrimp-like creatures, that are eaten by small fish, that are in turn eaten by large fish. Turn off the lights, wipe out the phytoplankton, and the entire food chain, this elaborate house of cards, collapses. Something similar is true on land.

The beings of Earth depend on one another. Life on Earth is an intricately woven tapestry or web. Yank out a few threads here and there, and you can't be sure whether that's all the damage you've done, or whether the whole fabric will now unravel.

Insects and other arthropods are the principal agents by which dead plants and animal excrement are cleaned up. Scarabs—the dung beetles identified with the sun god and worshipped by the ancient Egyptians—are specialists in waste management. They collect the nitrogen-rich animal excrement accumulating on the surface of our planet and transport this fertilizer down where the plant roots are. Some sixteen thousand beetles have been counted on a single fresh elephant pat in Africa; two hours later the pat was gone.[12] The Earth's surface would be very different (and very messy) without dung beetles and their like. In addition, the microscopic feces of mites and springtails are major constituents of the soil humus from which the plants grow. Animals then eat the plants. We live off each other's solid wastes as well.

Other inhabitants of the soil kill off the young plants. Here is Darwin's account of a little experiment he did to illustrate the hidden ferocity lurking just beneath the placid surface of a country garden:

[On] a piece of ground 3 feet long and 2 feet wide, dug and cleared, and where there could be no choking from other plants, I marked all the seedlings of our native weeds as they came up, and out of 357 no less than 295 were destroyed, chiefly by slugs and insects. If turf which has long been mown, and the case would be the same with turf closely browsed by quadrupeds, be let to grow, the more vigorous plants gradually kill the less vigorous though fully grown plants . . .[13]

Some plants provide food for specific animals; in turn, the animals act as agents for the sexual reproduction of the plants—in effect, couriers taking sperm from male plants and using it for artificial insemination of female plants. This is not quite artificial selection, because the animals are not much in charge. The currency these procurers are paid in is usually food. A bargain has been struck. Maybe the animal is a pollinating insect, or bird, or bat; or a mammal to whose furry coat the reproductive burrs adhere; or maybe the deal is food supplied by the plants in exchange for nitrogenous fertilizer supplied by the animals. Predators have symbionts that clean their coats or scales or pick their teeth in exchange for leavings. A bird eats a sweet fruit; the seeds pass through its digestive tract and are deposited on fertile ground some distance away: another business transaction consummated. Fruit trees and berry-bearing bushes often take care that their offerings to the animals are sweet only when the seeds are ready to be dispersed. Unripe fruit gives bellyaches, the plants' way of training the animals.

The cooperation between plants and animals is uneasy. The animals cannot be trusted; given a chance, they'll eat any plant in sight. So the plants protect themselves from unwelcome attention with thorns, or by producing irritants, or poisons, or chemicals that make the plant indigestible, or agents that interfere with the predator's DNA. In this endless slow-motion war, the animals then produce substances that disable these adaptations by the plants. And so on.

The beasts and vegetables and microbes are the interlocking parts, the gear train, of a vast, intricate and very beautiful ecological machine of planetary proportions, a machine plugged into the Sun. Pretty nearly, all flesh is sunlight.

Where the ground is covered with plants perhaps 0.1% of the sunlight is converted into organic molecules. A plant-eating animal saunters by and eats one of these plants. Typically the herbivore extracts about a tenth of the energy in the plant, or about one ten-thousandth of the sunlight that could, with 100% efficiency, have been stored in the plant. If the herbivore is now attacked and eaten by a carnivore, about 10% of the available energy in the prey will wind up in the predator. Only one part in a hundred thousand of the original solar energy makes it to the carnivore. There are no perfectly efficient engines, of course, and we expect losses at each stage in the food chain. But the organisms at the top of the food chain seem inefficient to the point of irresponsibility.*

A vivid image of the interconnection and interdependence of life on Earth was provided by the biologist Clair Folsome, who asks you to imagine what you would see if all the cells of your body, flesh and bones, were magically removed:

> What would remain would be a ghostly image, the skin outlined by a shimmer of bacteria, fungi, round worms, pin worms and various other microbial inhabitants. The gut would appear as a densely packed tube of anaerobic and aerobic bacteria, yeasts, and other microorganisms. Could one look in more detail, viruses of hundreds of kinds would be apparent throughout all tissues.

And, Folsome stresses, any other plant or animal on Earth, under the same dispensation, would reveal a similar "seething zoo of microbes." [14]

———

A biologist from some other solar system, in an unblinking examination of the teeming lifeforms of Earth, would surely note that they are all made of almost exactly the same organic stuff, the same molecules almost always performing the same functions, with the same genetic codebook in use by almost everybody. The organisms on this planet are not only kin; they live in intimate mutual contact, imbibing each other's wastes, dependent on one another for life itself, and sharing the same fragile surface layer. This conclusion is not ideology, but

---

* In principle the ecological machine could continue as long as the Sun continues to shine, estimated at another 5 billion years. It's hard not to wonder—we carnivores at the apex of the food chain, the beneficiaries of a process with a thousandth of a percent efficiency—if there might not be some more efficient way for us to harness the Sun.

reality. It depends not on authority, faith, or special pleading by its proponents, but on repeatable observation and experiment.

The beings of our planet are imperfectly linked and coordinated; and there is certainly nothing like a collective intelligence of all the life on Earth—in the sense that all the cells of a human body are subject, within stringent constraints, to a supervening volition. Still, the alien biologist might be excused for lumping together the whole biosphere—all the retroviruses, mantas, foraminifera, mongongo trees, tetanus bacilli, hydras, diatoms, stromatolite-builders, sea slugs, flatworms, gazelles, lichens, corals, spirochetes, banyans, cave ticks, least bitterns, caracaras, tufted puffins, ragweed pollen, wolf spiders, horseshoe crabs, black mambas, monarch butterflies, whiptail lizards, trypanosomes, birds of paradise, electric eels, wild parsnips, arctic terns, fireflies, titis, chrysanthemums, hammerhead sharks, rotifers, wallabies, malarial plasmodia, tapirs, aphids, water moccasins, morning glories, whooping cranes, komodo dragons, periwinkles, millipede larvae, angler fish, jellyfish, lungfish, yeast, giant redwoods, tardigrades, archaebacteria, sea lilies, lilies of the valley, humans, bonobos, squid and humpback whales—as, simply, Earthlife. The arcane distinctions among these swarming variations on a common theme may be left to specialists or graduate students. The pretensions and conceits of this or that species can readily be ignored. There are, after all, so many worlds about which an extraterrestrial biologist must know. It will be enough if a few salient and generic characteristics of life on yet another obscure planet are noted for the cavernous recesses of the galactic archives.

Chapter 8

SEX AND DEATH

[S]ex endows the individual with a dumb and powerful instinct, which carries his body and soul continually towards another; makes it one of the dearest employments of his life to select and pursue a companion, and joins to possession the keenest pleasure, to rivalry the fiercest rage, and to solitude an eternal melancholy. What more could be needed to suffuse the world with the deepest meaning and beauty?

GEORGE SANTAYANA,
The Sense of Beauty (1896)[1]

Death is the great reprimand which the will to live, or more especially the egoism which is essential to this, receives through the course of nature; and it may be conceived of as a punishment for our existence. It is the painful loosening of the knot which the act of generation had tied . . .

ARTHUR SCHOPENHAUER,
The World as Will and Idea, Supplements[2]

Fireflies out on a warm summer's night, seeing the urgent, flashing, yellow-white phosphorescence below them, go crazy with desire; moths cast to the winds an enchantment potion that draws the opposite sex, wings beating hurriedly, from kilometers away; peacocks display a devastating corona of blue and green and the peahens are all aflutter; competing pollen grains extrude tiny tubes that race each other down the flower's female orifice to the waiting egg below; luminescent squid present rhapsodic light shows, altering the pattern, brightness, and color radiated from their heads, tentacles, and eyeballs; a tapeworm diligently lays a hundred thousand fertilized eggs in a single day; a great whale rumbles through the ocean depths uttering plaintive cries that are understood hundreds or thousands of kilometers away, where another lonely behemoth is attentively listening; bacteria sidle up to one another and merge; cicadas chorus in a collective serenade of love; honeybee couples soar on matrimonial flights from which only one partner returns; male fish spray their spunk over a slimy clutch of eggs laid by God-knows-who; dogs, out cruising, sniff each other's nether parts, seeking erotic stimuli; flowers exude sultry perfumes and decorate their petals with garish ultraviolet advertisements for passing insects, birds, and bats; and men and women sing, dance, dress, adorn, paint, posture, self-mutilate, demand, coerce, dissemble, plead, succumb, and risk their lives. To say that love makes the world go around is to go too far. The Earth spins because it did so as it was formed and there has been nothing to stop it since. But the nearly maniacal devotion to sex and love by most of the plants, animals, and microbes with which we are familiar is a pervasive and striking aspect of life on Earth. It cries out for explanation.

What is all this in aid of? What is this torrent of passion and obsession about? Why will organisms go without sleep, without food, gladly put themselves in mortal danger for sex? Some beings, among them good-sized plants and animals such as dandelions, salamanders, some

lizards and fish, can reproduce without sex. For more than half the
history of life on Earth organisms seem to have done perfectly well
without it. What good is sex?

What's more, sex is expensive. It takes formidable genetic programming to wire in seductive songs and dances; to manufacture sexual
pheromones; to grow heroic antlers used only in defeating rivals; to
establish interlocking parts, rhythmic motions, and mutual zest for
sex. All this represents a drain on energy resources that could just as
well be used for something of more obvious short-term benefit to the
organism. Also, some of what the beings of Earth do or endure for sex
endangers them directly: The displaying peacock is much more vulnerable to predators than if he were inconspicuous, fearful, and dun-
colored. Sex provides a convenient and potentially deadly channel for
the transmission of disease. All these costs must be more than offset
by the benefits of sex. What are those benefits?

———

Embarrassingly, biologists don't fully understand what sex is for. In
this respect the situation has hardly changed since 1862 when Darwin
wrote

> We do not even in the least know the final cause of sexuality; why
> new beings should be produced by the union of the two sexual ele-
> ments . . . The whole subject is as yet hidden in darkness.

Through 4 billion years of natural selection, instructions have been
honed and fine-tuned—more elaborate, more redundant, more fool-
proof, more multiply capable instructions—sequences of As, Cs, Gs,
and Ts, manuals written out in the alphabet of life in competition with
other similar manuals published by other firms. The organisms become
the means *through* which the instructions flow and copy themselves,
by which new instructions are tried out, *on* which selection operates.
"The hen," said Samuel Butler, "is the egg's way of making another
egg." It is on this level that we must understand what sex is for.

We do understand much of the molecular machinery of sex. To
begin with, let's consider some of those microbial beings that routinely
do what many people would consider impossible—reproducing without sex*: Once every generation their nucleic acids faithfully copy

---

* In vitro fertilization is of course still sex.

themselves out of the A, C, G, and T molecular building blocks they manufacture for the purpose. The two functionally identical DNAs then each take half the cell and run—a little like a property settlement in a divorce. Sometime later, the process repeats itself. Each generation is a dreary repetition of the one before, and every organism is the spitting image—nearly identical, down to the last mitochondrion and flagellar propulsion system—of its single parent. If the organism is well-adapted and the environment repetitive and static, this arrangement might work well. The monotony is broken, rarely, by mutation. But mutation, as we've stressed, is random and much more likely to do harm than good. All subsequent generations will be afflicted unless, improbably, there's a compensating mutation down the line. The pace of evolution under such circumstances must be slow, as indeed seems to be reflected in the fossil record between 3.5 and about 1 billion years ago—until the invention of sex.

Now, instead of slow, random change in the genetic materials, imagine that you could in one step glue onto part of the existing messages a long, complex set of new instructions—not merely a change in one letter of one word of the DNA, but whole volumes of consumer-tested manuals. Imagine the same kind of reshuffling occurring in subsequent generations. This is a dumb idea if you're ideally adapted to an unchanging or very marginal environment; then any change is for the worse. But if the world you must adapt to is heterogeneous and dynamic, evolutionary progress is better served when reams of new genetic instructions are made available in each generation than when all there is to deal with is an occasional conversion of an A into a C. Also, if you can reshuffle genes, you or your descendants can get out of the trap set by the accumulation, generation after generation, of deleterious mutations.[3] Harmful genes can quickly be replaced by advantageous ones. Sex and natural selection work as a kind of proofreader, replacing the inevitable mutational errors by uncontaminated instructions. This may be why the eukaryotes diversified—into the separate evolutionary lines that would lead to protozoa (like paramecia), plasmodia (like those that cause malaria), algae, fungi, all land plants and all animals—just around the time that eukaryotes hit upon sex.

Some modern organisms—ranging from bacteria to aphids to aspens—sometimes reproduce sexually and sometimes asexually. They can go either way. Others—dandelions, for example, and certain whiptail lizards—have recently evolved from sexual to asexual forms, as seems clear from their anatomy and behavior: Dandelions produce flowers and nectar that are useless for their current reproductive style; no matter how busy the bees are, they cannot be agents for dandelion fertilization. In the whiptail lizards, everyone is female and the hatchlings have no biological fathers. But reproduction still requires heterosexual foreplay—the formality of copulation with males of other, still sexual, lizard species, even though they cannot impregnate these females, or a ritual pseudocopulation with other females of the same species.[4] Apparently, we are observing these dandelions and lizards so soon after their evolution from sexual to asexual beings that there has been insufficient time for the scripts and props of sex to have withered away. Perhaps there are circumstances when it's wise to reproduce sexually and others when it isn't; certain beings may prudently cycle from one state to the other, depending on the external environment. This option is, however, unavailable to us. We are stuck with sex.

Today a reshuffling of genetic instructions, similar to what happens in sex, occurs—oddly—in infection: A microbe enters a larger organism, evades its defenses, and insinuates its nucleic acid onto that of its host. There's an intricate machinery in the cell, idling and ready to go, which reads and replicates preexisting sequences of A, C, G, and T. The machinery's not good enough, though, to distinguish foreign nucleic acids from native ones. It's a printing press for instruction manuals, and it will copy anything when its buttons are pushed. The parasite pushes the buttons, the cell's enzymes are issued new instructions, and hordes of newly minted parasites are spewed out, itching for more subversion.

Occasionally, the dead manage to have sex and generate offspring. When a bacterium dies, its contents are spilled into the surroundings. Its nucleic acids don't know much about the death of the bacterium and even as they slowly fall to pieces, the fragments remain for a time functional—like the severed leg of an insect. Should such a fragment be ingested by a passing (and intact) bacterium, it may be incorporated into the resident nucleic acids. Perhaps it is used as an independent record of what undamaged instructions should say, helpful in repairing DNA altered by oxygen. Maybe this extremely rudimentary form of sex arose along with the Earth's oxygen atmosphere.

Bizarre chimerical gene combinations happen more rarely—for example, between bacteria and fish (not only are there bacterial genes in fish today; there are also fish genes in bacteria), or baboons and cats. They seem to have been brought about by a virus attaching itself to the DNA of a host organism, reproducing with and accommodating to the host over the generations, and then shaking loose to infect another species while carrying some of the original host's genes with it. Cats are known to have acquired a baboon virogene somewhere on the shores of the Mediterranean Sea 5 to 10 million years ago.[5] Viruses are looking more and more as if they are peripatetic genes that cause disease only incidentally. But if genetic exchanges can occur today in such widely divergent organisms, it must be far easier for them to occur, by accident, in organisms of the same or very closely related species. Perhaps sex started out as an infection, becoming later institutionalized by the infecting and infected cells.

Two distant relatives, members of the same species, each in the process of replication, find their nucleic acid strands, one from each, laid down, cozily, alongside one another. A short segment of one very long sequence might be, say,

   . . . ATG AAG TCG ATC CTA . . .

and the corresponding segment of the other

   . . . TAC TTC GGG CGG AAT . . .

The long nucleic acid molecules both break apart at the same place in the sequence (here, just after AAG in the first molecule and TTC in the second), whereupon they recombine, each picking up a segment of the other:

   . . . ATG AAG GGG CGG AAT . . .

and

   . . . TAC TTC TCG ATC CTA . . .

Because of this genetic recombination, there are two new sequences of instructions and therefore two new organisms in the world—not exactly chimeras, since they come from the same species, but nevertheless each constituting a set of instructions that may never before have coexisted in the same being.

A gene, as we've said, is a sequence of perhaps thousands of As, Cs, Gs, and Ts which codes for a particular function, usually by synthesizing a particular enzyme. When DNA molecules are severed, just prior to recombination, the cut occurs at the beginning or the end of a gene, and almost never in its middle. One gene may have many functions. Important characteristics of the organism—height, say, aggressiveness, coat color, or intelligence—will generally be the consequence of many different genes acting in concert.

Because of sex, different combinations of genes can now be tried out, to compete with the more conventional varieties. A promising set of natural experiments is being performed. Instead of generations patiently waiting in line for a lucky sequence of mutations to occur—it might take a million generations for the right one, and the species might not be able to wait that long—the organism can now acquire new traits, new characteristics, new adaptations wholesale. Two or more mutations that don't do much good by themselves, but that confer an enormous benefit when working in tandem, might be acquired from widely separated hereditary lines. The advantages (for the species, at least) seem clear, if only the costs were bearable. Genetic recombination provides a treasure trove of variability on which natural selection can act.[6]

Another proposed explanation for the persistence of sex, wonderful in its novelty, invites us to consider the age-old arms race between parasitic microbes and their hosts. There are more disease microorganisms in your body at this moment than there are people on Earth. A single bacterium reproducing twice an hour will leave a million successive generations during your lifetime. With so many microbes and so many generations, an immense number of microbial varieties are available for selection to operate on—especially selection to overcome your body's defenses. Some microbes change the chemistry and form of their surfaces faster than the body can generate new model antibodies; these tiny beings routinely outwit at least some parts of the human immune system. For example, an alarming 2% of the plasmodium parasites that cause malaria significantly change their shapes and styles of stickiness each generation.[7] In light of the formidable adaptive powers of disease microorganisms, a real danger would arise if we humans were genetically identical, generation after generation. Very quickly, the blur of evolving pathogens might have our number. A variety that outsmarts our defenses might click into place. But if our DNA is reassorted every generation, we have a much better chance of keeping ahead of the potentially deadly infestation of disease microbes.[8] In this highly regarded hypothesis, sex provides essential confusion to our enemies and is the key to health.

———

Because females and males are physiologically different, they sometimes pursue different strategies, each to propagate its own hereditary line; and these strategies, while of course not wholly incompatible, introduce a certain element of conflict in the relations between the sexes. In many species of reptiles, birds, and mammals, the female produces only a small number of eggs at a time, perhaps only once a year. It then makes evolutionary sense for her to be discriminating in her choice of mates, and devoted to nurturing the fertilized eggs and the young.

The male, on the other hand, with plentiful sperm cells—up to hundreds of millions per ejaculation and the capability of many ejaculations a day in a healthy young primate—can often better continue *his* hereditary line through numerous and indiscriminate matings, if he can pull it off. He may be much more ardent and eager, and at the same time much more likely to drift from partner to partner—cajoling, displaying, intimidating, and impregnating as many females as possible. Moreover, since there are other males with identical strategies, a male can't be sure that a particular fertilized egg or hatchling or cub is his; why should he spend time and effort nurturing and raising a youngster that might not even carry his genes? The investment might benefit his rival's descendants and not his own. Better to be off fertilizing more females.

This is by no means an invariable pattern, though; there are species in which the female is eager to mate with many males, and there are species in which the male plays a major, even a primary, role in raising the young. Over 90% of the known species of birds are "monogamous"; so are 12% of the monkeys and apes, to say nothing of all the wolves, jackals, coyotes, foxes, elephants, shrews, beavers, and miniature antelope.[9] However, monogamous doesn't mean sexually exclusive; in many species in which the male helps raise the children and provides care for their mother, he also is sneaking out for a little sex on the side; and she is often receptive to other males. Biologists call it a "mixed mating strategy," or "extra-pair copulation." As much as 40% of the young reared by "monogamous" bird pairs are revealed by DNA fingerprinting to have been sired by extramural encounters, and numbers almost as large may apply to humans. Still, the motif of nurturing females, who are choosy about their sex partners, and males given to sexual adventure and many partners is very widespread, especially among the mammals.

———

There's a good deal of plumbing, odor signaling, and other machinery in higher organisms to get the genes of one organism in contact with those of another, so the molecules can lie down next to one another and recombine. But that's mere hardware. The central sexual event, from bacteria to humans, is the exchange of DNA sequences. The hardware serves the purpose of the software.

In its beginning, all sex must have been fumbling, confused, haphazard, the microbial equivalent of bedroom farce. But the advantages that sex confers on future generations seem to be so great that, provided the costs were not too high, selection for improved sexual hardware must soon have been up and running, along with whatever new software was required to stiffen a resolve for sexual congress. Passionate organisms, other things being equal, leave more descendants than those of more tepid dispositions. Unenlightened on the selective advantage of new DNA combinations, organisms nevertheless developed an overwhelming compulsion to trade their hereditary instructions. Like hobbyists who exchange comic books, postage stamps, baseball cards, enameled pins, foreign coins, or celebrity autographs, they didn't think it out; they just couldn't help themselves. Trade is at least a billion years old.

Two paramecia may conjugate, as it's called, exchange genetic material, and then drift apart. Recombination does not require gender. There aren't boy bacteria and girl bacteria, and bacteria do not have sex—do not recombine segments of their DNA—with every act of reproduction. Sexual plants and animals do. However you bring it about, recombination means that every new being has two parents rather than only one. It means that members of the same species— and, except during courtship, the members of most species are solitary and asocial—must arrange a centrally important act that can only be performed in pairs. The two genders might have slightly different goals and strategies, but sex calls, as an absolutely minimum requirement, for cooperation.

Once so powerful an impetus is let out into the world, it might lead, through slow and natural stages, to other kinds of cooperation. Sex brings an entire species together—not just by protecting one another from the cumulative build-up of dangerous mutations, not just by providing new adaptations to a changing environment, but also in the sense of an ongoing, collective enterprise, cross-linking different hereditary lines. This is very different from the asexual practice, where there are many parallel lines of descent, the organisms nearly identical within each line, generation upon generation, and no close relatives between lines.

When sex becomes central to reproduction, the attractiveness of each sex to the other, and the drama of choosing among rivals is moved to center stage. Associated themes include sexual jealousy; real and mock fighting; careful noting of the identities and whereabouts of potential sexual partners and rivals; coercion and rape—all of which in turn lead swiftly, as Darwin pointed out, to the evolution of strange and wonderful appendages, color patterns, and courting behavior that humans often find beautiful, even in members of distantly related species. Darwin thought this sexual selection might be the origin of the human aesthetic sense. Here is a twentieth-century biologist on what sexual selection has brought forth in birds:

> crests, wattles, ruffs, collars, tippets, trains, spurs, excrescences on wings and bills, tinted mouths, tails of weird or exquisite form, bladders, highly coloured patches of bare skin, elongated plumes, brightly hued feet and legs . . . The display is nearly always beautiful[10]

—especially to the bird of the opposite sex who chooses sexual partners partly on the basis of their good looks. Fashions in beauty then spread rapidly through the population, even if the style isn't a bit of good in, say, evading predators. Indeed, they spread even if the lifetimes of those who adopt them are thereby considerably shortened, provided the benefit for future generations is sufficiently large. One promising explanation of the showy displays of male birds and fish to the females of their species is that all this is to assure her of his health and prospects.[11] Bright plumage and shiny scales demonstrate the absence of an infestation of ticks or mites or fungi, and females—unsurprisingly—prefer to mate with males unburdened by parasites.

———

The sockeye salmon exhaust themselves swimming up the mighty Columbia River to spawn, heroically hurdling cataracts, in a single-minded effort that works to propagate their DNA sequences into future generations. The moment their work is done, they fall to pieces. Scales flake off, fins drop, and soon—often within hours of spawning—they are dead and becoming distinctly aromatic. They've served their purpose. Nature is unsentimental. Death is built in.

This is very unlike the far less dramatic asexual reproduction of beings like paramecia, where, pretty closely, remote descendants are genetically identical to their distant ancestors. The ancient organisms can with some justice be described as still alive. With all its manifold advantages, sex brought something else: the end of immortality.

Sexual organisms do not generally reproduce by fission, by splitting in two. The big macroscopic sexual organisms reproduce by making special sex cells, often the familiar sperm and egg, that assemble the genes of the next generation. These cells survive just long enough to accomplish their task, and are hardly able to do anything else at all. In sexual beings, the parent does not evenhandedly distribute its body parts and transmute into two offspring; rather, the parent eventually dies, leaving its world to the next generation, which in its time also dies. Individual asexual organisms die by mistake—when they run out of something, or when they experience a lethal accident. Sexual organisms are *designed* to die, preprogrammed to do so. Death serves as a poignant reminder of our limitations and frailties—and of the bond with our ancestors who, in a way, died that we might live.

The more active the enzymes devoted to DNA proofreading and repair in big multicellular organisms, the longer the life span tends to be. When these enzymes—themselves of course synthesized under the control of the organism's DNA—become sparse or inactive, replication errors proliferate and are compounded, and the individual cells increasingly try to implement nonsense instructions. By relaxing the extreme fidelity of its replication, DNA can arrange, at the appropriate moment, for its own death, and that of the organism doing its bidding.

Where sex mandates the death of the individual organism, it provides life to the hereditary line and the species. Still, no matter how many consecutive generations have been recorded of nearly identical asexual beings, eventually the accumulation of deleterious mutations destroys the clone. Eventually, there is a generation where all the individuals are smaller and more feeble, and then you can hear extinction knocking. Sex is the way out. Sex rejuvenates the DNA, revivifies the next generation. There's a reason we rejoice in it.

A billion years ago, a bargain was struck: the delights of sex in exchange for the loss of personal immortality.[12] Sex and death: You can't have the former without the latter. Nature, she drives a hard bargain.

———

The first living things had no parents. For about 3 billion years, everyone had one parent, and was pretty close to immortal. Now, many beings have two parents and are unambiguously mortal. There are, so far as we know, no lifeforms that regularly have three parents or more*—although it doesn't seem much more difficult, in terms of plumbing and allure, to arrange than two. The variety of genetic recombination would be correspondingly greater. And the ability to recognize an error in the message (as the deviant sequence when the three are intercompared) would be much improved. Perhaps on some other planet . . .

On hearing the love call of the male, the female cowbird promptly adopts a come-hither posture, unmistakably indicating her readiness for copulation. Mature female cowbirds raised in isolation will adopt this posture upon hearing the male's serenade for the very first time. The male, if he's raised in isolation, if he's never heard the cowbird love song in his life, still knows it by heart. The musical score, and

---

* Although strands from two different dead bacteria might, on rare occasions, be incorporated by a live bacterium.

information on how to appreciate it, are encoded in their DNA. Perhaps on hearing it the female, at least a little, falls in love with him. Perhaps, on seeing her fetching response to his music, the male, at least a little, falls in love with her.

In contrast to parental care and kin selection, which are so prominent among the birds and mammals, many frogs and fish eat their young. Cannibalism is a commonplace—not just in extraordinary circumstances such as overcrowding or famine, but under normal, everyday conditions: The little ones are plentiful, they've gone through all the effort of fattening themselves up into convenient and nutritious packages, only a few need to survive to continue the hereditary line, and an affectionate family life that might exert a restraining influence is lacking. But parental care is not restricted to the birds and mammals. It pops up here and there among fish and even invertebrates. Dung beetle mothers, who have laid their eggs in the "brood balls" they've skillfully rolled out of animal feces, dote on their young. And Nile crocodiles, whose powerful jaws can bite a human in two, walk about carefully carrying their little hatchlings, who peer out from between the mothers' teeth "like sightseers on a bus."[13]

Even if it is merely genetic sequences working out their self-interest, something that an outside observer might interpret as love has been building in the kingdom of the animals, especially since the extinction of the dinosaurs. With the origin of the primates, it begins its full flowering. It works to bind a species together, in effect to fashion something approaching a common loyalty.

The primacy of reproduction, the sense that the next generation is all, or nearly all, that matters, is made most clear in those many species that promptly die, both sexes, in huge numbers, immediately after conception has occurred and precautions have been taken to safeguard the fertilized eggs. In other species, including our own, the parents play a vital role in protecting and educating the young, and so for them there is life after copulation. Otherwise, the parental generation would have served its purpose, and been hustled off before it came into competition for scarce resources with its own progeny.

The adaptive value of getting DNA strands together has been so substantial that vast changes have been worked in anatomy, physiology, and behavior to accommodate the needs of these molecules. While cooperation was present long before sex—in stromatolite colonies, say, or in the symbiotic relationships of chloroplasts and mito-chondria with the cell—sex has introduced a new kind of cooperation, common endeavor, and self-sacrifice into the world. In the differing sexual strategies of male and female, sex has also introduced a novel creative tension—one that cries out for reconciliation and compro-mise—as well as a potent new motive for competition. Our own spe-cies is as good an example as any of the nearly determining role of sex —not just the sex act itself, but all the attendant preparation, conse-quences, associations, and obsessions—in establishing much of the personality, character, agenda, and drama of life on Earth.

ON IMPERMANENCE

Only
for sleep we come,
for dreams.

Lie! It is a lie.
We come to live on Earth.

As a weed we become
each springtime,

swell green, our hearts
open,

the body makes a few flowers
and drops away withered somewhere.

Poems of the Aztec Peoples[14]

Chapter 9

WHAT THIN PARTITIONS...

How instinct varies in the grovelling swine,
Compar'd, half-reasoning elephant, with thine!
'Twixt that, and reason, what a nice barrier,
Forever sep'rate, yet forever near!
Remembrance and reflection how ally'd!
What thin partitions sense from thought divide!

ALEXANDER POPE,
Essay on Man[1]

M ost people would rather be alive than dead. But why? It's hard to give a coherent answer. An enigmatic "will to live" or "life force" is often cited. But what does that explain? Even victims of atrocious brutality and intractable pain may retain a longing, sometimes even a zest, for life. Why, in the cosmic scheme of things, one individual should be alive and not another is a difficult question, an impossible question, perhaps even a meaningless question. Life is a gift that, of the immense number of possible but unrealized beings, only the tiniest fraction are privileged to experience. Except in the most hopeless of circumstances, hardly anyone is willing to give it up voluntarily—at least until very old age is reached.

A similar puzzlement attaches to sex. Very few, at least today, have sex for the conscious purpose of propagating the species or even their own personal DNA; and such a decision for such a purpose, coolly and rationally entered into, is exceedingly rare in adolescents. (For most of the tenure of humans on Earth, the average person did not live much beyond adolescence.) Sex is its own reward.

Passions for life and sex are built into us, hardwired, pre-programmed. Between them, they go a long way toward arranging for many offspring with slightly differing genetic characteristics, the essential first step for natural selection to do its work. So we are the mostly unconscious tools of natural selection, indeed its willing instruments. As deeply as we can go in assessing our own feelings, we do not recognize any underlying purpose. All that is added later. All the social and political and theological justifications are attempts to rationalize, after the fact, human feelings that are at the same time utterly obvious and profoundly mysterious.

Now imagine us with no interest at all in "explaining" such matters, no weakness for reason and contemplation. Suppose you unquestioningly accepted these predispositions for surviving and reproducing, and spent your time solely in fulfilling them. Might that be something like the state of mind of most beings? Every one of us can recognize these two modes coexisting within us. A moment of introspection is often all it takes. Religious writers have described them as our animal and spiritual states. In everyday speech, the distinction is between feeling and thought. Inside our heads there seem to be two different ways of dealing with the world, the second, in the sweep of evolutionary time, arisen in earnest only lately.

——

Consider the world of the tick.[2] Plumbing aside, what must it do to reproduce its kind? Ticks often have no eyes. Males and females find each other by aroma, olfactory cues called sex pheromones. For many ticks the pheromone is a molecule called 2,6-dichlorophenol. If C stands for a carbon atom, H for hydrogen, O for oxygen, and Cl for chlorine, this ring-shaped molecule can be written $C_6H_3OHCl_2$. A little 2,6-dichlorophenol in the air and ticks go wild with passion.[3]

After mating, the female climbs up a bush or shrub and out onto a twig or leaf. How does she know which way is up? Her skin can sense the direction from which light is coming, even if she cannot generate an optical image of her surroundings. Poised out on the leaf or twig, exposed to the elements, she waits. Conception has not yet occurred. The sperm cells within her are neatly encapsulated; they've been put in long-term storage. She may wait for months or even years without eating. She is very patient.

What she's waiting for is a smell, a whiff of another specific molecule, perhaps butyric acid, which can be written C_3H_7COOH. Many mammals, including humans, give off butyric acid from their skin and sexual parts. A small cloud of the stuff follows them around like cheap perfume. It's a sex attractant for mammals. But ticks use it to find food for prospective mothers. Smelling the butyric acid wafting up from below, the tick lets go. She drops from her perch and falls through the air, legs akimbo. If she's lucky, she lands on the passing mammal. (If not, she falls to the ground, shakes herself off, and tries to find another bush to climb.)

Clinging to the fur of her unsuspecting host, she works her way through the thicket to find a less hairy spot, a patch of nice warm bare skin. There, she punctures the epidermis and drinks her fill of blood.*

The mammal may feel a sting and rub the tick off, or intently comb through its hair and pick it off. Rats may spend as much as one-third their waking hours grooming themselves. Ticks can draw a great deal of blood, they secrete neurotoxins, they carry disease microbes. They're dangerous. Too many of them on a mammal at the same time can lead to anemia, loss of appetite, and death. Monkeys and apes meticulously search through each other's fur; this is one of their principal cultural idioms. When they find a tick, they remove it with their precision grip and eat it. As a result, they are remarkably free from such parasites in the wild.

If the tick has avoided the hazards of grooming, and has become engorged with blood, she drops heavily to the ground. Thus fortified, she unseals the chamber with the stored sperm cells, lays the fertilized eggs in the soil (perhaps ten thousand of them) and dies—her descendants left to continue the cycle.

Note how simple are the sensory abilities required of the tick. They may have been feeding on reptile blood before the first dinosaurs evolved, but their repertoire of essential skills remains fairly meager. The tick must be crudely responsive to sunlight so she knows which way is up; she must be able to smell butyric acid so she knows when to fall animalward; she must be able to sense warmth; she must know how to inch her way around obstacles. This is not asking much. Today we have very small photocells easily able to find the sun on a cloudless day. We have many chemical analytic instruments that can detect small amounts of butyric acid. We have miniaturized infrared sensors that sense heat. Indeed, all three such devices have been flown on spacecraft to explore other worlds—the Viking missions to Mars, for example. A new generation of mobile robots being developed for planetary exploration is now able to amble over and around large obstacles. Some progress in miniaturization would be needed, but we are not very far from being able to build a little machine that could duplicate—indeed far surpass—the central abilities of the tick to sense the

---

* It's not the taste of the blood that attracts her, but the warmth. If she drops onto a butyric acid–scented toy balloon filled with warm water, she will readily puncture it and, an inept Dracula, gorge herself on tap water.

outside world. And we certainly could equip it with a hypodermic syringe. (Harder for us to duplicate just yet would be its digestive tract and reproductive system. We are very far from being able to simulate from scratch the biochemistry of a tick.)

What would it be like inside the tick's brain? You would know about light, butyric acid, 2,6-dichlorophenol, the warmth of a mammal's skin, and obstacles to clamber around or over. You have no image, no picture, no vision of your surroundings; you are blind. You are also deaf. Your ability to smell is limited. You are certainly not doing much in the way of thinking. You have a very limited view of the world outside. But what you know is sufficient for your purpose.[4]

———

There's a thump on the window and you look up. A moth has careened headlong into the transparent glass. It had no idea the glass was there: There have been things like moths for hundreds of millions of years, and glass windows only for thousands. Having bumped its head against the window, what does the moth do next? It bumps its head against the window again. You can see insects repeatedly throwing themselves against windows, even leaving little bits of themselves on the glass, and never learning a thing from the experience.

Clearly there's a simple flying program in their brains, and nothing that allows them to take notice of collisions with invisible walls. There's no subroutine in that program that says, "If I keep bumping into something, even if I can't see it, I should try to fly around it." But developing such a subroutine carries with it an evolutionary cost, and until lately there were no penalties levied on moths without it. They also lack a general-purpose problem-solving ability equal to this challenge. Moths are unprepared for a world with windows.

If we have here an insight into the mind of the moth, we might be forgiven for concluding that there isn't much mind there. And yet, can't we recognize in ourselves—and not just in those of us gripped by a pathological repetition-compulsion syndrome—circumstances in which we keep on doing the same stupid thing, despite irrefutable evidence it's getting us into trouble?

We don't always do better than moths. Even heads of state have been known to walk into glass doors. Hotels and public buildings now affix large red circles or other warning signs on these nearly invisible barriers. We too evolved in a world without plate glass. The difference between the moths and us is that only rarely do we shake ourselves off and then walk straight into the glass door again.

Like many other insects, caterpillars follow scent trails left by their fellows. Paint the ground with an invisible circle of scent molecule and put a few caterpillars down on it. Like locomotives on a circular track, they'll go around and around forever—or at least until they drop from exhaustion. What, if anything, is the caterpillar thinking? "The guy in front of me seems to know where he's going, so I'll follow him to the ends of the Earth"? Almost always, following the scent trail gets you to another caterpillar of your species, which is where you want to be. Circular trails almost never occur in Nature—unless some wiseacre scientist shows up. And so this weakness in their program almost never gets caterpillars into trouble. Again we detect a simple algorithm and no hint of an executive intelligence evaluating discordant data.

When a honeybee dies it releases a death pheromone, a characteristic odor that signals the survivors to remove it from the hive. This might seem a supreme final act of social responsibility. The corpse is promptly pushed and tugged out of the hive. The death pheromone is oleic acid [a fairly complex molecule, $CH_3(CH_2)_7CH = CH(CH_2)_7COOH$, where $=$ stands for a double chemical bond]. What happens if a live bee is dabbed with a drop of oleic acid? Then, no matter how strapping and vigorous it might be, it is carried "kicking and screaming" out of the hive.[5] Even the queen bee, if she's painted with invisible amounts of oleic acid, will be subjected to this indignity.

Do the bees understand the danger of corpses decomposing in the hive? Are they aware of the connection between death and oleic acid? Do they have any idea what death is? Do they think to check the oleic acid signal against other information, such as healthy, spontaneous movement? The answer to all these questions is, almost certainly, No. In the life of the hive there's no way that a bee can give off a detectable whiff of oleic acid other than by dying. Elaborate contemplative machinery is unnecessary. Their perceptions are adequate for their needs.

Does the dying insect make a special last effort to generate oleic acid, to benefit the hive? More likely, the oleic acid derives from a malfunction of fatty acid metabolism around the time of death, which is recognized by the highly sensitive chemical receptors in the survivors. A strain of bees that had a slight tendency to manufacture a death pheromone would do better than one in which decomposing, disease-ridden dead bodies were littering the hive. And this would be true even if no other bee in the hive were a close relative of the recently departed. On the other hand, since they are all close relatives, special manufacture of a death pheromone can be understood perfectly well in terms of kin selection.

———

So here's a bejeweled insect, elegantly architectured, prancing among the dust grains in the noonday sun. Does it have any emotions, any consciousness? Or is it only a subtle robot made of organic matter, a carbon-based automaton packed with sensors and actuators, programs and subroutines, all ultimately manufactured according to the DNA instructions? (Later, we will want to look more closely at what "only" means.) We might be willing to grant the proposition that insects are robots; there's no evidence, so far as we know, that compellingly argues the contrary; and most of us have no deep emotional attachments to insects.

In the first half of the seventeenth century, René Descartes, the "father" of modern philosophy, drew just such a conclusion. Living in an age when clocks were at the cutting edge of technology, he imagined insects and other creatures as elegant, miniaturized bits of clockwork—"a superior race of marionettes," as Huxley described it,[6] "which eat without pleasure, cry without pain, desire nothing, know nothing, and only simulate intelligence as a bee simulates a mathematician" (in the geometry of its hexagonal honeycombs). Ants do not have souls, Descartes argued; automatons are owed no special moral obligations.

What then are we to conclude when we find similar very simple behavioral programs, unsupervised by any apparent central executive control, in much "higher" animals? When a goose egg rolls out of the nest, the mother goose will carefully nudge it back in. The value of this behavior for goose genes is clear. Does the mother goose who has been incubating her eggs for weeks understand the importance of retrieving one that has rolled away? Can she tell if one is missing? In fact, she will retrieve almost anything placed near the nest, including ping-pong balls and beer bottles. She understands something, but, we might say, not enough.

> If a chick is tied to a peg by one leg, it peeps loudly. This distress call makes the mother hen run immediately in the direction of the sound with ruffled plumage, even if the chick is invisible. As soon as she catches sight of the chick, she begins to peck furiously at an imaginary antagonist. But if the fettered chick is set before the mother hen's eyes under a glass bell, so that she can see it but not hear its distress call, she is not in the least disturbed by the sight of him.
>   . . . The perceptual cue of peeping normally comes indirectly from an enemy who is attacking the chick. According to plan, this sensory cue is extinguished by the effector cue of beak thrusts, which chase the foe away. The struggling, but not-peeping chick is not a sensory cue that would release a specific activity.[7]

Male tropical fish show fighting readiness when they see the red markings of other males of their species. They also get agitated when they glimpse a red truck out the window. Humans find themselves sexually aroused by looking at certain arrangements of very small dots on paper or celluloid or magnetic tape. They pay money to look at these patterns.

So now where are we? Descartes was prepared to grant that fish and poultry are also subtle automatons, also soulless. But then what about humans?

Descartes was here treading on dangerous ground. He had before him the chastening example of the aged Galileo, threatened with torture by the self-styled "Holy Inquisition" for maintaining that the Earth turns once each day, rather than the view, clearly expressed in the Bible, that the Earth is stationary and the heavens race around us once each day. The Roman Catholic Church was quite prepared to coerce conformity—to intimidate, torture, and murder to force people to think as it did. At the very beginning of Descartes's century, the Church had burned the philosopher Giordano Bruno alive because he thought for himself, spoke out, and would not recant. And here, the proposal that animals are clockwork automatons was a far riskier and theologically more sensitive matter than whether the Earth turns—touching not peripheral but central dogmas: free will, the existence of the soul. As on other issues, Descartes walked a fine line.

We "know" we are more than just a set of extremely complex computer programs. Introspection tells us that. That's the way it feels. And so Descartes, who attempted a thorough, skeptical examination of why he should believe anything, who made famous the proposition *Cogito, ergo sum* ("I think, therefore I am"), granted immortal souls to humans, and to no one else on Earth.

But we, who live in a more enlightened time, when the penalties for disquieting ideas are less severe, not only may, but have an obligation to, inquire further—as many since Darwin have done. What, if anything, do the other animals think? What might they have to say if properly interrogated? When we examine some of them carefully, do we not find evidence of executive controls weighing alternatives, of branched contingency trees? When we consider the kinship of all life on Earth, is it plausible that humans have immortal souls and all other animals do not?

The moth doesn't need to know how to fly around the pane of glass, or the goose to retrieve eggs but not beer bottles—again because glass windows and beer bottles have not been around long enough to have been a significant factor in the natural selection of insects and birds. The programs, circuits and behavioral repertoires are simple when no benefit accrues from their being complex. Complex mechanisms evolve when the simple ones will not do.

In Nature, the goose's egg-retrieval program is adequate. But when the goslings hatch, and especially just before they're ready to leave the nest, the mother is delicately attuned to the nuances of their sounds, looks, and (perhaps) smells. She has learned about her chicks. Now, she knows her own very well, and would not confuse them with someone else's goslings, however similar they may seem to a human observer.

In species of birds where mix-ups are likely, where the young may fledge and mistakenly land in a neighboring nest, the machinery for maternal recognition and discrimination is even more elaborate. The goose's behavior is flexible and complex when rigid and simple behavior is too dangerous, too likely to lead to error; otherwise it *is* rigid and simple. The programs are parsimonious, no more complex than they need be—if only the world does not produce too much novelty, too many windows and beer bottles.

Consider our prancing insect again. It can see, walk, run, smell, taste, fly, mate, eat, excrete, lay eggs, metamorphose. It has internal programs for accomplishing these functions—contained in a brain of mass, perhaps, only a milligram—and specialized, dedicated organs for carrying the programs out. But is that all? Is there anyone in charge, anyone inside, anyone controlling all these functions? What do we mean by "anyone"? Or is the insect just the sum of its functions, and nothing else, with no executive authority, no director of the organs, no insect soul?

You get down on your hands and knees, look at the insect closely, and you see it cock its head, triangulating you, trying to get a sense of this immense, looming, three-dimensional monster before it. The fly strides unconcernedly; you lift the rolled-up newspaper and it quickly buzzes off. You turn on the light and the cockroach stops dead in its tracks, regarding you keenly. Move toward it and it scampers into the woodwork. We "know" such behavior is due to simple neuronal subroutines. Many scientists get nervous if you ask about the consciousness of a housefly or a roach. But sometimes you get an eerie feeling that the partitions separating programs from awareness may be not just thin, but porous.

We know the insect decides who to eat, who to run away from, who to find sexually attractive. On the inside, within its tiny brain, does it have no perception of making choices, no awareness of its own existence? Not a milligram's worth of self-consciousness? Not a hint of a hope for the future? Not even a little satisfaction at a day's work well done? If its brain is one millionth the mass of ours, shall we deny it one millionth of our feelings and our consciousness? And if, after carefully weighing such matters, we insist it is still "only" a robot, how sure are we that this judgment does not apply as well to us?

We can recognize the existence of such subroutines precisely because of their unbending simplicity. But if instead we had before us an animal brimming over with complex judgments, branched contingency trees, unpredictable decisions, and a strong executive program, would it seem to us that there is more here than just an elaborate, exquisitely miniaturized computer?

The honeybee scout returns to the hive from a foraging expedition and "dances," rapidly crawling in a particular, fairly complex pattern over the honeycomb. Pollen or nectar may adhere to her body, and she may regurgitate some of her stomach contents for her eager sisters. All this is done in complete darkness, her motions monitored by the spectators through their sense of touch. Given only this information, a swarm of bees then flies out of the hive in the proper direction to the proper distance to a food supply they've never visited as effortlessly as if this was their daily, familiar commute from home to work. They partake of the meal described to them. All this occurs more often when food is scarce or the nectar especially sweet.[8] How to encode the location of a field of flowers into the language of dance, and how to decode the choreography is knowledge present in the hereditary information stored inside the insect. Maybe they are "only" robots, but if so these robots have formidable capabilities.

When we characterize such beings as only robots, we are also in danger of losing sight of the possibilities in robotics and artificial intelligence over the next few decades. Already, there are robots that read sheet music and play it on a keyboard, robots that translate pretty well between two very different languages, robots that learn from their own experiences—codifying rules of thumb never taught to them by their programmers. (In chess, for example, they might learn that it is generally better to position bishops near the center than near the periphery of the board, and then teach themselves circumstances in which an exception to this rule is warranted.) Some open-loop chess-playing robots can defeat all but a handful of human chess masters. Their moves surprise their programmers. Their completed games are routinely analyzed by experts who speculate about what the robot's "strategy," "goals," and "intentions" must have been. If you have a large enough pre-programmed behavioral repertoire and if you are able to learn enough from experience, don't you begin to appear to an outside observer *as if* you're a conscious being making voluntary choices—whatever may or may not be going on inside your head (or wherever you keep your neurons)?[9]

And when you have a massive collection of mutually integrated programs, capability for learned behavior, data-processing prowess, and means of ranking competing programs, might it not start feeling, on the inside, a little bit like thinking? Might our penchant for imagining someone inside pulling the strings of the animal marionette be a peculiarly human way of viewing the world? * Could our sense of executive control over ourselves, of pulling our own strings, be likewise illusory—at least most of the time, for most of what we do? How much are we really in charge of ourselves? And how much of our actual everyday behavior is on automatic pilot?

Among the many human feelings that, although culturally mediated, may be fundamentally preprogrammed, we might list sexual attraction, falling in love, jealousy, hunger and thirst, horror at the sight of blood, fear of snakes and heights and "monsters," shyness and suspicion of strangers, obedience to those in authority, hero worship, dominance of the meek, pain and weeping, laughter, the incest taboo, the infant's smiling delight at seeing members of its family, separation anxiety, and maternal love. There is a complex of emotions attached to each, and thinking has very little to do with any of them. Surely, we can imagine a being whose internal life is nearly wholly composed of such feelings, and nearly devoid of thought.

———

The spider builds her web near our porch light. The fine, tough thread reels out from her spinneret. We first notice the web glistening with tiny droplets after a rainstorm, the proprietor repairing a damaged circumferential strut. The elegant, concentric, polygonal pattern is carefully stabilized with a single guy thread extending to the cowl of the lamp itself, and another to a nearby railing. She repairs the web even in darkness and foul weather. At night, when the light is on, she sits at the very center of her construction, awaiting the hapless insect who is attracted by the light and whose eyesight is so poor that the web is quite invisible. The moment one becomes entangled, news of this event travels to her in waves along the threads. She rushes down a radial strut, stings it, quickly wraps it in a white cocoon, packaging it for future use, and rushes back to her command center—composed, a marvel of efficiency, not even, as far as we can see, a little out of breath.

* One promising finding in artificial intelligence is the discovery that distributed data processing—many small computers working in parallel without much of a central processing unit—does very well, by some standards better than the largest and fastest lone computer. Many little minds working in tandem may be superior to one big mind working alone.

How does she know to design, construct, stabilize, repair, and utilize this elegant web? How does she know to build it near the lamp, to which the insects are attracted? Did she scamper all over the house tallying the abundance of insects in various potential campsites? How could her behavior be pre-programmed, since artificial lights have been invented much too recently to be taken account of in the evolution of spiders?

When spiders are given LSD or other consciousness-altering drugs, their webs become less symmetrical, more erratic, or, we might say, less obsessive, more freeform—but also less effective in catching insects. What has a tripping spider forgotten?

Maybe its behavior is entirely pre-programmed in its ACGT code. But then, couldn't much more complex information be locked away in a much longer, much more elaborate code? Or maybe some of this information is learned from past adventures in spinning and repairing webs, immobilizing and eating prey. But then look how small that spider's brain is. How much more sophisticated behavior might emerge out of the experience of a much larger brain?

The web is anchored opportunistically to a local geometry of lamp cowling, metal railing, and wood siding. That could not per se have been pre-programmed. There must have been some element of choice, of decision making, of connecting a hereditary predisposition to an environmental circumstance never before encountered.

Is she "only" an automaton, unquestioningly performing actions that seem to her the most natural thing in the world—and being rewarded, her behavior reinforced by an ample supply of food? Or might there be a component of learning, decision making, and self-consciousness?

Adopting high standards of engineering precision, she spins her web now. She reaps the reward later, maybe much later. She patiently waits. Does she know what she's waiting for? Does she dream of succulent moths and foolish mayflies? Or does she wait with her mind a blank, idling, thinking of nothing at all—until the telltale tug sends her scurrying down one of the radial struts to sting the struggling insect before it frees itself and escapes? Are we really sure she doesn't have even a faint and intermittent spark of consciousness?

We would guess that some rudimentary awareness flickers in the most humble creatures, and that with increasing neuronal architecture and brain complexity, consciousness grows. "When a dog runs," said the naturalist Jakob von Uexküll, "the dog moves his legs; when a sea urchin runs, the legs move the sea urchin." [10] But even in humans, thinking is often a subsidiary state of consciousness.

If it were possible to peer into the psyche of a spider or a goose, we might detect a kaleidoscopic progression of inclinations—and maybe some premonitions of conscious choice, actions selected from a menu of possible alternatives. What individual nonhuman organisms may perceive as their motivations, what they feel is happening inside their bodies, is for us one of the nearly inaudible counterpoints to the music of life.

When an animal goes out to seek food, it often does so according to a definite pattern. A random search is inefficient, because the path would turn back on itself many times; the same places would then be examined again and again. Instead, while the animal may dart off to left and right, the general search pattern is almost always progressive forward motion. The animal finds itself on new ground. The search for food becomes an exercise in exploration. A passion for discovery is hardwired. It's something we like to do for its own sake, but it brings rewards, aids survival, and increases the number of offspring.

Perhaps animals are almost pure automatons—with urges, instincts, hormonal rushes, driving them toward behavior which in turn is carefully honed and selected to aid the propagation of a particular genetic sequence. Perhaps states of consciousness, no matter how vivid, are as Huxley suggested, "immediately caused by molecular changes in the brain substance." But from the point of view of the animal, it must seem—as it does with us—natural, passionate, and occasionally even thought out. Perhaps a flurry of impulses and intersecting subroutines at times feels something like the exercise of free will. Certainly the animal cannot much have an impression of being impelled *against* its will. It voluntarily chooses to behave in the manner dictated by its contending programs. Mainly, it's just following orders.

So when the days become long enough, it feels an unfocused restlessness, something like spring fever. It hasn't thought through conception, gestation, the optimum season for the birth of the young and the continuance of its genetic sequences; all that is far beyond its abilities. But from the inside it may well feel as though the weather is intoxicating, life is tempestuous, and moonlight becomes you.

———

We do not mean to be patronizing. The depth of understanding exhibited by our fellow creatures is of course limited. So is ours. We also are at the mercy of our feelings. We too are profoundly ignorant about what motivates us. Some of those beings have, as familiar aspects of their everyday lives, sensibilities wholly absent in humans. Other beings have different tastes and appreciations of the outside world—"To a worm in horseradish, the horseradish seems sweet," as an old Yiddish folk adage has it. Beyond that, the horseradish worm lives in a world of smells, tastes, textures, and other sensations unknown to us.

Bumblebees detect the polarization of sunlight, invisible to uninstrumented humans; pit vipers sense infrared radiation and detect temperature differences of 0.01°C at a distance of half a meter; many insects can see ultraviolet light; some African freshwater fish generate a static electric field around themselves and sense intruders by slight perturbations induced in the field; dogs, sharks, and cicadas detect sounds wholly inaudible to humans; ordinary scorpions have microseismometers on their legs so they can detect in pitch darkness the footsteps of a small insect a meter away; water scorpions sense their depth by measuring the hydrostatic pressure; a nubile female silkworm moth releases ten billionths of a gram of sex attractant per second, and draws to her every male for miles around; dolphins, whales, and bats use a kind of sonar for precision echo-location.

The direction, range, amplitude, and frequency of sounds reflected back to echo-locating bats are systematically mapped onto adjacent areas of the bat brain. How does the bat perceive its echo-world? Carp and catfish have taste buds distributed over most of their bodies, as well as in their mouths; the nerves from all these sensors converge on massive sensory processing lobes in their brains, lobes unknown in other animals. How does a catfish view the world? What does it feel like to be inside its brain? There are reported cases in which a dog wags its tail and greets with joy a man it has never met before; he turns out to be the long-lost identical twin of the dog's "master," recognizable by his odor. What is the smell-world of a dog like? Magnetotactic bacteria contain within them tiny crystals of magnetite—an iron mineral known to early sailing ship navigators as lodestone. The bacteria literally have internal compasses that align them along the Earth's magnetic field. The great churning dynamo of molten iron in the Earth's core—as far as we know, entirely unknown to uninstrumented humans—is a guiding reality for these microscopic beings. How does the Earth's magnetism feel to them? All these creatures may be automatons, or nearly so, but what astounding special powers they have, never granted to humans, or even to comic book superheroes. How different their view of the world must be, perceiving so much that we miss.

Each species has a different model of reality mapped into its brain. No model is complete. Every model misses some aspects of the world. Because of this incompleteness, sooner or later there will be surprises —perceived, perhaps, as something like magic or miracles. There are different sensory modalities, different detection sensitivities, different ways the various sensations are integrated into a dynamic mental map of . . . a snake, say, in full hunting slither.

But Descartes was unimpressed. He wrote to the Marquis of Newcastle:

> I know, indeed, that brutes do many things better than we do, but I am not surprised at it; for that, also, goes to prove that they act by force of nature and by springs, like a clock, which tells better what the hour is than our judgment can inform us.[11]

———

As life evolved, the repertoire of feelings expanded. Aristotle thought that "in a number of animals we observe gentleness or fierceness, mildness or cross-temper, courage or timidity, fear or confidence, high spirit or low cunning, and, with regard to intelligence, something equivalent to sagacity."[12] Emotions that Darwin argued are manifested by at least some mammals other than humans—chiefly dogs, horses, and monkeys—include pleasure, pain, happiness, misery, terror, suspicion, deceit, courage, timidity, sulkiness, good temper, revenge, selfless love, jealousy, hunger for affection and praise, pride, shame, modesty, magnanimity, and a sense of humor.[13]

And at some point, probably long before the first humans, a new set of emotions—curiosity, insight, the pleasures of learning and teaching—also slowly emerged. Neuron by neuron, the partitions began to go up.

ARE ANIMALS MACHINES?
FOUR VIEWS

A Seventeenth-Century View: Descartes:

[A]s you may have seen in the grottoes and the fountains in royal gardens, the force with which the water issues from its reservoir is sufficient to move various machines, and even to make them play instruments, or pronounce words according to the different disposition of the pipes which lead the water . . .

The external objects which, by their mere presence, act upon the organs of the senses; and which, by this means, determine the corporal machine to move in many different ways, according as the parts of the brain are arranged, are like the strangers who, entering into some of the grottoes of these waterworks, unconsciously cause the movements which take place in their presence. For they cannot enter without treading upon certain planks so arranged that, for example, if they approach a bathing Diana, they cause her to hide among the reeds; and if they attempt to follow her, they see approaching a Neptune, who threatens them with his trident; or if they try some other way, they cause some other monster, who vomits water into their faces, to dart out; or like contrivances, according to the fancy of the engineers who have made them. And lastly, when the *rational soul* is lodged in this machine, it will have its principal seat in the brain, and will take the place of the engineer, who ought to be in that part of the works with which all the pipes are connected, when he wishes to increase, or to slacken, or in some way to alter their movements . . .

All the functions which I have attributed to this machine (the body), as the digestion of food, the pulsation of the heart and of the arteries; the nutrition and the growth of the limbs; respiration, wakefulness, and sleep; the reception of light, sounds, odours, flavours, heat, and such like qualities, in the organs of the external senses; the impression of the ideas of these in the organ of common sense and in the imagination; the retention, or the impression, of these ideas on the memory; the internal movements of the appetites and the passions; and lastly, the external movements of all the limbs, which follow so aptly, as well as the action of the objects which are presented to the senses, as the impressions which meet in the memory, that they imitate as nearly as possible those of a real man: I desire, I say, that you should consider that these functions in the machine naturally proceed from the mere arrangement of its organs, neither more nor less than do the movements of a clock, or other automaton, from that of its weights and its wheels; so that, so far as these are concerned, it is not necessary to conceive any other vegetative or sensitive soul, nor any other principle of motion, or of life.[14]

An Eighteenth-Century View: Voltaire:

What a pitiful, what a sorry thing to have said that animals are machines bereft of understanding and feeling, which perform their operations always in the same way, which learn nothing, perfect nothing, etc.!

What! that bird which makes its nest in a semi-circle when it is attaching it to a wall, which builds it in a quarter circle when it is in an angle, and in a circle upon a tree; that bird acts always in the same way? That hunting-dog which you have disciplined for three months, does it not know more at the end of this time than it knew before your lessons? Does the canary to which you teach a tune repeat it at once? Do you not have to spend a considerable time in teaching it? Have you not seen that it has made a mistake and that it corrects itself?

Is it because I speak to you, that you judge that I have feeling, memory, ideas? Well, I do not speak to you; you see me going home looking disconsolate, seeking a paper anxiously, opening the desk where I remember having shut it, finding it, reading it joyfully. You judge that I have experienced the feeling of distress and that of pleasure, that I have memory and understanding.

Bring the same judgment to bear on this dog which has lost its master, which has sought him on every road with sorrowful cries, which enters the house agitated, uneasy, which goes down the stairs, up the stairs, from room to room, which at last finds in his study the master it loves, and which shows him its joy by its cries of delight, by its leaps, by its caresses. [15]

A Nineteenth-Century View: Huxley:

Consider what happens when a blow is aimed at the eye. Instantly, and without our knowledge or will, and even against the will, the eyelids close. What is it that happens? A picture of the rapidly-advancing fist is made upon the retina at the back of the eye. The retina changes this picture into an affection of a number of the fibres of the optic nerve; the fibres of the optic nerve affect certain parts of the brain; the brain, in consequence, affects those particular fibres of the seventh nerve which go to the orbicular muscle of the eyelids; the change in these nerve-fibres causes the muscular fibres to alter their dimensions, so as to become shorter and broader; and the result is the closing of the slit between the two lids, round which these fibres are disposed. Here is a pure mechanism, giving rise to a purposive action, and strictly comparable to that by which Descartes supposes his waterwork Diana to be moved. But we may go further, and inquire whether our volition, in what we term voluntary action, ever plays any other part than that of Descartes' engineer, sitting in his office, and turning this tap or the other, as he wishes to set one or another machine in motion, but exercising no direct influence upon the movements of the whole . . .

Descartes pretends that he does not apply his views to the human body, but only to an imaginary machine which, if it could be constructed, would do all that the human body does; throwing a sop to Cerberus unworthily; and uselessly, because Cerberus was by no means stupid enough to swallow it . . .

. . . [W]hat living man, if he had unlimited control over all the nerves supplying the mouth and larynx of another person, could make him pronounce a sentence? Yet, if one has anything to say, what is easier than to say it? We desire the utterance of certain words: we touch the spring of the word-machine, and they are spoken. Just as Descartes' engineer, when he wanted a particular hydraulic machine to play, had only to turn a tap, and what he wished was done. It is because the body is a machine that education is possible. Education is the formation of habits, a superinducing of an artificial organisation upon the natural organisation of the body; so that acts, which at first required a conscious effort, eventually became unconscious and mechanical. If the act which primarily requires a distinct consciousness and volition of its details, always needed the same effort, education would be an impossibility.

According to Descartes, then, all the functions which are common to man and animals are performed by the body as a mere mechanism, and he looks upon consciousness as the peculiar distinction of the *"chose pensante,"* of the "rational soul," which in man (and in man only, in Descartes' opinion) is superadded to the body. This rational soul he conceived to be lodged in the pineal gland, as in a sort of central office; and here, by the intermediation of the animal spirits, it became aware of what was going on in the body, or influenced the operations of the body. Modern physiologists do not ascribe so exalted a function to the little pineal gland, but, in a vague sort of way, they adopt Descartes' principle, and suppose that the soul is lodged in the cortical part of the brain—at least this is commonly regarded as the seat and instrument of consciousness.

. . . [T]hough we may see reason to disagree with Descartes' hypothesis that brutes are unconscious machines, it does not follow that he was wrong in regarding them as automata. They may be more or less conscious, sensitive, automata; and the view that they are such conscious machines is that which is implicitly, or explicitly, adopted by most persons. When we speak of the actions of the lower animals being guided by instinct and not by reason, what we really mean is that, though they feel as we do, yet their actions are the results of their physical organisation. We believe, in short, that they are machines, one part of which (the nervous system) not only sets the rest in motion, and co-ordinates its movements in relation with changes in surrounding bodies, but is provided with special apparatus, the function of which is the calling into existence of those states of consciousness which are termed sensations, emotions, and ideas. I believe that this generally accepted view is the best expression of the facts at present known.

. . . It is quite true that, to the best of my judgment, the argumentation which applies to brutes holds equally good of men; and, therefore, that all states of consciousness in us, as in them, are immediately caused by molecular changes of the brain-substance. It seems to me that in men, as in brutes, there is no proof that any state of consciousness is the cause of change in the motion of the matter of the organism. If these positions are well based, it follows that our mental conditions are simply the symbols in consciousness of the changes which take place automatically in the organism; and that, to take an extreme illustration, the feeling we call volition is not the cause of a voluntary act, but the symbol of that state of the brain which is the immediate cause of that act. We are conscious automata . . .[16]

A Twentieth-Century View: James L. and Carol G. Gould:

In considering the issue of mental experiences in animals, we have begun to wonder if the implicit assumption that humans are almost wholly conscious and aware (and hence fully competent to evaluate our cognitively less sophisticated animal brethren) is correct. Could it be that the degree to which conscious thinking is involved in the everyday lives of most people is greatly overestimated? We know already that much of our learned behavior becomes hardwired: despite the painfully difficult process of learning the task originally, who has to concentrate consciously as an adult on how to walk or swim, tie a shoe, write words, or even drive a car along a familiar route? Certain linguistic behavior, too, falls into such patterns. Michael Gazzaniga, for instance, tells the story of a former physician who suffered from a left (linguistic) hemisphere lesion so serious that he could not form even simple three-word sentences. And yet, when a certain highly touted but ineffective patent medicine was mentioned, he would launch into a well-worn and perfectly grammatical five-minute tirade on its evils. This set piece had been stored on the undamaged right side (along with the usual collection of songs, poetry, and epigrams) as a motor tape requiring no conscious linguistic manipulation to deliver.

. . . Indeed, what evidence is there that those sublime intellectual events known as "inspiration" involve any conscious thought? Most often our best ideas are served up to us out of our unconscious while we are thinking or doing something perfectly irrelevant. Inspiration probably depends on some sort of repetitive and time-consuming pattern-matching program which runs imperceptibly below the level of consciousness searching for plausible matches.

It strikes us that a skeptical and dispassionate extraterrestrial ethologist studying our unendearing species might reasonably conclude that *Homo sapiens* are, for the most part, automatons with overactive and highly verbal public relations departments to apologize for and cover up our foibles.[17]

Chapter 10

THE NEXT-TO-
LAST REMEDY

When all the world is overcharged with inhabitants,
then the last remedy of all is war . . .

THOMAS HOBBES,
Leviathan, II, 30[1]

Once organisms get really good at sex, once they evolve the plumbing and the passion for it, there gets to be a danger: So many competent, DNA-exchanging beings may be born that they will improvidently gobble up all the food or nutrients or prey, and then almost everyone, including their close relatives, will die. This must have occurred innumerable times in the history of life.

Take a being as modest as a bacterium, weighing in at a trillionth of a gram, and let it reproduce with no impediments. In the second generation there will be two bacteria; in the third generation, four; in the fourth generation, eight; and so on. If we imagine that none of those offspring die, then in 100 generations they will collectively weigh as much as a mountain; in 135 generations, as much as the Earth; in 150 generations, as much as the Sun; and in 185 generations, as much as the Milky Way galaxy.

Of course, such prodigious increases in mass are arithmetic exercises only. They could never occur in the real world. For one thing, the replicating microbes would soon run out of food. Your descendants cannot weigh as much as a mountain if there's not a mountain's worth of food to eat—much less an Earth's worth or a Sun's or a galaxy's. There is only so much food available. Thus, your descendants will quite soon be in competition with one another for scarce resources. But because of the enormous power of exponential reproduction, an organism with even a slight advantage in finding or utilizing food rapidly supplants the competition (or at least its descendants do). Fast reproducers generate large populations, and competition for resources; they provide the raw material for a natural selection that efficiently magnifies small differences in fitness, differences that might be too small or subtle for even the most skilled naturalist to notice. This was the central argument of Darwin's unpublished 1844 manuscript on evolution, and of his article in the *Proceedings of the Linnaean Society* of London for 1858.[2]

So what happens in fact when there's too much crowding? Some responses seem to serve a larger purpose. Sibling shark embryos fight to the death in utero. In many nonhuman mammals, brothers and sisters of the same litter compete for access to nipples; often, there is a least competent infant, unsuccessful in elbowing its way to a nipple —the runt of the litter, who becomes progressively weaker with each failed attempt to nurse. The Virginia opossum has thirteen teats and, generally, more than thirteen pups per litter. Only those who regularly get to a teat live. Such competitions weed out the weak. Those species with more teats than pups permit weakling and unaggressive young-sters to reach adulthood. If they are unlikely to compete successfully as adults and pass their genes on, their mother has, from the point of view of her genes, been wasting her time nursing such pups. Those mothers with fewer teats or more pups have a selective advantage. Concern about cruelty and suffering doesn't, so far as we know, enter into it.

Cities aside, we humans routinely experiment on crowding animals into confined enclosures. The institutions responsible are called zoos; some are much more pernicious than others. A well-known problem of zoos is that many of the inmates are somehow less able to "breed in captivity"; another problem is sustained and violent conflict, usually between males of the same species. Zookeepers have learned that if they wish to maintain their "inventories," they must often separate the males. Experiments have also been performed in the laboratory to study overcrowding. In all of these cases it's important to remember the artificiality of the circumstances. An option available in the wild is unachievable in captivity: No matter what the provocation, a caged animal cannot flee conflict and make a new start somewhere else.

Norway rats have been bred in scientific laboratories since the middle nineteenth century. Artificial selection has elicited—partly through unconscious choices by laboratory personnel—a strain of rats that is calmer, tamer, less aggressive, more fertile, and with significantly smaller brains than their wild ancestors. All this is a convenience for those experimenting on rats.[3]

In a now-classic experiment,[4] the psychologist John B. Calhoun let Norway rats reproduce in an enclosure of fixed size until the number of occupants, and therefore the population density, was very high. He made sure, however, to provide everyone with enough to eat. What happened?

As the population increased, a range of unusual behavior was noted. Nursing mothers became somehow distracted, rejecting and abandoning their infants, who would wither away and die. Despite the surplus of ordinary food, the bodies of the newborn would be greedily eaten by passersby. An adult female in heat or estrus would be pursued relentlessly, not by one, but by a pack of males. She had no hope of escape, or even sanctuary. Obstetrical and gynecological disorders proliferated, and many females died giving birth, or from complications soon after. When crowded together, the rats lost their inclination or ability to build nests for themselves and their young; their desultory constructions were amateurish and ineffective.

Among the males Calhoun distinguished four types: the dominant, highly aggressive ones who, although "the most normal," would occasionally go "berserk"; the homosexuals who made sexual advances to adults and juveniles of both sexes (but, significantly, only to non-ovulating females): their invitations were generally accepted, or at least tolerated, but they were frequently attacked by the dominant males; a wholly passive population that "moved through the community like somnambulists" with nearly complete social disorientation; and a subgroup Calhoun calls the "probers," uninvolved in the struggle for status but hyperactive, hypersexual, bisexual, and cannibalistic.

If there were no differences between rats and people, we might conclude that among the consequences of crowding humans into cities—other things being equal—would be more outbreaks of street fighting and domestic violence, child abuse and neglect, soaring infant and maternal mortality, gang rape, psychosis, increased homosexuality and hypersexuality, gay bashing, alienation, social disorientation and rootlessness, and a decline in traditional domestic skills. It's suggestive, surely. But people are not rats.

Crowding in cats leads to a nightmarish tableau of incessant hissing and squalling, fur standing on end, remorseless fighting, and the designation of pariahs who are attacked by all. But people are not cats either.

Crowding in our nearer relatives, the baboons, can lead to bloodshed and social disorder at least on the scale of rats and cats, as we treat later. In many animals overcrowding also leads to increased susceptibility to disease, and smaller adult stature. But as vervet monkeys become more and more crowded together, the inmates begin studiously avoiding one another, inspecting with great interest the ground on which they sit and the motion of clouds in the sky above. In chimpanzees, crowding does tend to make everybody a little edgy. There is more aggression. But not *much* more. As the population density increases, chimps make concerted efforts at appeasing one another, at peacemaking.[5] They have neural machinery and a social idiom to compensate for overcrowding. Are we not more like chimps than like rats?

The rat response to overcrowding, even at its most pathological, might be viewed as making sense in a remorseless evolutionary way. If the population density becomes too high, then mechanisms are set into motion to reduce it. Huge numbers of socially disinterested adults, illness, increased homosexuality, and soaring infant and maternal mortality, all serve this purpose. Eventually, the population crashes, overcrowding is relieved, and the next generation is back to business as usual—until the population pressures build up again. Some of the behavioral responses to high population density in Calhoun's rats, and in many other species, might be looked on not as barbarous and unfeeling, but as a calamitous necessity, the capability for which has been painstakingly evolved.

We've phrased this in terms of group selection, but interpretations in the idiom of kin selection are also possible. We could, instead, have stressed that overcrowding is, almost invariably in Nature, a prelude to famine, so it makes a desperate kind of sense to abandon or eat nursing infants, or to cease building nests for the young, or to arrange that babies be stillborn or not conceived at all.[6]

In many animals—howler monkeys, for example—high population density leads to takeovers by alien males and the wholesale slaughter of resident infants. This behavior is especially vivid in animals where dominant males keep harems or try to prevent other males from reproducing.[7] But is it fundamentally due to overcrowding, or to the evolutionary strategy of the new dominant male? It benefits the proliferation of his complement of genes to remove all distractions from the females as quickly as possible, move them into ovulation (which killing their young accomplishes), and impregnate them before he's overthrown by the next usurper.* The more crowding there is, the more challenges from sexual rivals and the more such infanticides. Whether all of the anomalous behavior of Calhoun's rats can be understood in these ways is still unclear; but surely some of it can.

———

If, sympathizing with the rats, cats, and baboons in these experiments, we wished to help them, what could we do? We might be tempted to organize a jailbreak and return them to their natural environments. We would eliminate the overcrowding and—assuming the animals could fend for themselves—hope they would revert to their normal behavior and social organization. But then shouldn't evolution also have invented mechanisms for dispersing competing organisms so they're not in each other's way—especially the most flagrantly aggressive variety, usually the young adult males? This would be to the advantage of both the individual and the species.

In fact, Nature provides such a safety valve: Instead of staying on to fight to the death, the potential losers—those who estimate that they would be vanquished if they continued fighting, or those who judge that the probable benefits of fighting are not worth the risk—may simply pick up and leave. There is an escape clause in their contract, a get-out-of-jail-free card, which precipitously reduces the incidence of mutilation and murder. A few formalities and they're gone. But lock them up in a zoo or a laboratory apartment house for rats and all possibility of escape is denied them. That's when they go crazy.

Some kind of mutual repulsion is needed, like that provided by electrical charges of the same sign or polarity. When two electrons are far from one another, they hardly feel each other's influence. But bring them close together and a powerful force of electrical repulsion

---

* A very nice test of these ideas are the observations by the animal behavior expert Stephen Emlen. He thought to examine jacanas, birds in which the usual sex roles are reversed: Males do all the parenting and the females compete vigorously for something like a harem of males. Those females who don't possess a harem don't reproduce, so the dominant females are often challenged by lower-ranking females. When a takeover attempt succeeds, the incoming female routinely destroys the eggs and kills the chicks. She then sexually solicits the males, who now have no young to distract them—and so are able to attend to propagating the genetic sequences of the incoming female. The genetic strategy of infanticide is situational, not gender-based.

is brought into play, the force being stronger the closer together the electrons are. Something similar is true for magnets. Opportunistic animals able, under favorable conditions, to reproduce exponentially need a similar mutual repulsion, increasing quickly as the animals are brought into systematic close contact. There is such a force in Nature: intraspecific aggression, aggression within, internal to, a given species.

Most competition in animals is with members of the same species. How could it be otherwise? They have almost precisely the same habitat, the same tastes in food, the same erotic aesthetic, the same nesting and sleeping places, the same foraging and hunting grounds. If the animals are spread out, there's enough food and other resources for everyone, while they can still remain near enough so they can find each other when it's time to mate. If they're crowded together, conflict escalates and even the strongest animals run an increased risk of lethal combat.

Spreading out is accomplished by aggression, but aggression is not the same as violence and rarely goes as far as violence.[8] Often it's enough to announce menacingly to all within earshot that this is your territory and no intruders will be tolerated. You might patrol the frontiers, spraying your urine or depositing your feces in prominent, strategic locations—or leaving, through special scent glands and much dragging and rubbing, an aromatic token of your proprietary interest. If you're a grizzly bear, you might try marking a pine tree as high up as you can reach; when potential poachers grasp how big you must be to mark so high, they'll give you wide berth.

About 80% of the different orders of mammals are armed with specialized scent glands. Gazelles have them in front of their eyes, camels on their feet and neck, sheep on their bellies, some pigs on the wrist, chamois behind the horns, pronghorns on the jaw, peccaries on the back, musk deer in front of the genitals, and goats on the tail. Water voles rub their hind feet over their flank gland and rhythmically drum them on the ground. Gerbils and woodrats rub their bellies directly on the ground, secreting their scent mark from a ventral gland. Some animals have five or six different kinds of scent glands in various places on their bodies, each conveying a different chemical proclamation. Cats spray carefully titrated amounts of urine on the drapes and upholstery, in case some presumptuous alien feline might enter the living room and curl up before the fire. Rabbits meticulously deposit piles of feces, each pellet coated by the anal scent gland, at crossroads in the warren—like the altars of Hecate on the highways of ancient Greece.

Some animals mark others with these scents, and rats urinate on their partners' bodies—perhaps as a sign of proprietorship over individuals as well as territories. Animals can distinguish male and female, their own group or strain from others, age, individual identity, and the sexual receptivity of females, all by odor alone.[9] Scientists have begun to decipher the stock phrases of their chemical communications—maybe just "foreigners keep out: this means you," or "single male, well-bred, wishes to meet attractive single female . . . ," or "for a good time, follow this scent trail." Sometimes it seems to be something much more subtle. Animals are busy filling the olfactory communications channels with a richness and fineness of discrimination long ago lost to humans. With all our instruments, we have not yet learned how to reenter that world.

If, despite all your aromatic notices, someone invades your territory, it might be enough to make threatening gestures, or swoop down on him, or bare your teeth and growl. Clearly, claw-to-claw or talon-to-talon mortal combat each time there's a minor jurisdictional dispute is too costly for everybody—winner and loser. It's much better to disperse the population through bluff, deception, feints, and a vivid pantomime of what violence you will visit on the intruder should he persist in ignoring your restrained and reasonable warnings. Deterrence is the way these matters are arranged, by and large, on the planet Earth. Real violence lies at the extreme end of the spectrum of aggressive possibilities, a last resort, as Hobbes said. Nature almost always settles somewhere short of that.

To avoid misunderstandings, it's important to have evolved unambiguous conventions not only for what constitutes aggression, but also for what constitutes submission. Typical submissive gestures in mammals are the opposite of typical aggressive gestures[10]—averting the eyes so they look anywhere but at the adversary; absolute motionlessness; a kind of bowing in which the forelegs and head are lowered and the rump raised; hiding from view those body parts that are conspicuous in threat displays; and turning jugular artery or belly up, exposing vital organs to the adversary as if inviting evisceration. The pantomime is lucid: "Here is my belly, do with me as you will." It's followed almost always by a magnanimous gesture from the victor.* Different species have different hereditary conventions on what constitutes and symbolizes submission. Fighting is transformed into ritual; instead of bloody combat, there is an exchange of data.

Such aggression—most often between males of the same species in disputes over territory or females—is very different from predatory aggression, aggression against members of another species. The two modes share some features in common (baring the teeth, for example), but the one is mainly bluff and the other is in deadly earnest. They engage different parts of the brain. In rivalries of love, cats will hiss, spit, arch their backs, make their hair stand on end, raise their tails high, and dilate their pupils. (Note how many of these postures and gestures make the animal seem larger and more dangerous than it is.) They rarely do each other serious harm, though. A genetic propensity for attacking others of your species, and eliciting attacks from them, has a maladaptive side to it—even if you win every fight, you might be badly injured, or a minor cut might later become infected. Bloodless rituals and symbolic combat are far more practical.

Predatory aggression is just the opposite. Its early object is to come as close as possible to the victim before it realizes what's up. The cat will slink an inch at a time if it must, ears slicked back, hair tightly following the contours of the body, tail lowered. It stalks in absolute silence. Then the pounce, the kill, and dinner—all done with consummate delicacy and grace. No hissing and spitting here. *Intra*specific aggression is almost all show, display, intimidation, coercion, stagecraft. Only rarely does it end in mortal combat. *Inter*specific aggression, that's different. That's business. The prey may get away, but the predator's intent is murder. Few species systematically confuse the two modes of aggression.

Mock combat is a staple in the theater of intraspecific aggression; both parties go through the motions, but neither is seriously hurt. The deadly, needle-toothed piranha fish of South American rivers fight among themselves, or at least the males do, but never by biting: If there were biting, everyone could get hurt. Instead they push and

---

* Another aspect of the gestural vocabulary of appeasement is infantile behavior in adults, including begging. It's a little like human lovers using baby talk and calling each other "baby." They're applying a lexicon established in infancy to another purpose.

shove with their tail fins. They want to communicate aggression, but not to bloody the water. It's as if the combatants walk a fine line between cowardice and murder. Most often—crowded conditions may be another story—the line is walked with astonishing precision. But, as a reminder of how fine the line is, in many species intraspecific fighting is more likely when the animals are hungry. One kind of behavior spills over into the other.

The female blue heron hears the love screech of the male. There may be several males calling at once—to the wind, for all they know. She picks her heart's desire and settles on a branch nearby. The male immediately begins to court her. The moment she indicates interest and approaches him, though, he changes his mind, becomes unpleasant, shoos her away, or even attacks her. As soon as the discouraged female flies off, he screeches after her—"frantically," according to Nikko Tinbergen, the pioneering chronicler of blue heron life. If she gives him another chance and flies back, he may very well attack her again. Gradually, though, should the female's patience last that long, the fickle male's grumpiness subsides and he may actually be ready to mate. He is conflicted and ambivalent. Sex and aggression are mixed up in his mind, and the confusion is so profound that, if not for the patience of the female, this species might fail to reproduce itself. If ever there was an avian candidate for psychotherapy, the male blue heron is our nominee. But a similar confusion in the minds especially of males holds for many species, including reptiles, birds, and mammals. Some of the brain's neural circuitry for aggression seems dangerously cheek by jowl with the neural circuitry for sex. The resulting behavior is strangely familiar. But of course humans are not herons.

Often you can see the ambivalence, the tension between inhibiting and disinhibiting the aggressive machinery in the animal's behavior. It is literally "of two minds." A fighting cockerel, whose pecks and spurs are deadly, may in the midst of a confrontation turn aside and peck at a pebble on the ground, which after a moment it drops. In human as in animal behavior this is called "displacement." The aggressive feelings are transferred or displaced to someone or something else, so the passions can be discharged without causing real injury. The cockerel is not angry at the pebble, but the pebble is a handy as well as a safer target.

Some male tropical fish use their vivid coloration to keep other males away, that is, to protect territories and females. The females are, however, similarly decorated. During courtship the female, if attracted to the male, dispenses with her usual indications of submissiveness or readiness for escape and signals her amorous intent by a display to the male—a display, however, which is very similar to the male's own aggressive posture. In some species, the male becomes enraged (and probably a little confused); he responds by displaying his coloration broadside to her, beating his tail fearsomely, and charging her. But, as noted in a famous study by Konrad Lorenz, he does not actually attack her. (If he did, he would leave fewer offspring.) Instead, narrowly missing the female, he races on and attacks someone else, usually the male in charge of the next territory, who may have been minding his own business, browsing in the algae. Eventually things settle down. Our protagonist no longer attacks his neighbor or charges the female. The species continues. Here, instead of displacing aggression away from a formidable enemy to an inoffensive target, the displacement goes the other way around. This sort of redirection is widespread. Again, gestures, postures, and displays about sex are very close to those about violence. The two can get confused.

One wolf will greet another by placing its mouth around the other's muzzle. Many other mammals do likewise. Those taming wild animals may be startled when they are at the receiving end of such a greeting. The wolf stands on its hind legs, places its forelegs on the scientist's shoulders, and places its jaws around the scientist's head. This is just the wolf's way of being friendly. If you're an animal who doesn't know how to talk, a very clear signal is communicated: "See my teeth? Feel them? I could hurt you, I really could. But I won't. I like you." Once more, a very narrow line separates affection from aggression.

Chimpanzees engaged in what humans call horseplay put on a characteristic "playface" to show that their combat gymnastics are meant only as a game. Courtship displays in gulls have been described as "fear and hostility, or attack and fleeing tendencies, expressed . . . in a manner that denies them."[11]

In cranes there's an "appeasement ceremony" in which the male spreads his wings, exaggerates his size, raises his beak . . . and then, still in a threat posture, turns himself aside—presenting a vulnerable and very visibly marked part of his anatomy, perhaps the side or back of his head. The pantomime may be repeated several times and incorporate an attack on a piece of wood or something else handy. The message being communicated is clear: "I am big and threatening, but not toward you—toward the other, the other, the other."[12]

Smiling may have a similar origin. Baring one's teeth carries the message: "I think you're food," or at least "Watch out for me." But in the symbolic language of animals, this signal may be softened and altered: "Even if you *are* food, even though I'm well-equipped to eat you, you're safe with me." All over the world, in virtually every human culture, smiling signifies affection and good fellowship (with certain nuances conveying a touch of nervousness and deference). All over the world, in nearly every human culture, in civilian as well as military life—in handshakes, high fives, salutations among mounted Sioux, hails to Caesar and heils to Hitler, upon greeting a superior officer or waving farewell—we humans offer our right hands in greeting, demonstrating while still at a safe distance that we are unarmed and therefore pose no threat. In a species given from its earliest days to clubs, knives, spears, and axes, this is information worth having.

———

With occasional exceptions, animals do not seem consciously to work out what to do in a given situation and then, weighing alternatives, opt for aggression. It's too slow a process to survive the hurly-burly of the biological world. Instead, the animal senses threat or prey, and a tenth of a second later it responds. A complex set of physiological reactions begins—adrenaline pours into the bloodstream, limbs begin to flex—reactions that are ordinarily sitting there in the animal on ready standby, awaiting the release signals.

In the neural architecture of mammals there is hardwired circuitry for aggression and predation. When a certain region in the brain of a solitary cat is electrically stimulated, she begins to stalk imaginary prey. Turn the current off and she stretches and licks her paws; the hallucination has vanished. Rats that do not look twice at a mouse will, when an electric current is made to pass through the appropriate parts of their brains, become crazed killers—dedicated, implacable mouse-murdering machines. The stimulated neural circuits are present for a reason; in the ordinary course of the animal's life, they will be excited by some cue from the outside world—a motion, a smell, a sound, causing an electrical stimulation—and the brain machinery for aggression or predation is set into motion. When given a juicy bone still covered with meat, even puppies as young as two weeks old will growl and bark. Dry dog food does not trigger the same hardwired and impassioned response. Humans have such machinery too. Sometimes a misfiring or miswired circuit can set it into motion with very little stimulus from the outside world, or even none at all.

It's as if all of us birds and mammals—but especially the males— are walking around wearing a control panel with a set of push buttons on it. The panels are prominently displayed, easy for others to get to (or even for us to get to—so we can pump ourselves up on our own, a skill of professional athletes). When pressed, the buttons disinhibit a set of powerful, passionate, and sometimes deadly responses that are ordinarily kept under tight controls. Put this way, it may seem odd that Nature has made the buttons so easy to push, so readily available, so vulnerable to exploitation.[13]

A cannibalistic species of firefly simulates the color and frequency of the come-hither flashes of another, country bumpkin species of firefly. The love buttons have been pushed on the naive insects; they see visions of sultry females where there is only a gaping mouth. To lure uninterested or recalcitrant females into mating, males of many species are often ready to press buttons designed for quite different purposes,

> such as feeding, defense, timidity in the face of aggression, or brood care. They may give a brief threatening lunge, cry like a baby, mimic an alarm call, hop on one leg as if wounded, or (as in peacocks) peck at the ground as if food has been found.[14]

Undeterred by scruple, they will use any method that works. In many cultures, young men try to press all available buttons for sex, perhaps offering wholly insincere promises of fidelity and devotion; or they taunt each other into fighting by casting aspersions on another's courage, say, or his mother's sexual behavior. The benefits of having these buttons so readily available must outweigh the risks. The inflexibility of such hair-trigger responses might be a cause for worry, though.

These behavior patterns also are encoded in the nucleic acids. Every deterrent flourish, every postural hint of submission, is meticulously written down in the ACGT language. That being the case, you might expect variations in the style or intensity of aggression from animal to animal within a given species, as is indeed the case. If you take a population of mice and breed the aggressive ones with each other and the peaceful ones with each other, eventually you produce two strains of markedly different temperament. This isn't due to pup-rearing practices, because the young of aggressive parents, when raised by peaceful mothers, are aggressive, and vice versa. It's a commonplace that through artificial selection dog breeders have produced nervous, high-strung, ferocious breeds—for example, rottweilers or pit bulls—and friendly, peaceful strains, often useless as watchdogs, such as cocker spaniels. In mouse and dog aggression, heredity often seems to take precedence over home environment. (It might be the other way around in humans, or the two influences might be evenly matched.)

———

Nearly all the social mammals are organized as groups of females (often relatives) with their offspring. Males, otherwise absent, are conspicuously present when the females are in heat. They may be busy dominating, fighting, or mating, but in terms of basic social structure and the bringing up of the young, they are often a shadowy presence. Usually, the young are raised by single mothers. Among the exceptions to this rule are chimpanzees, gorillas, gibbons, wild dogs, perhaps wolves. And, more than occasionally, humans.

In temperate and polar climates, there's a good reason for the young to be born in the spring—so they may have the rest of spring and all of summer and autumn to grow up before having to face the rigors of winter. If the gestation period is short (or alternatively around a year), then mating will also occur in the spring. To arrange for biological clocks to be built into animals, to stimulate the reproductive machinery at the right moment in springtime and to inhibit it at other times of the year, must have occupied great vistas of evolutionary time.

Natural selection has provided a wide range of visual, olfactory, auditory, and other cues to inform the normally uninterested males of the otherwise indetectable fact that ovaries are releasing eggs all around them. Sexual attention at other times is generally a wasted effort (it's used to bond male and female in species where both are needed to raise the young). So the female is designed with some internal calendar (perhaps triggered by the length of the day), and a series of signals and behaviors (alluring pheromones plus enticing postures, say). In the season of love, on cue, as if activated by some Cartesian clockwork, both sexes become mad with passion.

If mating is to occur in the spring, then the rivalry of males for the attentions of females should also peak in the spring. If the lives of deer depend in part on their speed and their ability to fight back when cornered by predators, then intraspecific tests of strength, speed, stamina, and strategy among stags are to the benefit of the genes of the victors as well as to the deer clan. This is ritualized combat, almost never to the death. The point of the exercise becomes instantly clear as the doe gives herself to the winner. A multitude of such dramas over many generations helps deer keep pace with hereditary improvements in, for example, the hunting skills of wolves.

In many predatory species, animals hunt together. Prey is flushed into ambush, or is exhausted by repeated feinting. Stragglers, usually the weak, the young, and the old, can be isolated. The predators may adopt a relay system, Group One performing feints only, and Group Two loping along to pick up the attack when Group One is exhausted. Cooperation makes hunting much more efficient, and the predators may now bring down animals much larger than they are.

Members of hunting packs have a kind of ethic: Whatever rivalries they bear are put aside during the hunt. For them too, "politics stops at the water's edge." There's a different set of social rules within the group than without. But it's an easy step from attacking animals of other species to attacking strangers of the same species. This is true of dogs and lions, which hunt in packs, and of ants and penguins, which do not. They behave as if special loyalty is owed to their group only; suspicion and hostility are due all others, even though they are fellow members of the same species. And this is not restricted to hunting packs. It's a fact of life among most sociable birds and mammals.

Ethnocentrism is the belief that our group (whichever it happens to be) is at the focus of everything good and true, the center of the social universe. *We* do things the way they were meant to be done. Xenophobia is the fear and hatred of strangers. *Their* behavior is wrongheaded or weird or abominable. They don't have the same respect for life that we do. And anyway they're out to get us. "Us against them," again. Ethnocentrism and xenophobia are extremely common among birds and mammals, although they do not constitute an invariable rule: Flocks of migrating birds, for example, are pretty much open to all comers of the same species.

If we're confronted by some stranger who means harm to both of us, then we're motivated to put aside whatever differences lie between us and together deal with the common enemy. Our chance—as individuals and as a group—of surviving an attack is greatly improved if we make common cause with our fellows. The existence of common enemies can work as a powerful unifying force. Common enemies make the social machinery purr. Those groups that incline to xenophobic paranoia might gain a cohesive advantage over groups that are initially more realistic and carefree. If you've exaggerated the threat, at least you've reduced internal tensions in your group; and if the external threat is more serious than you've privately estimated, your preparedness is higher. As long as the social costs stay within reasonable bounds, it may become a successful survival strategy. So there's a kind of contagion about xenophobia.

Even among animals that as adults have few natural enemies—dolphins, say, or wolves—the young are vulnerable. Special steps must be taken to protect them. Adult dolphins keep very close to the young. Wolf cubs are cautious and fearful in their first few months of life. Many nestlings beg for food with visual, not auditory, cues so as not to attract the unwelcome attention of predators. These measures are useful in dealing with both interspecific and intraspecific violence: Because so many group-living animals attack members of other groups who stray into their territory, the young have good reason to be wary of strangers.

Among the wildebeests, an African antelope hunted by many predators, the calf shakily stands within a few minutes after birth. Five minutes later it can follow its mother, and in twenty-four hours it can keep up with the herd. Wildebeests grow up fast. In other animals, of which humans are the most striking example, the young are born utterly helpless. If abandoned by their parents, they would perish in a few days, predators aside. A wildebeest mother need make few concessions to her young, apart from permitting them to suckle. Human mothers (and robin, wolf, and monkey mothers, among many others) must adopt a complex behavioral repertoire in order for there to *be a* next generation. In higher mammals, these special activities may last years or even decades—until the youngster is nearly fully grown. For so major an investment to be made, there must be a comparably major benefit. The long childhood of the higher mammals is connected with their larger brains and with the necessity that the young be taught. This frees the youngsters from the comparative inflexibility of having preprogrammed genetic knowledge only.

In many animals there's a period early in life during which profound and irreversible learning occurs, a time, for example, when a duckling will follow anything nearby that moves as if it were Mom—even if it's a bearded pioneer in animal behavior. This is called imprinting. Some imprinting goes on even before birth. Ducklings, before they hatch, memorize the voice of whoever is incubating them, and respond (by peeps from inside the egg). If it's a human who talks to the egg during incubation, that's the voice the duckling responds to after it's hatched. Imprinting may involve learning a call, a song, an odor, a shape, or a food preference, and is accompanied by deep emotional bonding. The information is implanted in the memory for a lifetime.

These sounds, smells, and sights are associated with food, warmth, love, and safety in an often-hostile world. Lambs, chicks, and goslings must reliably recognize and follow their perambulatory mothers; failure to do so is punishable by death. It's no wonder imprinting lasts for life. The predisposition to be imprinted is programmed in the DNA and subject to very strict constraints (in some cases imprinting can occur only in a specific one- or two-day period over an entire lifetime). But the specific information that is so indelibly etched is conditioned by environment and experience, and differs from animal to animal. In this way the youngster can learn, generally from its parents, wisdom too recent to have been inscribed in the latest edition of the nucleic acids.

An unfocused inclination towards ethnocentrism and xenophobia can be particularized as needed in each generation. The groups to which loyalty is owed and the ones deserving special hatred and contempt may change from generation to generation. Imprinting is a means for fitting general proclivities to practical politics, and is a form of education. The machinery stands ready for those who know how to use it. The young animals have a nearly eidetic memory. But they have no critical faculties. They'll believe anything—whatever they're taught. As the example of the parade of ducklings waddling adoringly after the ethologist reminds us, imprinting might lead, in unscrupulous higher animals, to misuse. The young are so ready to learn who to love and who to hate.

If the nipples and vaginas of nursing rats ("suckling dams," the scientists call them) are regularly swabbed with the scent of lemon, the male pups, when grown, are preferentially attracted to lemon-scented females—foreswearing the naturally aromatic, accessible, and nubile alternatives.[15] This odor imprinting suggests how powerfully early experiences can affect later sexual preference and performance. It's something like the line in the song that goes, "I want a girl just like the girl that married dear old Dad." But humans are not rats.

With long childhoods and efficient imprinting, animals can make wholesale changes in their behavior to adapt to a changing environment—taking only a few lifetimes instead of a geological age. In turn, this bonds mothers and offspring together still more closely. It creates something akin to love. It also means that different communities of the same species may have different patterns of behavior that are passed down the generations—even if the groups are, genetically, essentially identical. The strategy of long childhoods and early learning introduces a new element: culture.

———

Human life begins in a race of one against hundreds of millions. The stampeding sperm cells are competitive from the start. But the whole point of the rivalry is cooperation of the most intimate sort. The two cells wholly merge. They combine their genetic material. Two very different beings become one. The act of making a human being involves an almost bizarre mix of opposites—desperate competition against all odds, and cooperation so perfect that the partners' separate identities vanish. It would be inconsistent for beings who arise out of intense rivalry and begin in perfect cooperation to decry either.

"In the ways of Nature," said Marcus Aurelius, "there is no evil to be found."[16] Animals are aggressive not because they are savage, or bestial, or evil—those are words with very little explanatory power—but because such behavior provides food and defense against predators, because it spaces out the population and avoids overcrowding, because it has adaptive value. Aggression is a survival strategy, evolved to serve life. It coexists, especially in the primates, with compassion, altruism, heroism, and tender, self-sacrificing love for the young. These are also survival strategies. Eliminating aggression would be a foolish as well as an unachievable goal—it's built too deeply into us. The evolutionary process has worked to achieve the right level of aggression—not too much, not too little—and the right inhibitors and disinhibitors.

We emerge out of a turbulent mix of contradictory inclinations. It should be no surprise that in our psychology and our politics a similar tension of opposites should prevail.

Chapter 11

DOMINANCE AND SUBMISSION

When we no longer look at an organic being as
a savage looks at a ship, as something wholly
beyond his comprehension; when we regard
every production of nature as one which has
had a long history; when we contemplate every
complex structure and instinct as the summing
up of many contrivances, each useful to the
possessor, in the same way as any great
mechanical invention is the summing up of the
labour, the experience, the reason, and even
the blunders of numerous workmen; when we
thus view each organic being, how far more
interesting—I speak from experience—does the
study of natural history become!

CHARLES DARWIN,
The Origin of Species [1]

Order. Hierarchy. Discipline.

BENITO MUSSOLINI,
proposed national slogan [2]

The two pit vipers slither toward one another in silence, forked tongues flickering. Slowly they entwine in a languorous embrace. They raise themselves higher and higher off the ground. The glistening coils ebb and flow. Like some macroscopic echo of their underlying microscopic reality, they form a double helix.

Once, observers concluded that this is a reptilian courtship dance. They neglected to capture the snakes, though, and determine their sexes. When this is done, both snakes turn out to be male. So what are they doing? Since homosexual embraces are known throughout the animal kingdom, it still might be a courtship dance—except that it usually ends with one snake toppling the other to the ground, no overt sexual acts having transpired. Instead, this hypnotizing serpentine ritual seems to be a competition, like arm wrestling, played by strict rules. No combatant has ever been bitten or even injured, so far as we know. When the duel ends, whoever has been forced over accepts defeat and slithers away.

Is this contest about access to females? Sometimes there's no female in evidence, urging her champion on, or available as a reward for the victor. At the least, this is a struggle over hierarchy, over who's the top viper—which does not exclude the possibility that the encounter is homosexual as well: Male competition for dominance expressed in homosexual metaphor is a theme widespread among the animals.

Losing the struggle is apparently a blow to the snake's self-confidence. He seems morose and demoralized, unable many days later to defend himself against even weakling rivals. Here's one mechanism by which struggles for dominance later convert into mating success: A female viper, on meeting a lone male, will mimic male behavior and raise herself up as if preparing for this sportive combat. If, still despondent from his last defeat, he does not with sufficient vigor rise to the occasion, she looks elsewhere for a mate.[3] Almost without exception, the females manage to mate with the winners.[4]

Among pit vipers,[5] a male will take one or more sexually receptive females under his "protection" and do what he can to discourage the approach of other males. He will defend or compete for specific territories, especially those that contain resources important for the next generation of vipers. The most celebrated American pit viper, the prairie rattlesnake, does not mate as it comes out of hibernation in the spring, but waits until the late summer when a male must make a considerable effort to track down a female.

In contrast, the garter snakes of Manitoba hibernate in enormous dens of perhaps ten thousand individuals, the proverbial snake pit. In springtime, the females are sexually receptive as they emerge, one at a time, from the den. And a good thing, too: Waiting impatiently is a gang of several thousand males, who pounce on each female as she exits, forming a writhing, orgiastic, but largely infecund "mating ball." Competition among the males is fierce, both pre- and post-coitus; after mating, the victor will insert a vaginal plug so no rivals can succeed if he has failed to impregnate the object of his affections. Even among snakes there is a core of basic behavior—including dominance, territoriality, and sexual jealousy—that humans have no trouble recognizing.

———

With very few exceptions, animal societies are not democracies. Some are absolute monarchies, some fluid oligarchies, some—especially on the female side—hereditary aristocracies. Dominance hierarchies exist in almost all, except for the most solitary, species of birds and mammals. There's a rank order based mainly on strength, size, coordination, courage, bellicosity, social intelligence. Sometimes you can predict, just by looking, who's dominant: the stag with the most points on his antlers, say, or that large, spectacularly well-muscled gorilla with the silver back. In other cases it's someone you wouldn't have guessed, someone without imposing physical stature, someone whose leadership qualities may be apparent to the animals you're observing but not to you.

The dominant animal—as determined in ritualized or occasionally in earnest combat—is called "alpha" after the first letter of the Greek alphabet. After alpha comes beta, then gamma, delta, zeta, eta . . . and so on down to omega, the last letter of the Greek alphabet. Most often, alpha lords it over beta, who makes appropriate indications of submission; beta over gamma; gamma over delta; and so on down the hierarchy.* The alpha male might exhibit dominance behavior in the male hierarchy 100% of the time, the omega male or males 0% of the time, with those in between showing intermediate frequencies.

Apart from the dubious intrinsic satisfaction of intimidating others, high rank often carries with it certain practical benefits—the privilege of dining first and from the choicest morsels, say, or the right to have sex with whomever strikes your fancy. The most passionate enthusiasts of dominance hierarchies are almost always the males, although loosely parallel female dominance hierarchies occur in many species. Males generally dominate all females and all juveniles. Among the comparatively rare species in which females sometimes dominate males are the vervet monkeys, the very same who keep their cool when overcrowded.

While privileged access to desirable females is not the invariable accompaniment of high rank, it is a frequent benefit. In a population of mice, the top third of the hierarchy was responsible for 92% of the inseminations. In a study of elephant seals, the bulls in the top 6% of the dominance hierarchy impregnated 88% of the cows.[6] High-ranking males often work hard to prevent lower-ranking males from inseminating the females. Females sometimes act to incite rivalry among the males.[7] If the dominant males are going to father almost all the children, then clearly there's a major selective advantage to being a dominant male. Whatever inherited qualities predispose to accomplishing, maintaining, and enjoying dominance will swiftly be established throughout the population—or at least among the males. Social and individual constitutions will be reconfigured by evolution to this end. Indeed, there seem to be parts of the brain in charge of dominance behavior.[8]

Promotion in rank does not usually occur because of community

---

* Alpha also dominates gamma and those below gamma; beta dominates delta and those below delta, and so on. Since more animals submit than dominate, it might with greater justice be called a submission hierarchy than a dominance hierarchy. But we humans are transfixed by dominance and often, at least in the West and setting religion aside, a little repelled by submission. Vast libraries are written on "leadership" and virtually nothing on "followership."

social work or fighting off invaders. Promotion comes mainly from combat within the group—mainly ritualized, sometimes real. Darwin clearly understood how natural selection might bring this about:

> The law of battle for the possession of the female appears to prevail throughout the whole great class of mammals. Most naturalists will admit that the greater size, strength, courage, and pugnacity of the male, his special weapons of offence, as well as his special means of defence, have been acquired or modified through that form of selection which I have called sexual. This does not depend on any superiority in the general struggle for life, but on certain individuals of one sex, generally the males, being successful in conquering other males, and leaving a larger number of offspring to inherit their superiority than do the less successful males.[9]

If you're a second lieutenant in the hierarchy and wish to be promoted, you challenge your first lieutenant; he would challenge his captain; he his major; and so on, up the ladder. In this respect at least, animal dominance hierarchies and human military hierarchies differ. Perhaps certain dog-eat-dog corporate hierarchies provide a better parallel. In the case of a successful challenge, the two animals sometimes exchange status, silver bars for gold. Animals weakened by disease, injury, or age are generally broken to the ranks.

"This town ain't big enough for the both of us" isn't the way dominance hierarchies usually work. Faced with a testy alpha male, you have another option besides fight or flight. You can submit. Almost everybody does. Subordinate males ingratiate themselves to those at the top of the hierarchy through incessant bowing and scraping. From their proximity to power those next in rank tend to gain access to food and to females, the leavings of the alphas. Sometimes dominant males are so busy with their police functions that those lower in the hierarchy can arrange sexual trysts that never would have been permitted had the alphas been less preoccupied. Surreptitious fertilization of females when the alpha male isn't looking is called "kleptogamy." "Stolen kisses" has something of the same flavor. So being alpha is only one strategy for males to continue their lines. Being beta or gamma with an inclination for kleptogamy is also a strategy. There are others.

An unambiguous, well-defined dominance hierarchy minimizes violence. There's plenty of threat, intimidation, and ritual submission, but not much bodily harm. Violence does occur when the rank order is uncertain or is in a state of flux. When young males attempt to establish their place in the hierarchy, or when there's a struggle at the top for alpha status, then there can be serious injuries, even death by combat. But if you don't mind constantly subordinating yourself to those of superior rank, dominance hierarchies provide a peaceful and ritualized environment with few surprises. Perhaps this is part of the appeal for those humans drawn to the religious, academic, political, police, and corporate hierarchies, and the military establishment in peacetime. Whatever inconveniences hierarchy may impose are offset by the resulting social stability. The price may be paid in anxiety— anxiety about offending those of higher rank, being perceived as insufficiently deferential, forgetting yourself, committing lèse-majesté.

In maintaining the dominance hierarchy, all conflicts (chiefly ritual or symbolic combat) are between animals who know each other well. But xenophobic intraspecific aggression is different, occurring between animals with no perceived bonds, relationships, or even familiarity. It's an encounter with strange-smelling aliens, and the circumstance most likely to lead to casualties and deaths.

When an unfamiliar mouse arrives, rats drop what they're doing and attack it—dominant rats attacking the intruder's back and frequently mounting it in the process, while subordinate rats attack the intruder's flanks and rarely mount it. Each in his own way.[10] Among mice living in small groups, those at the top of the hierarchy tend to be most active in scuffling, intimidating, and fighting, in reacting to novelty, and in fathering baby mice. They also have sleeker coats than the subordinate males. But when it comes to fighting mice of another group,[11] suddenly democratic forms come into play and the subordinates fight side-by-side with the alphas.*

---

* The very recent history of human warfare provides a contrast: The alphas—generally old men—sequester themselves in safety, often where the young women are, and dispatch the subordinates—generally young men—out to fight and die. In no other species have alpha males gotten away with such cushy arrangements for themselves. It does require at least implicit cooperation between the alphas of rival groups, but this can often be arranged. Apart from the social insects, no other species has been clever enough to invent war. It is an institution optimally configured to benefit the alphas.

The simplest geometry of a dominance hierarchy is linear or straight-line. This is what we've been describing. The private defers to the corporal, the corporal to the sergeant (and if you look more closely, there are various hyperfine grades of privates, corporals, and sergeants), the sergeant to the second lieutenant, and so on, up through first lieutenant, captain, major, lieutenant colonel, colonel, brigadier general, major general, lieutenant general, plain old general, and general of the army or field marshal. The military establishments of different nations have different names for the various ranks, but the basic idea is the same: Everyone knows his rank. A currency of deference is offered by subordinate to superior. Homage is paid.

Linear hierarchies are a mode of social organization readily observable in domestic fowl, which is where the phrase "pecking order" originates. It's especially clear-cut among the hens. (In mammals the pecking order is often the chief fact of male social life.) Again, the alpha hen pecks beta and everybody lower; beta pecks gamma and everybody lower; and so on down the hierarchy to poor omega, who has no one at all to peck. The high-ranking males try to sexually monopolize the hens, but sometimes they fail. Cocks dominate hens except on rare occasions; the word "henpecked" refers to the exceptions and comes from everyday observation of barnyard life.

With large populations a linear rank order is rare; instead, little triangular loops break out in which delta dominates epsilon, epsilon dominates zeta, but zeta in addition to dominating eta also dominates delta, or maybe even someone higher up the hierarchy.[12] This leads to a social complexity that may be opposed by die-hard conservative chickens.

How does the dominance hierarchy get established? When two chickens are introduced to each other, there is usually a brief squabble —involving much clucking, squawking, pecking, and feathers flying. Or else one chicken takes a good look at the other and submits without a fight, as is usually the case when an immature chicken is confronted by a healthy adult. Among vigorous hens, the winner is the better fighter, or the better bluffer. A home-court advantage is reported: A hen is more likely to win the fight in her own yard than in her adversary's. Aggressiveness, bravery, and strength play their roles. After a single instance of dominance combat, the relationship between the two hens is often frozen; the higher-rank has the right to peck the lower-rank without fear of retribution. Flocks in which high-ranking hens are regularly removed and replaced by total strangers fight more, eat less, lose weight, and lay fewer eggs. In the long view, the pecking order is in the interest of the chickens.[13]

"Playing chicken" is an American male adolescent game of 1950s vintage in which each threatens the other to see who will flinch first. The most familiar example involves automobiles speeding directly toward one another; he who swerves first may gain his life (and, incidentally, save that of his rival) but lose his status. Calling it "playing chicken" recognizes its deep evolutionary origins. *Being* chicken, in the same youth culture, means being fearful of performing a risky or heroic action. Again, the behavior of subordinates in the barnyard dominance hierarchy is evoked; again, the choice of words betrays if not real knowledge at least a suspicion of the animal roots of the practice.

Another way in which our awareness of animal dominance hierarchies has insinuated itself into the language and proves useful in describing our own behavior is the use of the phrases "top dog" for the alpha male and "underdog" for everyone else. When we say we're for the underdog in sports or politics or economics, we're revealing an awareness of dominance hierarchies, their injustice and their shifting fortunes.

There are monarchical social systems in which everyone is dominated by the alpha male or the few highest-ranking males, and hardly any aggression occurs in the rest of the group. The dominant male spends a considerable amount of his time calming outraged subordinates and adjudicating disputes. Sometimes justice is a little rough, but often merely a bark or grimace will suffice. In such systems especially, dominance hierarchies carry with them social stability. The males of many species have evolved potent weaponry. Life would be a lot more dangerous if every time two piranha males, or two lions, or two stags, or two elephant bulls had a difference of opinion, it was a fight to the death. The dominance hierarchy—with relative status fixed for considerable periods of time, and the institutionalization of ritualized rather than real combat in settling serious disputes—is a key survival mechanism. Not only is there a genetic advantage for the dominant male, but also for everyone else. *Pax dominatoris*. Even if you have to take a lot of abuse, even if you sometimes resent the brass, it's safe, maybe even comfortable, in such a system—where everyone knows his place.

So what kind of selection is this? Is it simple individual selection for the alpha male, with the benefit for other males being only incidental? Is it kin selection, because the lower-ranking males are not-too-distant relatives of the alpha? Is it group selection, because such a group, structured and stabilized by a dominance hierarchy, is more likely to survive than one in which combat to the death is the norm? Are these categories separable and distinct?

The alpha might be of a mind to attack an offending inferior, but if the latter makes the species' characteristic submission gestures, the former feels obliged to spare him. They have not sat down and agreed on a moral code, no tablets have been carried down from the mountain, but the postural and gestural inhibitions to violence work very much like a moral code.

One of the most spectacular examples of dominance behavior in groups—known among animals as different as birds, antelopes, and (perhaps) midges—is called the lek:

> [L]eks are tournaments, held before and during the breeding season, day after day, when the same group of males meet at a traditional place and take up the same individual positions on an arena, each occupying and defending a small territory or court. Intermittently or continuously they spar with their neighbours one at a time, or display magnificent plumage, or vocal powers, or bizarre gymnastics . . . Though they have territories, yet they have a hierarchy with the top-ranking males typically placed in the middle and ungraded lesser aspirants ranged outside. Females come to these arenas in due course to be fertilized, and normally they make their way through to one or other of the dominants in the centre.[14]

Perhaps spring break at Ft. Lauderdale or Daytona Beach is one of the more conspicuously lekish human institutions.

Among reptiles, amphibians, and even crustaceans, dominance behavior is common.[15] The varanids (such as the komodo dragon) are very good at ritualized and stereotyped intimidation displays. They rattle or lash their tails, rear up on their hind legs, inflate their throats, and, if their rival has not yet submitted, attempt to wrestle him to the ground. In crocodiles, dominance is established by slapping the head into the water, roaring, lunging, chasing, and biting, pretend or real. When interrupted in his mating embrace, a male frog croaks; the deeper his croak, the greater his implied size when disengaged, and the more diffident is the would-be intruder. A toothless, brightly colored Central American frog, genus *Dendrobates*, intimidates intruders by performing a vigorous sequence of push-ups. But among the skinks, in which aggression is released seasonally when the heads of the males turn bright red, the virtues of intimidation by bluff are often lost sight of, and the two rivals tear into each other without so much as a preliminary throat swelling. When hermit crabs introduce themselves, they devote a few seconds to taking each other's measure—by stroking one another with their antennae; the smaller then promptly submits to the larger.[16] Stalk-eyed flies do the same; the more dominant individuals are the ones with the more widely separated eyes.

It's rare that any male starts out as an alpha. Generally you have to work your way up through the ranks. But in the intervals between your challenges it would be a mistake to be too disruptive. Even for the very ambitious a talent for subordination and submission is needed. Also, it's hard to predict who will achieve high-ranking status. Sometimes greatness is thrust upon unsuspecting animals by the course of events. Accordingly, everyone needs to be able to rise to the occasion. If you're in a linear hierarchy, you must know how to dominate the animals below you and submit to those above. An inclination for both dominance and submission must beat within the same breast. Complex challenges make for complex animals.

———

Nothing we've said so far indicates anything about female preference. What if she finds the alpha male arrogant, boorish, taking too much for granted? Or just plain ugly? Does she have the right to refuse? At least among hamsters, this is not an option.

Here's an experiment[17] done on Syrian hamsters by the psychologist Patricia Brown and her colleagues: To begin, males, matched for size and body weight, were allowed to interact with one another in pairs to establish dominance. Chasing and biting were among the behaviors counted as dominant; defensive postures, evasions, raised tails, and full cowering submission were counted as subordinate traits. The dominants accounted for over ten times more aggressive acts than an equal number of subordinate animals; the subordinate animals tallied ten times more submissive acts than those judged dominant. It never took more than an hour for a pair of hamsters to decide who was dominant and who was subordinate.

Now although these males knew how to fight, they'd never had a sexual experience. Each of them was made to wear a little leather harness attached to a tether, which, like a dog's leash, limited how far he could roam. Next, an ovulating female was released; she could have access to the tethered males, but beyond a certain point their leashes would prevent them from following her or offering unwelcome attentions. Whatever sexual contact might be in the offing would be on her terms.

We imagine her, steely-eyed, slowly looking the males over head to tail in their kinky leather outfits. Because the earlier dominance conflict was largely ritual, there were no injuries to betray which was the subordinate animal. Each male was in its own partitioned area, so they could not see one another and betray to the female their relative status through gestures of dominance or submission. Would she, despite the absence of signs apparent to the human observers, select the dominant male? Or would she find some other trait more attractive? The females were not hesitant or demure. In less than five minutes, every one of them presented herself for copulation to one of the males. In every case it was the dominant male. Prior familiarity was not required. Somehow she knew. There were no questions asked about his education, family, financial prospects, or the gentleness of his disposition. Every female was eager for sex with the dominant male.

How could the female know? The answer seems to be that she could smell dominance. There is literally a chemistry between them, the odor of power. The dominant males give off some effluvium, some pheromone that subordinate males do not.[18]

"I'm a celebrity. That's what celebrities do," offered one-time heavyweight boxing champion Mike Tyson in explaining his scattershot propositioning of virtually every contestant at a beauty pageant. Former U.S. Secretary of State Henry Kissinger, not known for his looks, explained a beautiful actress's attraction to him in these words: "Power is the greatest aphrodisiac."

Dominant males preferentially copulate with attractive females. The females are as accommodating as they can be. They crouch down, they raise their hindquarters, they lift their tails out of the way. (We're back to hamsters.) In Brown's rodents-in-motorcycle-jackets experiment, during the first half hour of mating the number of "intromissions" by dominant males averaged 40; those subordinate males able to score at all (usually after the dominants were done) averaged a measly 1.6 for the half hour.

Now suppose you grow up in a society in which such behavior is the community standard. Wouldn't you tend to conclude that the animal who mounts and who makes repeated pelvic thrusts is the dominant partner, while the animal who crouches, who is receptive and passive, is subordinate in rank? Would it be surprising if this powerful symbol of dominance and submission were generalized in the gestural and postural vocabulary of the status-obsessed males?

Before the invention of language, animals need clear symbols to communicate with one another. There's a well-developed non-verbal language, which we've already described, including "My belly's up and I surrender" and "I could bite you but I won't, so let's be friends." It would be very natural if everyday reminders of status in the rank hierarchy were established by brief ceremonial mountings of males by males. He who mounts is dominant; he who is mounted is subordinate. No "intromission" is required. Such symbolic language is in fact widespread, and we will discuss it in greater detail in later chapters. It may have little or no overt sexual content.

Under natural conditions, ordinary Norway rats—the same common variety whose social structure collapsed in Calhoun's overcrowding experiments—arrange themselves into social hierarchies. A dominant might approach a submissive animal, sniff and lick its anogenital area, and mount it from the rear, holding on with the forepaws. The submissive animal might elevate its hindquarters so as to indicate its eagerness to be mounted. Male aggression in maintaining the dominance hierarchy includes banging flanks, rolling over and kicking, pinning the opponent with the forepaws, and boxing—the two animals actually stand toe-to-toe and let loose with left jabs and right uppercuts. Under normal conditions it's rare that anyone is injured.

Even among lobsters the aggressive posture is to stand upright—

indeed, on their toes (or at least the tips of their claws). The submissive posture is flat on the ground, legs somewhat akimbo. The idea is to show that you can't (quickly) do any harm, even if you want to. Many gestures in a similar spirit can be found among humans. Police confronting possibly armed suspects will order them to raise their hands (so it's clear they're weaponless); or clasp their hands behind their necks (ditto); or lean at a high inclination angle against a wall (so their hands must support them); or lie prone. Submissive words are well and good ("I didn't mean nothin', honest"), but a police officer putting his or her life on the line requires a firmer postural guarantee.

In almost all higher mammals copulation occurs with the male entering the female's vagina from behind. The female crouches down to assist the male in mounting her. She may make special motions to aid his entry, and those motions, like the bump and grind, become part of the symbolic language of enticement. The reason for the crouch is partly to present a favorable geometry for entry, but it also indicates that she has no intention of going anywhere. She's not about to run away. Something similar can be seen in many other species. A male beetle come a-courting taps on the female's carapace—in different beetle species, drumming with his feet, his antennae, his mouthparts, or his genitals—and she is instantly immobilized.[19] The strange attractiveness to men of grotesquely deformed small feet (in China for nearly a millennium), and of very high heels (throughout the modern West), as well as traditional, constraining women's clothing[20] and the fetish of female helplessness in general, may be a human manifestation of the same symbolism.

In many species the alpha male systematically threatens any other male attempting to mate with *any* female in the group, especially when conception is possible. Because of clandestine impregnations by subordinate males—kleptogamy—in which the females are often willing partners, the alpha does not always succeed; but he's highly motivated to try. This is true within female dominance hierarchies as well. In domestic fowl, for example, the alpha female tends to attack any female that so much as walks up to an adult male during the breeding season. In gelada baboons, in which there is a female dominance hierarchy, high-ranking females do not, on average, mate more frequently during ovulation than do the lower-ranking females; but the lower-ranking females rarely give birth. Something about their inferior status diminishes their fertility. Perhaps they are advertising ovulation when in fact no egg is released, or maybe they have many spontaneous abortions. But whatever it is, their low status prevents them from having babies. In marmosets, subordinate females tend to suppress their ovulations, but when they are freed from the female dominance hierarchy, they quickly become pregnant.[21] Thus, genes contributing to high status in the female hierarchy—large stature, say, or superior social skills—get preferentially passed on to the next generation. This tends to stabilize a hereditary aristocracy.

In cattle and many other animals, the alpha male may try to gather around him a harem of females and chase away the other males, but his success is often limited. When the breeding time has passed, the males return to their solitary ways and the females (and young) resume their own social grouping. Among deer this is called a hind group and entails its own dominance hierarchy. Commonly, the leader of such communities is determined not by bluff, threat, or fighting ability, but by age: The oldest fertile female leads. (The same convention is adopted among all-female herds of African elephants; even when composed of hundreds of elephants, the social structure is extremely stable.) These groups seem to be organized around protection. When attacked, they form a diamond- or spindle-shaped pattern, with the alpha female in front and the beta bringing up the rear. If the pursuers are gaining, the beta female may valorously stop short and engage the leading predator. As the rest of the group makes its escape, the alpha and beta may then exchange sentry duty.

In skirmishes the advantages of the dominance hierarchy are clear. Even female mammals who evince little enthusiasm for individual dominance nevertheless arrange themselves into battle hierarchies in times of trouble. So dominance hierarchies have at least two functions, extremely useful both for individuals and for the group: They reduce dangerous and divisive fighting within the group (promoting what we might call political stability); and they are optimized for intergroup and interspecies conflict (providing what we might call military preparedness).

A third purported advantage of dominance hierarchies is that they preferentially propagate the genes of the alphas, those who are physically or behaviorally fit. We might imagine a common conditional strategy for everyone in the group that would go something like: "If

I'm big and strong, I intimidate; if I'm small and weak, I retreat." This benefits everyone one way or another, and the sole focus is on the "I."

Being human, we naturally feel some whiff of resentment when we imagine ourselves dropped into such a dominance hierarchy with its craven submissiveness and manifest cruelties. Being human, we might also imagine the pleasures of a well-run social machine in which everybody knows his place, in which nobody gets out of line and causes trouble, in which deference and respect to superiors is routinely shown. Depending on whether we come from a more democratic or a more authoritarian upbringing, schooling, or society, we might feel that the benefits of the dominance hierarchy outweigh any affront to freedom and dignity, or vice versa. But this discussion isn't yet about us. Humans are not red deer or hamsters or hamadryas baboons. For these species the cost-benefit analysis has been made. For them, law-and-order is the higher good. That there are innate individual rights and liberties of hamsters, needing institutional protection, is not a self-evident truth.

———

To play the hierarchy game, at the very least you must be able to remember who's who, to recognize rank, and to make the appropriate responses, dominant or submissive as circumstances dictate. The ranks are not fixed in time, so you must be able to reassess and revise facts of central importance. Dominance hierarchies bring benefits, but require thinking and flexibility. It's not enough to have inherited nucleic acid instructions on how to threaten and how to submit. You must be able to *apply* those behaviors appropriately to a changing array of acquaintances, allies, rivals, lovers—whose dominance status is situational and whose identity and current circumstances cannot possibly be encoded in the nucleic acids. As is also true for hunting and escape strategies or learning from parents, hierarchies require brains. Nevertheless, the instructions in the genes are often vastly more in control than whatever wisdom resides in the brain.

Early on, animals may not have been very adept at distinguishing *individuals*, contenting themselves with "If he gives off my favorite sex attractant, he's my guy." In interaction with predator and prey, or in the sexual adventuring of males who are not obliged to care for the offspring, there's no high premium on the niceties of individual recognition. Then you can get away with "They all smell the same to me" or "They're all the same in the dark." Then you can stereotype and there are few adaptive penalties you must pay. But as evolutionary time passes, finer distinctions must be made. It might be useful to know who the father of your child is so you can encourage him to play a role in raising and protecting it. It might be useful to know the exact position of all the other males in the dominance hierarchy if you wish to avoid daily conflicts about rank, or if you wish to advance up the ladder.

One of the many surprises in modern primate research is how readily the human observer—even if wholly insensitive to olfactory cues—can distinguish and recognize all the baboons in the troop, all the chimps in the band. If you spend a little time with them, they no longer all "look alike." It takes some motivation and a little thought, but it's well within our powers. Without such individual recognition, the greater part of the social life of higher animals, as of humans, remains hidden from us. With humans—because of language, dress, and behavioral eccentricities—individual recognition is much easier. Still, the temptation to divide humans and other species into a small number of stereotyped categories, rather than recognizing differences and judging individuals one at a time, remains deep within us.

Racism, sexism, and a toxic mix of xenophobias still powerfully influence action and inaction. But one of the proudest achievements of our own age is the developing global consensus—despite many false starts—that we're at last ready to leave behind this vestige of long ago. Many ancient voices speak within us. We are capable of muting some, once they no longer serve our best interests, and amplifying others as our need for them increases. This is cause for hope.

As for the larger issue of dominance and submission, the jury is still out. True, all but the pomp and costume of monarchy have, in the last few centuries, been swept off the world stage, and attempts at democracy seem fitfully to be breaking out planetwide. But the call of the alpha male and the compliant assent of the omegas remain the daily litany of human social and political organization.

ON   IMPERMANENCE

As for Man, his days are as grass; as a flower of the field, so he flourisheth.

For the wind passeth over it, and it is gone; and the place thereof shall know it no more.

Psalm 103, verses 15, 16, King James translation

Chapter 12

THE RAPE
OF CAENIS

Not the immortal gods can flee,
nor the men who live only a day.
Who has you within him
   is mad.

SOPHOCLES,
Antigone, 781 [1]

Over the Earth he flies
and the loud-echoing salt-sea.
He bewitches and maddens the heart
of the victim he swoops upon.
He bewitches the race of the mountain-hunting
lions and beasts of the sea,
and all the creatures that Earth feeds,
and the blazing sun sees—
and man, too—
over all you hold kingly power,
Love, you are the only ruler
over all these.

EURIPIDES,
Hippolytus, 1268 [2]

One of the myths of ancient Greece tells of Caenis, "loveliest of the maids of Thessaly," who, while walking alone on an isolated shore, was spied by Poseidon—god of the sea, elder brother of the king of the gods, and sometime rapist. Mad with lust, the god attacked her on the spot. Afterwards, he took pity, and asked what he might give in reparation. Manhood, was her answer. She wished to be transformed into a man—not just any kind of man, but one extravagantly male, a warrior and "invulnerable." Then she would never again be subjected to such a humiliation. Poseidon agreed. The metamorphosis was completed. Caenis became Caeneus.

Time passed. Caeneus fathered a child. With his sharp and expertly wielded sword he killed many. But the swords and spears of his adversaries could not penetrate his body. The metaphor here is not hard to fathom. Eventually Caeneus became so full of himself that he scorned the gods. He erected his spear in the marketplace and made the people worship it and sacrifice to it. He insisted, on pain of death, that they worship no other gods. The symbolism is again lucid.

Extreme arrogance, of which this is a fair example, was called by the Greeks *hubris*. It was almost exclusively a male trait. Sooner or later it would attract the attention and then the retribution of the gods—especially toward those humans insufficiently deferential to the immortals. The gods craved submission. When news of Caeneus's effrontery finally reached Zeus, whose desk was doubtless piled high with such casefiles, he ordered the centaurs—chimeras, half-man, half-horse—to execute his merciless judgment. Dutifully they attacked Caeneus, taunting him: "Do you not remember at what price you gained this false appearance of a man? . . . Leave wars to men." But the centaurs lost six of their number to Caeneus's swift sword. Their lances bounced off him "like a hailstone from a roof." Disgraced at being "conquered by an enemy but half-man"—a hollow complaint, coming from a centaur—they resolved to smother him with timber, destroying vast stands of trees "to crush his stubborn life with forests for our missiles." He had no special powers concerning breathing, and after a struggle they managed to subdue and then to suffocate him. When the time came to bury the body, they were amazed to find that Caeneus had reverted back to Caenis; the invincible warrior had become, once again, the vulnerable young woman.[3]

Perhaps poor Caenis had overdosed on the stuff that Poseidon used to effect the metamorphosis. There is a proper amount of whatever it is that makes one male, the ancient Greeks recognized, and too much or too little can get you into trouble.

———

The testicles of a sparrow are about a millimeter long and weigh about a milligram. (That's one of the reasons you never hear that someone's hung like a sparrow.) With testes intact, the scrappy birds enter into their mainly linear hierarchy, chase away other birds who invade their territory, and, if they're high-ranking, make successful overtures to fertile females. But reach under those feathers, remove those two tiny organs, and, after the bird has recovered, all of these traits are lost, or nearly so. Aggressive birds become submissive, territorial birds become complacent about intruders, passionate birds lose interest in sex. Now inject a certain steroid molecule into the sparrow and it regains its plucky enthusiasm for sex, aggression, dominance, and territoriality.

Shortly after castration, male Japanese quails stop strutting, crowing, and copulating. They also fail to elicit the interest of female Japanese quails. Treat them with that same steroid and they're back to strutting, crowing, and copulating, and the females find them irresistible once more. Castrate a young male fiddler crab and he will never develop his distinctive asymmetrical giant claw.

Humans have understood some of this for thousands of years. Captured warriors were castrated so they'd make no trouble. We still describe an ineffective leader as a "political eunuch." Chieftains and emperors castrated men so they could guard the harems without succumbing to temptation (or at least—the compromise sometimes reached—without impregnating any of the residents); and so their loyalties to the leader would not be adulterated by family ties or other distracting affections and obligations. It is remarkable that almost exactly the same molecule should produce such fundamental changes in behavior in sparrows, quail, crabs, and humans.

The steroid molecule that works these transformations like some wizard's potion is testosterone. Along with other, similar molecules, it's called an androgen. It's manufactured (from, of all things, cholesterol) mainly in the testicles,[4] enters the bloodstream, and induces a complex set of behaviors that we recognize as characteristically male. Here too, the connection is acknowledged in the language, as in the expression "He's got balls"—meaning he's shown exemplary courage and independence, he's not a coward or a sycophant.

In newly formed groups of male monkeys, the higher the rank in a forming dominance hierarchy, the more testosterone is found to be circulating in the blood. But when the hierarchy settles down to symbolic encounters, and betas are routinely submitting to alphas, the correlation vanishes.[5] The more testosterone an animal has, the farther away he's willing to roam to challenge and dominate potential rivals.[6] With high testosterone levels there's a cross-species tendency for dominance within the group to be extended to dominance over a piece of territory. The boss and the landlord become one.

In the brains of many animals are specific receptor sites to which the testosterone molecule and other sex hormones chemically bind, and which are in charge of hormone-induced behavior. There may be separate brain centers responsible for strutting, crowing, bullying, fighting, copulating, defending territory, and fitting into the dominance hierarchy; but each center has a button pushed by testosterone. The behavior is actuated once the testosterone migrates from the testicles through the blood to the brain. In the individual brain cells, the presence of testosterone activates otherwise untranscribed and ignored segments of the ACGT sequence, synthesizing a set of key enzymes. As with many hormones, testosterone is at the nexus of an array of positive and negative feedback loops that maintain the concentration of the molecule circulating in the blood.

Male animals don't just endure, but seem to delight in, testosterone-mediated scuffles, intimidation, and combat. Mice will learn to run a complex maze when the only reward or reinforcement is the opportunity to have a tussle with another male. There are abundant similar examples in our species. Activities that are central to leaving many offspring tend to be entered into with enthusiasm. Sex itself is the most obvious example. Aggression is in the same category.

Even among animals with very short gestation periods, such as mice, the delay between conception and birth is too long for the animal to associate cause and effect. To leave it to mice to figure out the connection between copulation and the creation of the next generation is to condemn their genes to extinction. Instead there must be an absolutely overwhelming need for sex and—as a means of reinforcement—a delight in partaking of it. This is just the DNA creatively demonstrating its control in the most overt and clear-cut way.

A deal has been struck: The animal will forgo food, will conform to extreme postural indignities, will put its very life at risk so its strands of DNA can join up with the strands from some other animal of the same species. In exchange, there will be a few moments of sexual ecstasy, one of the currencies in which the DNA pays off the animal that carries it around and nurtures it. There are many other examples of DNA-mediated delight in activities tending toward adaptive fitness —including parental love for children, joy in exploration and discovery, courage, camaraderie, and altruism, as well as the standard array of testosterone-driven traits making bosses and landlords.

Hormones similar to testosterone play a central role in the development of sexual organs and sexual behavior all the way down to the aquatic fungi. Steroids must have evolved very early to be so widely distributed today, perhaps going a fair way back to the invention of sex around a billion years ago.

This trans-species use of the same molecule for roughly the same sexual purpose has some bizarre consequences. For example, the chief sex pheromone in the pig is 5-alpha-androstenol—chemically similar to testosterone. It's mixed in with the boar's saliva (as testosterone is present in the spit of men). When a sow in heat smells this steroid on a slavering boar, she promptly adopts the come-hither mating posture. Oddly, truffles, the French culinary delight, produce exactly the same steroid and in a higher concentration than in boar spit. This seems to be why pigs are used by gastronomes to find and unearth truffles. (How strange it must seem to the sows, always falling in love with little black pieces of fungus, only to have them cruelly snatched away by humans.) Since truffles are fungi, in which steroids play key sexual roles, perhaps tormenting sows is just an accidental side-effect—or perhaps it serves the function of inciting pigs to dig so the spores are spread more widely and the Earth is covered with truffles.

Now in light of all this, what are we to make of the fact that 5-alpha-androstenol is also copiously produced in the underarm perspiration of men?[7] Long ago—before institutionalized hygiene, before the present perfumed and deodorized age—might it have played a part in human and prehuman courtship and mating behavior? (The noses of women, we cannot help noting, are often at the same level as the armpits of men.*) Might this have something to do with the willingness of the rich to spend exorbitant sums on tiny pieces of a nearly tasteless cork-like substance?

A genetically male embryo deprived of testosterone and other androgens will emerge with what look very much like female genitals. Conversely, the genitalia of a genetically female embryo subjected to high levels of testosterone and other androgens will be masculinized: If smaller amounts of the steroid are present, perhaps she's born with only a somewhat bigger clitoris; if larger amounts, her clitoris becomes a penis, and her labia majora fold over to become a scrotum. She may develop a normal-looking male penis and scrotum, although the scrotum will have no testicles within. (She'll also have nonfunctioning ovaries.) Such girls as they grow up are found to prefer guns and cars to dolls and kitchen supplies, boy to girl playmates, and enjoy roughhousing and the outdoors; they may also find women sexually more attractive than men.[8] (There's no evidence for the converse—for example, that most tomboys have excessive amounts of androgens.)

The difference between male and female, not genetically but on so fundamental a matter as which set of external genitalia you are to have, depends on how much male steroid you encountered in the first few weeks after conception. Leave that bit of developing embryonic tissue alone and it will become a female. Suffuse it with a little testosterone-like hormone and it will become a male.† The tissue is spring-

---

* One of the expert reviewers of this book complains, "I would struggle to help noticing . . .You don't need to be right at armpit level to smell. Consider any gym." But gyms are permeated with the accrued perspiration of many athletes over many years. Another expert reviewer notes that molecules like 5-alpha-androstenol are now marketed as alleged aphrodisiacs.

† Thus, Aristotle's contention[9]—echoed millennia later by Sigmund Freud—that "the female is, as it were, a mutilated male" is wrong. (Neither is a male a testosterone-

loaded to respond to the androgen (the word literally means "male maker"), which serves as a means of internal communication. There are buttons on the developing embryo that only androgens can push. Once they *are* pushed, substantial machinery, whose existence you might otherwise never have guessed, takes over and works mythic transformations.

Across widely different animal species, another class of sex hormones, the estrogens, curbs aggressiveness in females, and yet another, progesterone, increases the feminine inclination to protect and care for the young. (The words signify, respectively, something like estrus-maker and gestation-promoter.) Mother rats, as all mammals, are attentive to their offspring: They build and defend nests, nurse the pups, lick them clean, retrieve them when they wander away, and teach them. None of this behavior is evident in virgin females, though, who studiously ignore newborn pups, or even make some efforts to avoid them. But prolonged treatment with the female hormones progesterone and estradiol—bringing the hormone levels of virgins up to those typical of late pregnancy—results in the emergence of marked maternal behavior. Rats with high levels of estrogen are also less anxious and fearful and less likely to engage in conflict.[10]

These female hormones are produced mainly in the ovaries. But when we see a calm, competent, and loving mother, most of us are not driven to exclaim "Man, has she got ovaries!" The reason doubtless has something to do with the ready accessibility of testicles for accidental or experimental removal, dangling as they do in vulnerable external sacs *—quite differently situated than the ovaries, which are locked away for safekeeping within the vault of the body. But clearly ovaries must equally be counted as among the family jewels.

The female hormones control the estrus cycle—which culminates

---

altered female, although that's a little closer to the truth.) Women's bodies do synthesize estradiol, the most potent of the estrogens, from testosterone.

* In order, it is conventionally thought, to keep their temperature a few degrees lower than if they were situated inside the body. If the testicles were located within the warm abdomen, sperm cells, it is said, would be sparsely produced and men would be largely sterile. The benefits of external testicles outweigh the risks. But sparrows and scrappy songbirds carry their testes within; and yet, even at the elevated temperatures, their sperm cells seem to be spunky enough. Our understanding of why males of some species wear their testes outside, and others inside, seems incomplete.

when the females are ovulating and, usually, broadcasting olfactory and visual cues that they're available for mating. In many species this doesn't happen often and doesn't last long; cows, for example, are interested in sex for about six hours every three weeks. Cows don't date much. "For most species," writes Mary Midgley,[11] "a brief mating season and a simple instinctive pattern makes of it a seasonal disturbance with a definite routine, comparable to Christmas shopping." In a wide variety of mammals, from guinea pigs to small monkeys, mating outside of estrus is not only discouraged by the female, it's also made physically impossible by an organic chastity belt: The vagina is sealed by a membrane or plug grown specially for the purpose, or—even more decisive—it's fused shut.

In contrast, among most humans and some apes, sex is not only possible but is equally probable at virtually any phase of the cycle. Some humans monitor the cycle (by measuring small changes in body temperature) and then *avoid* sex around the time of ovulation. This Church-condoned contraceptive technique is the mirror image of the practice of most animals—who garishly advertise ovulation and avoid sex at all other times. It is a reminder of how far from our ancestors our culture has taken us, and what fundamental changes in us are possible.

For many animals the ovulation cycle is a few weeks in length. Not many species have periods almost exactly equal to the lunar cycle (the time from new moon to new moon). Whether this peculiarity of humans is more than a coincidence—and if so, why it should be—is unknown.

Mammals suckle their young, but only the females are appropriately endowed.* It's one of the few cases where the definition of a major

---

* Exceptions are, in a way, common. Male pigeons and doves routinely feed the young a regurgitated "crop milk," low in sugar, high in fat—just the opposite of the milk of mammals. The cock emperor penguin, after incubating the egg for forty days, generates a rich milk in his esophagus. When the chick hatches, this is its only food. It doubles its weight on Father's milk and is doing very well by the time the emperor penguin hen returns engorged with tiny shrimp. Both sexes of greater flamingos generate a kind of milk which is mixed with their blood and fed to the chicks in the first month of life; each parent provides about a tenth of a liter of this formula each day.[12] Many animals—wolves, for example—feed their young with regurgitated food, but this is very different from milk.

classification category in biology, or taxon, is determined by the characteristics of only one of the sexes. Giving milk is also hormonally mediated. Mother's milk is essential for the young, who are born helpless, unable to digest the adult diet. This is another reason that females spend more time with, and therefore have a greater investment in, the young. The males are generally more interested in other things—dominance, aggression, territoriality, many sex partners.

The connection between steroids and aggression applies with surprising regularity across the animal kingdom. Remove the principal source of sex hormones and aggression declines, not just among the mammals and birds, but in lizards and even fish. Treat castrated males with testosterone and the aggression returns. Give estrogen to intact animals and aggression diminishes, again across all these species. The repeated use of these same steroids for the same functions, turning aggression on and off, for so many different animals, is a testament both to their effectiveness and to their antiquity.

Aggression is adaptive, but only in controlled amounts. The repertoire of aggressive behavior is on call, awaiting only to be disinhibited. The steroids, their production titrated by the social environment and the biological clocks, do the disinhibiting. This being the case, why is it that males are so often more aggressive than females? If the females can generate a little less estrogen and a little more testosterone, can't they become as aggressive as males? Something like gender equality in aggression occurs in wolves, tree squirrels, laboratory mice and rats, short-tailed shrews, ring-tailed lemurs, and gibbons. In the southern flying squirrel, males are not territorial but females are, and most quarrels between the sexes are initiated by the females—and won by them.[13] The clear fact that males are more aggressive than females among us humans (where blood plasma testosterone is about ten times greater in men than in women) by no means commits the rest of the animal kingdom, or even the rest of the primates, to the same arrangement.

As anyone knows who has seen a pet tomcat drag himself home after an absence of a day or two—with an eye closed, an ear torn, his fur matted and bloody—testosterone exacts a price. What happens if you take a male animal—let's say, someone less combative than tomcats out for a night on the town—and equip him with an implant that keeps his testosterone blood levels high? When this is done to sparrows, hardy territorialists, there seems to be no significant increase in the sparrow murder rate. But when male cowbirds are implanted, their numbers markedly decrease;[14] many birds are now observed with unusually serious injuries, clearly obtained in combat with their fellows. Unlike sparrows, cowbirds establish dominance hierarchies but do not have core territorial refuges into which they can flee. Bluff can escalate into serious fighting if you're simultaneously charged up with testosterone and have no tradition of sanctuary. Another steroid deficit: Male birds with artificially high testosterone levels are less inclined to feed their hatchlings.[15] Macho males tend to neglect their family responsibilities.

Sex hormones are now manufactured by pharmaceutical companies, and widely used—legally and illegally. We can learn something about their role in Nature by asking why people use them. Anabolic steroids are molecules very like, but usually not identical to, testosterone. They're taken mainly by: (1) bodybuilders and athletes (who widely believe that certain feats of strength can be accomplished *only* by young men on steroids); (2) young men who wish to macho up, usually to attract women or other men; and (3) those who wish to disinhibit their meanness (nightclub bouncers, hit men in organized crime, prison guards, and so on).[16] The enhanced musculature does not come about through steroids alone; it also requires vigorous and systematic exercise. One of the side effects is facial and back acne. Anabolic steroids don't seem to grow hair. Large doses lead to dysfunction and atrophy of the testicles—perhaps the body's response to excessive testosterone titers; too much testosterone is socially sufficiently dangerous that a mechanism may have evolved so that tendencies toward excessive production aren't passed on to future generations.

Estrogen is taken by women, usually post-menopause or post-hysterectomy, to preserve sexual interest and lubrication, to slow loss of bone calcium, and to achieve a more youthful complexion. Bodybuilding and transsexual women may take anabolic steroids because they strikingly redistribute weight—from thighs to chest and biceps, for example. Transsexual men taking estrogen redistribute weight the other way, grow breasts, and feminize the nipples and areolae; there's also a general mellowing of temperament. Bearing in mind these consequences of taking sex hormones as an adult, and the much more profound influence they have on the embryo—actually determining which sexual organs will be present—it seems likely that far subtler changes in hormone levels might influence not just dominance, territoriality, aggression, care for the young, gentleness, anxiety level, and talent for conflict resolution, but also sexual appetite and preference.

———

Bulls, stallions, and roosters are made into steers, geldings, and capons because humans find their machismo inconvenient—the very same male spirit that the castrators likely admire in themselves. One or two skilled motions of the blade—or a deft bite by a reindeer-herding Lapp woman—and the testosterone levels are down to manageable proportions for the rest of the animal's life. Humans want their domestic animals to be submissive, easily controlled. Intact males are an awkward necessity; we want just enough of them to father a new generation of captives.

Something similar although less direct happens within the dominance hierarchy. From pit vipers to primates, the loser in ritual combat often experiences a steep decline in testosterone and related sex hormones, making him less likely to challenge the leadership at a later time, and therefore less likely to be injured. On a molecular level, he's learned his lesson. With fewer circulating steroids, he's now less ardent in his pursuit of females—at least when high-ranking males are around. This also is to the liking of the alphas. Again, decreases in testosterone levels following defeat are usually much more marked than any increases following victory.

Back to the testicles of sparrows: In a breeding area each little piece of territory has a male sparrow who will defend it against all comers.* Suppose a meddling ornithologist captures one of these territorial males and removes him from the territory. What happens? Other males from adjacent areas—many of them not previously able to defend a territory—move in. Of course they have to threaten and intimidate before they'll be taken seriously. So the general level of sparrow anxiety rises, both among the newcomers and among unreplaced spar-

---

* Well, against all *sparrow* comers. The dominance relationships in the same bit of forest within the communities of, say, owls, bears, raccoons, and humans are generally beneath the notice of sparrows.

rows in adjacent territories. Political tensions become high. If now we monitor the bloodstreams of the sparrows in the course of their disputes (which from our point of view, of course, seem petty, but to them it's Quemoy and Matsu), we find that everyone's testosterone level has risen—the newly introduced males who are trying to establish their territories, and the males of neighboring territories who are now required to do more in the way of defending than has been their recent practice. Something similar is true for many animals.

Those who have more testosterone, by and large, become more aggressive. Those who need more testosterone, by and large, generate it. Testosterone seems to play a vital role as both the cause and the effect of aggression, territoriality, dominance, and the rest of the "boys-will-be-boys" constellation of male behavioral traits. This seems to be true for widely differing species, including monkeys, apes, and humans.

In springtime, stimulated by the increasing day length, the testosterone level in male perching birds and songbirds (such as jays, warblers, and sparrows) goes up; they develop plumage, unveil a scrappy temperament, and begin singing. Males with larger repertoires breed earlier and produce more chicks. The repertoires of the most attractive males range up to dozens of distinguishable songs. Musical variety is the means by which more testosterone is converted into more birds.

When eggs are being laid, the male testosterone level remains high; they're protecting their mates. Once the females begin incubating the eggs and are uninterested in sexual advances, male testosterone levels fall. Suppose that the females are now given estrogen implants so they remain sexually alluring and receptive, despite their new maternal duties. Then the testosterone levels in the males remain high. As long as the female is sexually available, the male is inclined to be nearby and protective.[17]

These experiments suggest that an important selective advantage may accrue if a species breaks out of the estrus constraint. Continuous female sexual receptivity keeps the male around for all sorts of useful services. This is just what seems to have happened—maybe through a small adjustment in the DNA code for the internal estrogen clock—in our species.

Testosterone-induced behavior must be subject to limits and constraints. If it were carried to counterproductive lengths, natural selection would quickly readjust the concentration of steroids in the blood. Testosterone poisoning to the point of maladaptation must be very rare. In nectar-eating birds, bats, and insects it's possible to compare the energy expended in male steroid-driven defense against poachers with the energy that could be extracted from the flowers being guarded.* In fact, territoriality typically turns on only when the energy benefit exceeds the energy cost, only when there are so few delectable flowers to suck that it pays for you to expend the effort to chase away the competition. Nectar-eaters are not rigid territorialists. They won't fight all comers to protect a wasteland of stones. They make a cost-benefit analysis. Even in a rich garden of nectar-bearing flowers, often no territorial behavior is seen in the morning—because plentiful nectar has been accumulating at night when the birds were asleep. In the morning, there's enough to go around. Toward noon, though, when birds from far and wide have been feeding and the resource begins to get scarce, territoriality turns on.[18] Wings outstretched, beaks lunging, the locals drive away the intruders. Maybe they feel they've been nice guys long enough, but now they've had it up to *here* with these foreigners. Fundamentally, though, it's an economic, not a patriotic decision; practical, not ideological.

———

Many animals may do it, but at least among rats and mice it's well-demonstrated: Fear is accompanied by a characteristic odor, a fear pheromone, easily recognized by others.[19] Often, as soon as they sense you're afraid, your friends and relatives run away—useful for them, but not very helpful for you. It may even encourage the rival or predator who has prompted your fear in the first place.

In the heads of goslings and ducklings and chicks at the moment they peck their way out of the egg is, a classic experiment suggests, a rough knowledge of what a hawk looks like. No one has to teach it to them. Hatchlings know. They also know fear. Scientists make a very simple silhouette—cut out of cardboard, say: There are two projections which could be wings. They flank a body which is longer and rounded at one end and shorter and stumpy at the other. If the silhou-

* The question is similar to that posed by the artichoke: Are more calories burned in trying to get to its succulent heart than are afforded by eating the thing?

ette moves with the long projection first, it looks like a flying goose, wings spread, long neck preceding. Move the silhouette overhead, neck first, over the hatchlings and they go about their business. Who's scared of a goose? Now move the same silhouette stumpy end first— so it looks like a hawk with wings outstretched and long tail trailing— and there's a flurry of peeps and trepidation. If this experiment has been properly interpreted,[20] somehow, inside the sperm and the egg that made that chick, encoded in the ACGT sequence of their nucleic acids, there's a picture of a hawk.

Perhaps this inborn fear of raptors is akin to the fear of "monsters" that almost all babies manifest around the time they become toddlers. Many predators who are circumspect when a human adult is around would happily attack a toddler. Hyenas, wolves, and large cats are only a few of the predators that stalked early humans and their immediate ancestors. When the child begins to amble off on its own, it helps for it to know—in its marrow—that there are monsters out there. With such knowledge, it's much more likely to come running home to the grown-ups at the slightest sign of danger. Any mild predisposition in this direction will be resoundingly amplified by selection.*

In grown-up chickens there's a set of more organized and systematic responses, including specific auditory alarm calls that alert every chicken within hailing distance of the ominous news: A hawk is overhead. The cry announcing an aerial predator is distinctly different from that announcing a ground predator—a fox, say, or a raccoon. Since the bird sounding the alarm is also giving away its presence and location to the hawk, we might be tempted to consider it courageous, its behavior evolved through group selection. An individual selectionist might argue—how convincingly is another matter—that the cry works to stir other chickens into motion, whose scurrying might distract the hawk and save the bird that sounded the alarm.

Experiments by the biologist Peter Marler and his colleagues[21] show that, at least among cockerels, a propensity to make alarm calls de-

---

* Just as chicks seem to retain and refine this concern when they get to be adults, so do humans. The fear of non-human predators is another one of our readily available "buttons" that are easy to press in order to manipulate passionate behavior. Horror films are one, but hardly the most egregious, example.

pends very much on whether there's a companion nearby. With no other bird present, the cockerel may freeze or gaze up into the sky when seeing something like a hawk, but he doesn't cry out in alarm. He's more likely to sound the alarm if there's another bird within earshot; and, significantly, he's much more likely to cry out if his companion is another chicken—any chicken—rather than, say, a bobwhite. He's indifferent to plumage, though; chickens with very different color patterns are worthy of being warned. All that counts is that the companion be another domestic fowl. Maybe this is just sloppy kin selection, but it certainly edges toward species solidarity.

So is this heroism? Does the cockerel understand the danger he subjects himself to, and then, despite his fear, bravely cry out? Or is it more likely that squawking when there's a companion nearby but not when you're alone is a program in the DNA, and nothing more? See a hawk, see another chicken, cry out, and no agonizing moral struggle. When one of the combatants in a cockfight continues, although bleeding and blinded, to fight to the death, is he displaying "invincible courage" (as an English admirer of cockfighting has described it), or is this just a combat algorithm gotten out of hand, escaping the inhibition subroutines? Indeed, in humans does the hero have a lucid grasp of the danger, or is he or she merely following one of *our* preprogrammed subroutines? Most heroes report that they just did what came naturally, without much conscious thought.

The two sexes are not equally likely to produce alarm calls. In another study by Peter Marler and his colleagues,[22] cockerels cried out in alarm every time a hawk silhouette was presented; but hens made such calls only 13% of the time.* Castrated cockerels are much less likely to sound the alarm—except when they have testosterone implants, in which case the call rate goes back up. So testosterone plays a role not just in dominance hierarchies, sex, territoriality, and aggression, but also in providing early warning of predators, whether we hold the bearer to be hero or automaton.

———

* The sexes differ in other kinds of cries as well. For example, when a male comes upon some food he knows the female likes, he often generates a food call. But when the hen finds food, she does not call to the cock; indeed, she does not call at all, unless she has chicks. Hens without families prefer to dine alone.

Preadolescent female mice have a molecule in their urine that induces testosterone production in males who get a whiff. In turn, the males' urine now contains pheromones which, when sniffed by the immature female, quicken her sexual development. She matures early if there are males around, and late if there aren't—a positive feedback loop that saves unnecessary effort. (As you might expect, female mice who can't detect odors never come into heat.) What's more, normal pregnant females who sniff the urine from males of a different strain of mice spontaneously abort their pregnancies; they resorb the embryos back into their bodies and quickly come into heat.[23] This is convenient for the alien males. If the resident males don't like it, it's up to them to stop strangers from coming around with their abortion-inducing aromas.

In mice, as for many other animals, testosterone begins to be manufactured in earnest at puberty, and that's when serious aggression against other mice begins. In adult males, the more testosterone, the quicker will be the attack when a strange male appears at the territorial frontiers. Again, castrate the males and their aggressiveness declines. Again, deliver testosterone to the castrates and their aggressiveness increases. Male mice are given to "marking" their environment with tiny dribbles of urine—a practice they pursue with redoubled effort when other mice are around (or when they come upon some unfamiliar object, maybe a hairbrush). Because of embryo resorption, if the males are to leave progeny at all, they must be the chief urinators in their territory. Maybe marking is like nametags on luggage, "no trespassing" signs on private property, or heroic portraits of the national leader in public places. The doughty little mouse is singing "This land is my land" and "She belongs to me." Even when he's not physically present he wants passersby to take careful note of his proprietorship. As you might suspect, castrate the mouse and urinary marking declines strikingly; resupply testosterone and his compulsion to mark is rekindled.

Normal female mice are infrequent urinators. They are not inveterate markers. But what happens if anatomically normal female infants are jolted with testosterone? Then they begin marking often. (If a similar experiment is done in dogs, adult females who were given testosterone before birth adopt the urination posture of the males; they lift one leg and trickle the urine down the other—one more indignity visited at the hands of the scientists.) When female rats with ovaries surgically removed are supplied with testosterone, they become aggressive, alternating a masculine propensity for confrontation with distinctly feminine sexual behavior. But one thing about giving testosterone to normal females early in their lives: When they grow up, the males find them much less attractive.

While testosterone in the blood is intimately connected with the expression of aggression in male animals, it is by no means the whole story. There are, for example, molecules in the brain that repress aggression. Hereditary strains of rats that are unusually violent turn out to have less of these inhibitory brain chemicals than more peace-loving strains. Aggressive rats are calmed when there are more of these chemicals in their brains; peaceful rats are agitated when there is less of these chemicals. If you're a rat, busy watching violence in other rats—mice-killing, say—your level of inhibiting brain chemicals drops.[24] You're now more likely to be violent yourself, and not just toward mice. Your repressed aggressive tendencies have been disinhibited. And everybody else's. Hostility can then rapidly spread through your group, expressed differently by different individuals. Perhaps that's what happened with Calhoun's rats, so confined that aggression and despair spread in waves, reflected and amplified from multiple foci through the community. Violence is contagious.

In experiments performed by Heidi Swanson and Richard Schuster,[25] rats were given a complex cooperative task to learn, having to run together over specific floor panels in a particular sequence. If they succeeded, they were rewarded with sugar water; if they didn't, they found themselves racing around the experimental chamber for the fun of it. Nobody taught them what to do, or at least not directly. It was trial and error. The experiment was tried on pairs of males, pairs of females, pairs of castrated males, and pairs of castrated males with testosterone implants. Some of the rats had previously lived alone.

Here's how it turned out: Females, as well as male castrates, learned fairly quickly. Normal males and castrates with administered testosterone learned much more slowly. Males who had previously lived alone did still worse. Some pairs of previously solitary male rats—pairs with intact testicles as well as pairs of testosterone-jolted castrates—never learned at all.

For the solitary males this is just what you might expect: Because you live alone you have little experience in cooperating, so probably you're not going to do very well on a demanding test of cooperation.

But then, why should females who've been living alone be able to figure it out? The answer seems to be that if you're a solitary male, a loner, and you have to perform a complex task in coordination with someone else, testosterone makes you stupid. Every pair of males who ordinarily lived alone and couldn't figure out how to pass the test was engaged in violent combat. Communal living, by contrast, tended to calm them down.

Swanson and Schuster conclude that the learning deficits were not so much due to aggression *per se*, as to aggression in the context of the dominance hierarchy. Those who tended to be the winners in ritualized (or real) combat—almost always it was the same individuals —would strut and saunter with hair erect, threatening, feinting, and occasionally attacking. The subordinates would crouch, close their eyes, and either freeze for long periods or hide. But tendencies to strut or crouch or hide are not well suited for the gymnastic cooperation needed to get that sugar water.

Cooperation has strong democratic overtones. Extreme dominance/ submission hierarchies do not. The two are strongly incompatible. In these experiments, females intimidated others and fought as did the males, but today's winner was often yesterday's loser, and vice versa —unlike the males. Cowering and freezing were less common, and the female style of aggression didn't impede social performance as much as her male counterpart's.

The unfolding richness and complexity of testosterone-induced sexual behavior—dominance, territoriality and all the rest—is one means by which males compete to leave more offspring. It's not the only possibility. We've already mentioned selection at the level of competition among sperm cells, as well as those species in which the male leaves a vaginal plug when he's done to frustrate those who come after him. Male dragonflies attempt to undo the competition retroactively: Projecting from the male's penis is a whip-like prong that attaches itself to the mass of sperm previously deposited in the female. When he withdraws, he takes his rivals' semen with him. How much more direct the dragonflies are than the birds and mammals—our males violent, consumed with jealousy, spitting out threats and accusations, longing for exclusive sexual access to at least one female. The dragonfly male is spared much of this; he merely rewrites his mate's sexual history.

We've concentrated on aggression, dominance, and testosterone because they seem to be of central importance in understanding human behavior and social systems. But there are many other behavior-eliciting hormones fundamental for human well-being, including estrogen and progesterone in females. The fact that complex behavioral patterns can be triggered by a tiny concentration of molecules coursing through the bloodstream, and that different animals of the same species generate different amounts of these hormones, is something worth thinking about when it's time to judge such matters as free will, individual responsibility, and law and order.

Had Poseidon more carefully measured out whatever it was he gave to Caenis, the matter would not have come to Zeus' attention. Had Poseidon's own testosterone titer been lower, or had there been enforceable penalties against gods raping humans, Caenis might have lived a happy and blameless life. As it was, Caeneus was afflicted by hubris, surely; but only because of the rape and its aftermath. He was guilty of disrespect for the gods, but the gods had shown disrespect for her. There is not a hint that the piety of Thessaly would have been troubled had Poseidon left Caenis alone. She had been minding her own business, walking along the beach.

Chapter 13

THE OCEAN
OF BECOMING

Every valley shall be exalted, and every
mountain and hill shall be made low.

Isaiah 40:4

They will manage to cross the ocean of
becoming.

The Maitreyavyakarana (about 500 B.C.)[1]

L et's for a moment imagine your species is wildly successful. Through the slow evolutionary process it's become adapted with high precision to its environmental niche. You and all your fellows are now, perhaps even literally, fat and sassy. But, again, especially when you're so well adapted, any significant genetic change tends not to be in your best interest—just as a random change in some of the microscopic magnetic domains on an audio tape is unlikely to improve the music recorded there. You can't stop deleterious mutations from happening, just as you can't prevent a slow degradation of the recorded music, but those mutations are restrained from spreading through the species. Natural selection sifts through the population and quickly disposes of whatever doesn't work, or doesn't work as well. It is not considered an extenuating or mitigating circumstance that, by some remote accident, the mutation might be useful in the future. Darwinian selection is for the here and now. Summary judgment is rendered. With careful discrimination, the scythe of selection swings.

But now, let's imagine that something changes. A small world hurtling through space finds a blue planet smack in its path, and the resulting explosion sprays enough fine particles into the upper atmosphere to darken and cool the Earth; your lake then freezes over, or the savanna vegetation that sustains you shrivels and dies. Or the tectonic engine in the Earth's interior creates a new island arc and a flurry of volcanic explosions changes the composition of the air, so now more greenhouse gases are released into the atmosphere, the climate warms, and the tidepools and shallow lakes in which you have been luxuriously wallowing begin to dry up—or a dam of glacial ice is breached, creating an inland sea where your congenial desert habitat used to be.

Perhaps the change comes from a biological direction: The animals you eat are now better camouflaged, or defend themselves with greater obstinacy; or animals that eat you have become more adept at the hunt; or your resistance to a new strain of microorganism turns out to be poor; or some plant you habitually eat has evolved a toxin that makes you ill. There can be a cascade of changes—a relatively small physical alteration leading to adaptations and extinctions in a few directly affected species, and further biological changes propagating up and down the food chain.

Now that your world has changed, your once wildly successful species may be reduced to much more marginal circumstances. Now some rare mutation or an improbable combination of existing genes might be much more adaptive. The once-spurned hereditary information may now be given a hero's welcome, and we are reminded once more of the value of mutation and sex. Or, it may be, no new and more useful genetic information is generated fortuitously in the nick of time, and your species continues its downward drift.

Omnicompetent organisms do not exist. Breathing oxygen lets you be far more efficient in extracting energy from food; but oxygen is a poison for organic molecules, so arrangements for routine handling of oxygen by organic molecules are going to be expensive. The ptarmigan's white fur provides superb camouflage in the Arctic snows; but in consequence it absorbs less sunlight and greater demands are placed on its thermoregulatory system. The peacock's gorgeous tail makes him nearly irresistible to the opposite sex, but also provides a conspicuous luncheon advertisement for foxes. The sickle-cell trait confers immunity to malaria, but condemns many to debilitating anemia. Every adaptation is a trade-off.

Imagine designing a vehicle that drives off roads, flies through the air, and swims underwater. Such a machine, if it could be built at all, would perform none of its functions well. When we need to travel on "unimproved" land we build all-terrain vehicles, when beneath the water, submarines, and when through the air, airplanes. It's for good reason that these three kinds of vehicles, while roughly of similar shape, in fact tend not to look very much alike. Even so-called "flying boats" are not very seaworthy, nor are they very easy to fly.

Birds that are superb underwater swimmers, such as penguins, or highly capable runners, such as ostriches, tend to lose their ability to fly. The engineering specifications for swimming or running conflict with those for flying. Most species, faced with such alternatives, are forced by selection into one adaptation or the other. Beings that hold all their options open tend to be eased off the world stage. Overgeneralization is an evolutionary mistake.

But organisms that are too narrowly specialized, that perform exceedingly well but only in a single, restrictive environmental niche, also tend to become extinct; they are in danger of making a Faustian bargain, trading their long-term survival for the blandishments of a brilliant but brief career. What happens to them when the environment changes? Like barrelmakers in a world of steel containers, blacksmiths and buggy-whip tycoons in the time of the motorcar, or manufacturers of slide rules in the age of pocket calculators, highly specialized professionals can become obsolete virtually overnight.

If you're receiving a forward pass in American football, you must keep your eye on the ball. At the same time you must keep your eye on the opposition tacklers. Catching the ball is your short-term objective; running with it after you have it is your longer-term objective. If you worry only about how to outrun the defenders, you may neglect to catch the ball. If you concentrate only on the reception, you may be flattened the moment you receive the ball, and risk fumbling it anyway. Some compromise between short-term and longer-term objectives is called for. The optimum mix will depend on the score, the down, the time remaining, and the ability of the opposing tacklers. For any given circumstance there is at least one optimum mix. As a professional player you would never imagine that your job as a receiver is *solely* catching passes or *solely* running with the ball. You will have acquired a habit of quickly estimating the risks and the potential benefits, and the balance between short-term and long-term goals.

Every competition requires such judgments; indeed, they constitute a large part of the excitement of sport. Such judgments must also be made daily in everyday life. And they're a central and somewhat controversial issue in evolution.

The danger of overspecialization is that when the environment changes, you're stranded. If you're superbly adapted to your present habitat, you may be no good in the long term. Alternatively, if you spend all your time preparing for future contingencies—many of them remote—you may be no good in the short term. Nature has posed life a dilemma: to strike the optimum balance between the short-term and the long, to find some middle road between overspecialization and overgeneralization. The problem is compounded, of course, by the fact that neither genes nor organisms have a clue about what future adaptations are possible or useful.

Genes mutate from time to time, and because the environment is changing, it once in a great while happens that a new gene equips its bearer with improved means of survival. It is now more "fit" for its environmental niche. Its adaptive value, its potential to help the organism that bears it leave many viable offspring, has increased. If a particular mutation secures for its possessor a mere 1% advantage over those who lack it, the mutation will be incorporated into most members of a large, freely interbreeding population in something like a thousand generations [2]—which is only a few tens of thousands of years even for large, long-lived animals. But what if mutations conferring even so small an advantage occur too rarely; or what if several genes must all, improbably, mutate together, each in the right direction, in order to adapt to the new conditions? Then all members of the population may die.

Is there an evolutionary strategy by which individuals and the species can escape from this trap, some trick by which the extremes of overspecialization and overgeneralization can both be avoided? For major environmental catastrophes there may be no such strategy. The dinosaurs had proliferated into an impressive range of environmental niches, and yet not one of them survived the mass extinctions of 65 million years ago. For quick, but less apocalyptic environmental change there are several ways. It helps to reproduce sexually, as we've described, because recombination of genes greatly increases the overall genetic variety. It helps to occupy a large and heterogenous territory, and not be too specialized. And it helps if the population breaks up into many nearly isolated subgroups—as was first clearly described by the population geneticist Sewall Wright, who died almost a centenarian in 1987. What follows is a simplification of a complex subject, some aspects of which are under renewed debate.[3] But even if it were no more than metaphor, its explanatory power—for mammals, and especially for primates—is considerable.

———

The genes—the instruction manuals written down in the ACGT alphabet of DNA—are mutating. Some genes, in charge of important matters such as the business end of an enzyme, change slowly; indeed, they may change hardly at all in tens or even hundreds of millions of years—because such changes almost always make some molecular machine tool work more poorly, or not at all. Organisms with the mutated gene die (or leave fewer offspring) and the mutation tends not to be passed on to future generations. The sieve of selection strains it out. Other changes that do no damage—in an untranscribed nonsense sequence, or in the blueprints for *structural* elements involved in orienting the machine tool, say, or draping it over a molecular jig—can spread through future generations quickly, because an organism bearing the new mutation will not be eliminated by selection: In the code for structural elements, the particular sequence of As, Cs, Gs, and Ts hardly matters at all; what's needed are placeholders, *any* sequence that codes for the shape of a subcellular handle, say, never mind which amino acids the handle is made of. Changes in ACGT sequences that are ignored anyway also won't do any harm. Occasionally an organism hits the jackpot, and a favorable mutation will, in relatively few generations, sweep through the entire population; but the overall genetic change due to favorable mutations is slow, because they happen so rarely.

Some genes will be carried by almost all of the population; others will be present in only a tiny fraction of the population. But not even very useful genes will be carried by everyone, either because the gene is new and there hasn't been time enough for it to spread through the whole population, or because there are always mutations transforming or eliminating a given gene, even a beneficial one. If the absence of a useful gene isn't positively lethal, in a big enough population some organisms will always be without it. In general, any given gene is distributed through the population: Some individuals have it, and some don't. If you divide your species up into smaller, mutually isolated subpopulations, the percent of individuals that carry a given gene will vary from group to group.

There are around ten thousand active genes in a typical "higher" mammal. Any one of them may vary from individual to individual and group to group. A few are extinguished for a time or for all time. A few are spanking new and are being spread quickly through the population. Most are old-timers. How useful any given gene is (in the population of wolves or humans or whatever mammal we have in mind) depends on the environment, and that's changing too.

Let's follow one of those ten thousand genes. Maybe it's for extra testosterone production. But it could be any gene. The fraction of the population possessing this gene, relative to all possible alternative genes, is called the gene frequency.

Imagine now a set of isolated populations of the same species. Maybe they're troops of monkeys that live in adjacent, nearly identical mountain valleys, separated by impassable mountains. Whatever differences there are in the chances of survival or of leaving descendants in the two groups, it won't be because one is living in a more favorable physical environment.

Not all values of the gene frequency are equally adaptive. Instead, there's an optimum frequency in the population. If the gene frequency is too low, maybe the monkeys are insufficiently vigilant in defending themselves against predators. If it's too high, maybe they kill themselves off in dominance combat. When two isolated populations, in otherwise identical circumstances, have different constellations of active genes, their members will have different Darwinian fitness.

But the optimum frequency of this gene depends on the optimum frequency of other genes, as well as on the fluid and varying environment in which our monkeys must live. There might be more than one optimum frequency, depending on circumstances. The same is true for all ten thousand genes—their optimum frequencies all mutually dependent, all varying as the environment does. For example, a higher frequency of a gene for extra testosterone might be useful in dealing with predators and other hostile groups, provided genes for peacekeeping *within* the group were also more abundant. And so on. The optima interlace.

So a set of gene frequencies that once made your group superbly adapted may now constitute a marked disadvantage; and gene frequencies that once conferred only marginal fitness may now be the key to survival. What a disturbing concept of existence: Just when you're most in harmony with your environment, that's when the ice you're skating on begins to thin. What you should have been emphasizing, had you been able, is early escape from optimum adaptation— a deliberate fall from grace contrived by the well-adjusted, the elective self-humbling of the mighty. The meaning of "overspecialized" becomes clear. But this is a strategy, we well know from everyday human experience, that privileged populations are almost never willing to embrace. In the classic confrontation between short-term and long, the short-term tends to win—especially when there's no way to foretell the future.

Yes, they lack foresight. But how could they know? It's asking a great deal of monkeys to foresee future geological or ecological change. We humans, who with our intelligence ought to be much more capable prophets than monkeys, have difficulty enough foreseeing the future, and still more difficulty acting on our knowledge.[4] In military operations, ward-heeler politics, much of corporate strategy, and national response to the challenge of global environmental change, the short-term tends to predominate. So offhand, you might think that precautionary maintenance of a collection of gene frequencies that will be optimum for some *future* circumstance when no one is even aware of this fact is simply too difficult to arrange. You might think that there's a flaw in the evolutionary process, that life, under some circumstances, might get stranded.

What could possibly cause the gene frequency in different populations to drift to suboptimal values? Suppose the mutation rate went up because of some new chemical in the environment (belched up from the Earth's interior), or an increase in the flux of cosmic rays (perhaps from some exploding star halfway across the Milky Way). Then the gene frequencies in isolated populations diversify. You might even get a population that, by accident, winds up with the optimum frequencies needed to adapt to a future need. But that will be very rare. More likely, big changes will be lethal. So an increase in the mutation rate tends mainly to spread out the variation in gene frequencies, but not too much.

The population will, through mutation and selection together, tend to follow the changing circumstances, always working toward the optimum adaptation. If the external conditions vary slowly enough, the population might always be close to the optimal adaptation. Gene frequencies are always in slow motion. This gradual movement, driven by mutation and natural selection in a changing physical and biological environment, is just the evolutionary process outlined by Darwin; and Wright's continuously changing gene frequencies are a metaphor of natural selection.

———

Up to now each isolated subpopulation we've been considering has been large, comprising maybe thousands of individuals or more. But now, Wright's critical step: Let's think about small groups, with no more than a few dozen individuals. They tend to become closely inbred. After a few generations, who's available to mate with except relatives? So let's look at inbreeding for a moment before considering the evolutionary prospects of small populations.

Some human cultures have sex in private and eat in public, some do it the other way around; some live with their aged relatives, some abandon them, and some eat them; some institute rigid rules that even toddlers must obey, and some let children do almost anything they want; some bury their dead, some burn their dead, and some set them out for the birds to eat; some use cowrie shells for money, some use metal, some paper, and some do without money altogether; some have no gods, some have one god, some have many gods. But all of them abominate incest.

Incest avoidance is one of the few invariables common to the spectacular diversity of human cultures. Sometimes, though, exceptions were made for (who else?) the ruling class. Since kings were gods, or near enough, only their sisters were considered of sufficiently exalted status to be their mates. Mayan and Egyptian royal families were inbred for generations, brothers marrying sisters—the process mitigated, it is thought, by unsanctioned and unrecorded couplings with nonrelatives. The surviving offspring were not conspicuously more incompetent than the usual, run-of-the-mill kings and queens, and Cleopatra, Queen of Egypt—officially the product of many consecutive generations of incestuous matings—was gifted by many standards. The historian Plutarch described her as not incomparably beautiful; still,

> the contact of her presence, if you lived with her, was irresistible; the attraction of her person, joining with the charm of her conversation, and the character that attended all she said and did, was something bewitching. It was a pleasure merely to hear the sound of her voice, with which, like an instrument of many strings, she could pass from one language to another; so that there were few of the barbarian nations that she answered by an interpreter.

She was fluent not only in Egyptian, Greek, Latin, and Macedonian, but also in Hebrew, Arabic, and the languages of the Ethiopians, the

Syrians, the Medes, the Parthians, "and many others."[5] She's described as "the only human being except Hannibal who [ever] struck fear into Rome."[6] She also gave birth to several apparently healthy children—although they were not fathered by her brother. One of them was Ptolemy XV Caesar, son of Julius Caesar and titled King of Egypt (until murdered at age seventeen by the future Emperor Augustus). Cleopatra certainly does not seem to have exhibited marked physical or intellectual deficits, despite the alleged close relation of her parents.

Nevertheless, inbreeding produces a statistical genetic deficit that takes its toll chiefly in the deaths of infants and juveniles (and we don't have a good record of Mayan and Egyptian royal children who died at birth or were put to death in infancy). There is considerable evidence for this in many—but by no means all—groups of animals and plants. Even in sexual microorganisms, incest causes striking increases in the deaths of the young.[7] In incestuous unions in zoos, mortality in the offspring increased steeply for forty different species of mammals— although some were much more vulnerable to close inbreeding than others.[8] In successive brother-sister matings in fruit flies, only a few percent of the offspring survived by the seventh generation.[9] In baboons, matings between first cousins result in infants that die, within the first month of life, about 30% more often than in baboon matings where the parents are not close relatives.[10] Most normally outbred plants—corn, for example—deteriorate on consistent inbreeding. They become smaller, scrawnier, more withered. That's why we have hybrid corn. Many plants with both male and female parts are configured, as Darwin first noted, so they cannot easily have sex with themselves ("self-incompatibility" this ultimate incest taboo is called). Many animals, including the primates, have taboos that inhibit mating with close relatives.[11]

Purebred dogs are prone to deformities and crippling defects. The biologists John Paul Scott and John L. Fuller performed breeding experiments—that is, artificial selection—on five breeds of dogs:

> In our experiments we began with what were considered good breeding stocks, with a fair number of champions in their ancestry. When we bred these animals to their close relatives for even one or two generations, we uncovered serious defects in every breed.

. . . [C]ocker spaniels [are] selected for a broad forehead with prominent eyes and a pronounced "stop," or angle between the nose and forehead. When we examined the brains of some of these animals during autopsy, we found that they showed a mild degree of hydrocephaly; that is, in selecting for skull shape, the breeders had accidentally selected for a brain defect in some individuals. Besides all this, in most of our strains only about 50 per cent of the females were capable of rearing normal, healthy litters, even under nearly ideal conditions of care.

Among other dog breeds, such defects are quite common.[12]

Similar genetic deficits are found in the limited data available on human incest in modern times. The increased infant death rate resulting from first cousin marriages[13] is only about 60%. But in a Michigan study[14] in the middle 1960s, eighteen children from brother-sister and father-daughter matings were compared with a control group of children from non-incestuous matings. Most of the children of incest (eleven out of eighteen) died within their first six months, or showed serious defects—including severe mental retardation. No history of such defects was found in the parents or their families. The remaining children seemed normal in intelligence and in all other respects, and were recommended for adoption. None of the children in the control group died or was institutionalized. Compared to brother-sister and father-daughter matings in other animals, though, these mortality and morbidity rates seem high; perhaps incestuous unions that produce abnormal children were more likely to come to the attention of the scientists making the study.

The dangers of repeated inbreeding seem so clear that we can safely conclude that unsanctioned sexual unions, impregnations of Queens of Egypt by someone other than the Pharaoh, occurred among Cleopatra's immediate ancestors. Even a few sibling matings in consecutive generations would probably have led to death, or at least to a Cleopatra very different from the vital individual history reveals to us. But one generation of outcrossing goes far to cancel the previous inbreeding.

Inbreeding is a particular danger in very small groups, because in them it can hardly be avoided. If a new nonlethal mutation occurs in one individual, it either gets lost—because, for example, its bearer has no offspring—or it's not many generations before it's in nearly everybody, even if it's slightly maladaptive. So now most males in the population have, say, a little too much testosterone; the conflicts and the distractions of conflict are taking their toll, and the youngsters are not being cared for as they should. The population has wandered from optimum adaptation; if inbreeding is intense, it may be that eventually none of the members of the group leaves offspring.

If inbreeding weren't so risky, you might think that small populations are the way to get to gene frequency constellations that are not now especially adaptive, but that will be so at some time in the future. If the population is small, then new mutations or new combinations of letters and sequences in the genetic code can propagate through the entire population in only a few generations. New random experiments in biology are being conducted that could not occur in large populations. As a result, almost always, the group goes hurtling away from optimum adaptation. But comparatively rare genes and gene combinations can be tried out so quickly in a small population that it can swiftly cover a lot of ground in the possible range of gene frequencies.

What's happening here is described as "accidents of sampling," which have much more profound consequences in small populations than in large ones: Imagine you're flipping a coin. Your chance of getting one head in one trial or flip is clearly 50%, one chance in two. The coin has only a head and a tail, and it has to turn up one side or the other. With two flips, the full menu of equally possible outcomes is: two tails, a head and a tail, a tail and a head, or two heads. So your chance of getting two consecutive heads is one in four, or, equivalently, one-quarter, or $\frac{1}{2} \times \frac{1}{2}$. With three flips, the chance that they're all heads is one chance in eight ($\frac{1}{2} \times \frac{1}{2} \times \frac{1}{2}$), or one in 2^3. You can flip ten heads in a row once in about a thousand trials ($2^{10} = 1024$). (If we'd witnessed only that trial, we might think you're phenomenally lucky.) But a hundred heads in a row will take about a billion billion trillion trials (2^{100} roughly equals 10^{30})—which is the same as forever.

In small populations major accidents of sampling are inevitable; in large populations they are nonexistent. Were a national opinion poll to query three people only, there would be little reason to believe the results—that is, to think these three opinions adequately sampled the opinions of most citizens. One of the individuals polled might, by accident, be a Libertarian or a Vegetarian, a Trotskyite or a Luddite, a Coptic or a Skeptic—all with interesting perspectives, but none an accurate reflection of the general population. Now imagine that the opinions of these three were somehow proportionately amplified to become the opinions of the population of the United States as a whole; a major transformation in national attitudes and politics would have been worked. The same can be true genetically when a few individuals from a large population establish a new and isolated community.

Accidents of sampling happen when the population sampled is very small. In many elections, when the pollsters sample five hundred or a thousand randomly chosen people, the results repeatedly prove to be representative of the nation as a whole.* With five hundred or a thousand truthful random samplings, the findings are accurate to within a few percent. (The variation expected is the square root of the sample size.) If you ask a large number of randomly selected people, you will reliably sample the average*; if you ask only a few, you may sample atypical or fringe opinions. Pollsters would gladly sample smaller populations; it would save them money. But they dare not—the errors would be too large, the sampled opinions too unrepresentative.

As in opinion polls, so it is in the genetics of populations: With a small enough group, substantial deviations † from the average can be sampled and become established. With mutually isolated small groups, many different sets of gene frequencies get tried out—most maladaptive, but a few, fortuitously, poised for the future. This is called genetic drift.

Or suppose that your name is Theodosius Dobzhansky and that you live in New York City. Even if you have ten sons, your name will continue to be "rare and outlandish" so long as you continue to reside in the big city. But move the family to a small town, have many descendants, and Dobzhansky will eventually become a common and

---

* Except when what is expressed in the privacy of the voting booth is too shameful to be admitted to the pollster.

† The pejorative flavor that attaches to the word "deviant"—which only means different from the average—suggests the nearly irresistible social pressures, in almost all human societies, to fit in with the crowd. The word "egregious," meaning exceptionally bad, is Latin for separated out from the herd. Again, the equation of different with bad —sensible for well-adapted populations in the short-term but dangerous in changing times and in the long-term.

unremarkable name. Similarly, any extraordinary hereditary predisposition in the Dobzhansky genes will affect only a tiny fraction of the population while you're in New York, but might in a few generations become a major genetic feature of the citizenry of the town.[15]

Is there any way to preserve the accidents of sampling inherent in small groups, while avoiding the slow deterioration intrinsic to incest? Imagine that each group is significantly inbred, but that outbreeding is sometimes indulged in. Individuals from largely isolated subpopulations occasionally find each other and mate, enough to mitigate the more severe genetic consequences of incest. Different constellations of genes will be established in each subpopulation by genetic drift. Each small group will have a different set of hereditary propensities. They will not all, therefore, be optimally adapted to current circumstances. Now that the environment has changed, none of them may be. Being far from optimally adapted, their lives will be hard. Not one of these groups will be as well off as it was earlier. Many groups will die out. Now, though, when the environmental crisis comes, a few of these smaller populations will find themselves, by accident, advantageously situated, "preadapted."

The trick is to combine the accidents of sampling of small groups (so at least one group will be by chance fortunately poised for the next environmental crisis) with the stability of large groups (so once the new, desirable adaptation is hit upon, it is spread to a substantial population). Because the lucky group—with newly optimal gene frequencies—is also in genetic contact with other groups, its new constellation of adaptive genes is passed on. Other groups acquire the new capabilities, the new mix of traits, the new adaptations; and simultaneously the most dangerous consequences of inbreeding are avoided.

Here then is a trial-and-error mechanism through which a large population can explore the mix of possible gene frequencies. When the adaptations that formerly led to our success now become only marginally useful, we have a way out. Dividing a species into many quite small, fairly inbred populations, but allowing occasional interbreeding among these populations, is the solution Sewall Wright proposed. It avoids both traps, overspecialization and overgeneralization.[16] And to the extent that major evolutionary steps occur relatively quickly in small, semi-isolated populations, the relative paucity of intermediate forms in the fossil record—one of the problems that plagued Darwin—would be explained.[17]

———

No organisms have ever sat down and decided, as a matter of conscious species-wide evolutionary policy, to divide themselves up into small populations, amplify accidents of genetic sampling, and at the same time avoid the more flagrant forms of incest. But, as always happens in the evolutionary process, any species that, by accident, makes appropriate arrangements preferentially reproduces. If enough evolutionary experiments are tried over the immense vistas of time available in the history of life, then very improbable adaptations—in group size, say, or in the balance between inbreeding and outbreeding —can be institutionalized. Here we are talking about the evolution of a mechanism to guarantee continuing evolution, a second-order or meta-evolutionary development.[18]

What would it feel like from the inside if you were a member of a species that had, through natural selection, made arrangements for genetic drift? You would enjoy living in small groups. You would hate crowds. For accidents of sampling to work on an appropriate time scale, a group might have to comprise no more than one hundred or two hundred individuals, and—according to Wright—would probably be best with only a few dozen members. Groups of six to eight or fewer tend to be unstable; they're too vulnerable to being wiped out by predators or flood or disease, a different example of accidents of sampling. You would conceive a passionate loyalty to the group, something like intense family feeling, superpatriotism, chauvinism, ethnocentrism. (Especially because most members of your group are close relatives, you might when necessary be moved to something like altruistic or even heroic actions on their behalf.) You would also need to avoid any merger of your group with another, because much bigger groups would inhibit accidents of sampling. So it would be helpful if you conceived a passionate hostility to other groups, a vivid sense of their deficiencies, something like xenophobia or jingoism.

Those other groups are, of course, composed of individuals of the same species as you. They look almost exactly like you. To fan the flames of xenophobia, you must examine them with minute attention and exaggerate whatever differences can be discerned, always to their disadvantage. They have slightly different heredities and slightly different diets, so they don't smell quite the same as you and yours. If your olfactory powers are sufficiently finely tuned, maybe their scents will render them grotesque, hateful, odious.

It would be even better if you could *establish* some distinctions. If differences in dress and language are unavailable—having not yet been invented, for example—differences in behavior, posture, or vocalizations would be helpful. Anything that can distinguish your group from the others could work to keep hatreds high and resist merger. Other groups, conveniently, are similarly disposed. These nonhereditary differences between one group and another—even arbitrary differences, only distantly connected with any adaptive advantage, but serving to preserve group independence and coherence—are called, collectively, culture. At a rudimentary level many animals have it.[19] Cultural diversity helps preserve genetic drift.

At the same time, avoiding too much inbreeding and guaranteeing at least occasional outbreeding are essential. So you would feel a revulsion about incest, or at least about the most consanguineous matings. Wherever possible, this revulsion would be reinforced by your copying the attitudes of your fellows, by culture. There would be an incest taboo (relaxed perhaps if the population is reduced to only a few survivors). Outbreeding might be officially proscribed—perhaps, among humans, by young men attacking males from other groups who, even accidentally, wander into the neighborhood, or by fathers mourning, as if dead, daughters who run off with foreigners. But despite the pervasive ethnocentrism and xenophobia, now and then you would find members of other, hostile groups unaccountably attractive. Surreptitious matings would occur. (This is, more or less, the theme of *Romeo and Juliet*, Rudolph Valentino's *The Sheik*, and a vast industry of books on romance, targeted at women.)

A promising survival strategy, in short, is this: Break up into small groups, encourage ethnocentrism and xenophobia, and succumb to the occasional sexual temptations provided by the sons and daughters of enemy clans. Devise your own culture: The more your species is capable of learned behavior, the greater the differences that can be established between one group and another. Behavioral differences eventually lead to genetic differences, and vice versa. Incomplete isolation—just the right mix of aloofness and sexual abandon with other groups—generates diversity. And diversity is the raw material on which selection operates.

There seems to be, then, a reason—at the heart of population genetics and evolution—for small semi-isolated groups as the substructure of larger populations, for xenophobia, ethnocentrism, territoriality, incest avoidance, occasional outbreeding, and migration away from the most successful communities. These mechanisms work especially for those species that find themselves in a swiftly changing environment, biologically or physically. Archaebacteria, ants, and horseshoe crabs have not much been in this category; birds and mammals have. So next time you hear a raving demagogue counseling hatred for other, slightly different groups of humans, for a moment at least see if you can understand his problem: He is heeding an ancient call that—however dangerous, obsolete, and maladaptive it may be today—once benefitted our species.

A solution has been found to the problem of how to arrange for gene frequencies to respond quickly to a volatile, changing environment. And the solution seems eerily familiar. After a journey into an abstract world of population genetics and gene frequencies, we turn a corner and suddenly find ourselves gazing at something that looks very much like . . . ourselves.

Chapter 14

GANGLAND

Brought face to face with these blurred copies of himself, the least thoughtful of men is conscious of a certain shock, due perhaps, not so much to disgust at the aspect of what looks like an insulting caricature, as to the awakening of a sudden and profound mistrust of time-honoured theories and strongly-rooted prejudices regarding his own position in nature, and his relations to the under-world of life; while that which remains a dim suspicion for the unthinking, becomes a vast argument, fraught with the deepest consequences, for all who are acquainted with the recent progress of the . . . sciences.

T. H. HUXLEY
Evidence as to Man's Place in Nature[1]

The Big Guy, he gets respect. He walks by, folks bow. Stick out their hands. Most times, he'll touch you. Hands stretch out, Big Guy touch 'em, one after the other. You feel real good. He looks you in the eye and it's like, you gotta do what he wants. I can't stand it when he looks at me like that. It makes me feel so good, I gotta look down at my feet.

He's crazy about me. The Big Guy, he'd as soon fuck me as look at me. Truth is, he'll fuck anything that moves. With him you don't try "I'm not in the mood" or "I got a headache"—all that gets you is hurt and he still gets what he wants. Forget that. You have to give in anyway. So whatever he's in the mood for, you're in the mood for. Lucky I really like it with the Big Guy. But who wouldn't? Anyway, he don't care what I do on my own time, long as I don't get knocked up.

A lot of the guys, they don't get much respect. They're not much fun to make it with. You got to do it anyway, though. They give you the look and you don't come running, they beat the shit out of you. Those guys, all they're interested in is one thing. One time, when the Big Guy is away, I won't do it and this guy, he picks up a big rock. Huge. He means business, so I have to let him. They're all like that. You don't come across, they get real pissed off. Those little guys, they think they're so big. They think they're hot stuff. They think they can have anyone they like.

When the Big Guy's around, sometimes he lets 'em and sometimes he don't. When he's away on a trip, or when his back is turned, we give the boys a little if we like 'em. You never know, one of 'em might be high rank some day. One of 'em might be the new Big Guy some day. But when the Big Guy's watching, if he don't want us to, we don't even look at the boys. We know what to do. We know our place.

Guys take a lot of stroking. Sometimes what they need is petting or kissing. Sometimes they need more. After, they're not so grumpy. You come across right away, the guys are nice to you, know what I mean?

Before I had my kid, I make it with ten, fifteen guys, one after the other. They can't wait to get on me.

The Big Guy, sometimes when he gets outta hand, all I gotta do is stroke him a little, and it's like he can't remember what was getting him so hot and bothered. The Big Guy, he's real nice to me. One time my kid's watching us in the act and tries to stop us. He climb on, hitting Big Guy with his little fists. Big Guy, he don't touch him. He think it's funny. He don't hurt my kid. He don't hurt me.

Buddy and Squint, they get lots of respect too. Not as much as Big Guy, but almost. Squint's the Big Guy's brother. He's got a thing for me, too. Squint takes the patrols out at night, far away, near the end of our turf. There's a gang that hangs out on the other side. They're the Strangers. Sometimes they raid us. We don't like Strangers. Our guys see Strangers, they go crazy. Strangers come here, they get what they deserve. We catch 'em, we tear 'em apart. Our patrols, they're out there protecting us and our kids. From Strangers.

One time everybody was tense. You could smell trouble. Me and the kid, we was scared. We was hugging each other real tight. Some Strangers come tearing through. Looking for sex and trouble. Rampage. Well, the Big Guy, he give 'em trouble. He come down on 'em hard. Before Buddy and Squint could help or anything, Big Guy stomp 'em real good. Those Strangers, they run away fast. They stay a little longer, they'd be dead. Best part was, even before the dust settled, they come 'round—Big Guy and Buddy and Squint—to me and the kid and all the others. They make sure we know everything's all right. Big Guy put his hand on my shoulder. He touch my cheek. He gimme a kiss. Big Guy, he's all right.

———

I like a little ass, same as the next guy. But what I really like is combat. You're out on patrol, you gotta be real quiet. You gotta be ready for action. Strangers could be anywhere. Anything could happen at night. Night's the most exciting.

We catch some Strangers, they've had it. One time Squint come on a Stranger mother holding her kid. He take the little brat by a leg and smash its head on the rocks. That'll teach Strangers to come around. Days later I seen her again, real sad, carrying that dead baby like it's still alive. But that's the way it goes. Strangers mess with our turf, they get what's coming.

Big Guy, he don't go out on patrols no more. In the old days, before Big Guy take over, it'd be him and me and Squint on patrol. That was great. Those Strangers, they come over here to steal our turf and fuck our females. Some of ours, the younger ones, they don't mind so much—they got a thing about quickies with Strangers. But us guys, we mind. Strangers, they ain't like us. We don't watch our step, they pick us off one by one.

They're fast and they're quiet. When we can't catch 'em, sometimes we throw rocks. I'm real good with rocks. I get high up somewhere and they don't see me, I cripple 'em with rocks, I break their ass. I hurt 'em and they can't hurt back. Them Strangers, they better not mess with me.

You gotta be careful, though. Old Boss, the boss before Big Guy, he was off chasing Strangers once. Soon as he was gone, some of the guys take his girlfriend—you know, the one he went off on a honeymoon with. They take her into the bushes. They try to cop a fuck on the side. She don't mind. Boss come back, he don't get so much respect like before. You really like a female, it gets you into trouble. Especially if you want to be a boss. It turn out OK for him, though. After Big Guy take over, Old Boss, he just spend all his days fucking. His hair's gray now, but he's happy.

Sometimes one of those Stranger females, she sashays over here, all young and sassy, looking for a little action—a real piece of ass, you know? Myself, I'd rather fuck 'em than kill 'em. But some of the guys, they get carried away. We don't like Strangers here. Still, sometimes she'll suck up to one of the guys and before you know it he sorta ease her into the gang.

In our gang everybody knows their place. Females especially. They do what they're told. Or else. Sometimes they make believe like they don't want it, but I know what they really want. Sometimes you gotta slap 'em around a little. Mostly you give 'em a look and right away, they're shaking their ass, they got that smile, their eyes are staring, they're moaning. Most of the time they beg for it.

Us guys, we don't want the Big Guy to get nervous. We show respect. So we let him climb all over us. It's not real; it's just for show. We suck up to the Big Guy. I'm high up, but on this I'm like the rest. He's my boss. If some tight-assed young guy don't wanna show respect, he better change his mind or he don't last long.

Big Guy, he's really something. I seen him fight off two, three, lots of Strangers, all at once, all by himself. One time he save a little kid that fall into the water. Would've drowned for sure. Big Guy, he's got balls.

After Big Guy it's pretty much what I say goes. I'm high up. Besides Big Guy, hardly anybody gets on me. 'Course, I need help now and then from the other guys. I spend a lot of time stroking them. But that's OK. You should see some of the guys my kid brother has to let get on him. Sometimes if Big Guy's pissed off, you can calm him down just by touching his cock. Sometimes you gotta do more. It just means you're cool.

When there's enough to eat and there's no Strangers around, everybody chills out. Guys get calm. In the early afternoon they all get sleepy, you know, and take naps. Not much trouble then. Too much calm, though, you get itchy for patrol.

I come up through the ranks. I don't get to be number two by accident. When I start out, I'm not grown yet, nobody give me respect. I want respect so bad back then. When I get big enough, some of the other kids, then some of their mothers and sisters, they start giving me respect. Then all the females. Then I gotta start working myself up with the guys. It was hard. Sometimes I gotta beg food from them. Meat especially. Sometimes, when they give me a little piece, I'd grab it all and run. They'd get real pissed off. It wasn't easy then. Now it's different. Now everybody give me respect. Even Squint, sometimes. Even the Big Guy, sometimes.

We get on good. I help him, he help me. He scratch my back, I scratch his, know what I mean? I'm real close to him, closer than anybody except maybe Squint. But one time he got mad at me for not showing enough respect. He think he's gonna teach me some manners. We have a big fight. Lots of other guys join in. More fights break out. More guys jump on. Maybe they're helping their brother, or maybe they're nervous about Big Guy and me fighting. Guys who're fighting ask for help from guys who're just watching. Pretty soon everybody's fighting.

But Big Guy, he don't look at nobody else but me. And he whip my ass. Then he start calming everybody down. I had to respect him. That was like a real Boss. Still, he beat me in front of everybody. One of these days I'm gonna make my move. He's been good for me. But I want him off me. Someday I'm gonna be all over him.

Right now, though, Big Guy and Squint and me, we gotta stick together. Some of the young guys are getting restless. They want to stick it to us. I know what those guys are like. When they see us they suck up to us. They show respect. But inside, they think "Up yours." They think, "My time's gonna come." Well, my time's gonna come first.

———

One thing I wouldn't let even the Big Guy mess with. That's my kid. That's where I draw the line. No one messes with him. When we're out together, scrounging for something to eat, and I see my kid looking up at me, I know I'd sooner die than let anybody hurt him. He feels like that about me too. When the guys—even top guys—threaten me, my kid come over and try to protect me. They respect him for it. 'Course, just like every other kid around here, all he's really got is his mother. If I don't protect him, who will? When he was little he'd eat stuff that make him sick. I gotta stop him. I gotta show him what's good to eat. He really need me then. He still does, more than he know. Sometimes the guys babysit and they seem to like him. But you can't trust guys.

One of the young guys wanna fuck his mother. She don't want to. One of these days he gonna hurt her bad. He can fuck his sister, but he should leave his mother alone. When the mood comes over the guys, though, they can't help themselves. They go crazy. They act like animals.

Sometimes guys go so crazy, they beat a kid to death for nothing, just for being there. A guy, he gets to be a pain in the ass, he gets chewed out by some big shot. So he go look for somebody to kick around, some nobody—some female, some kid. When guys get pissed off it's no good for anybody—least of all females and kids. You work real hard to get them calm.

One time my sister's kid, he musta got sick or something. All of a sudden he can't move his legs no more. He can't walk. He just drag himself along by his hands. He look real weird. First, folks look away. None of the guys come 'round to babysit no more. Later, they hassle him. Then they attack him. Then they kill him off, snap his neck. I was sad for my sister.

My kid, all he lives for is to be in the gang, get respect, go out on patrol. He's too little now, but his time'll come. He'd do anything for a pat from the Big Guy. Me, too. I love it when the Big Guy touches my hand.

And he stop the young guys from fighting. He's got a look that says "Up your ass." Most of the time he just flash that look and the guys, they calm down. Grown-ups, they know how far they can go. They make lots of threats. Except for Strangers, though, nobody gets hurt much. But real young guys, they don't know the difference. After they get to a certain age they can hurt each other bad. I don't want my kid hurt by some asshole who don't know his own strength. Big Guy puts a stop to that.

And he takes care of me. The Big Guy—or Buddy, but I know the Big Guy put him up to it—sometimes goes 'round handing out food. Meat especially. Meat's not so easy to come by. They always give me and the kid some. They give it mostly to the good-looking females, like me, to make sure we'll come across. But I'd do it for free, anytime he wants. A lot of folks beg for more when they hand the food out. Not me. I don't have to.

When the guys leave me alone, I spend all my time with my sister, my girlfriends, my grown-up daughter. We watch out for each other. We give each other respect. I'd be nowhere without them.

One time when I was young—before anybody fucked me except for play—I got fed up. I wasn't getting no respect. I was off by myself taking a walk and I see this cute guy. He don't see me. He's a Stranger —you can tell right away—but he's real cute. Then all of a sudden he's gone. After, I keep thinking about him. Maybe all Strangers are as cute as him. Maybe Strangers gimme respect. So I go to check 'em out.

It's a long walk and I don't wanna run into our patrols. But I get there OK. Pretty soon I find a guy. A Stranger guy. I don't think he's the same one I seen the first time, but he's real cute, too. I give him a look, and I can see he's eager. There's two females there, though, his kind, and they're not happy to see me like he is. They come at me, yelling and scratching and biting, and I run back home. It's a long way. When I get here, it don't seem anybody notice I'm gone—except for Mom, of course. She gimme a big hug. I miss Mom.

Chapter 15

MORTIFYING REFLECTIONS

When he bethought him of the first beginning
of all things, he was filled with a yet more
overflowing charity, and would call the dumb
animals, howsoever small, by the names of
brother and *sister*, forasmuch as he recognized
in them the same origin as in himself.

ST. BONAVENTURA
The Life of St. Francis[1]

We are astonished to see how slight and how
few are the differences, and how manifold and
how marked are the resemblances.

CHARLES BONNET
Contemplation de la Nature
(1781), on comparing apes and humans.[2]

Early in the fifth century B.C., Hanno of Carthage set sail into the western Mediterranean with a fleet of sixty-seven ships, each with fifty oars, carrying altogether thirty thousand men and women. Or at least this is what he claimed in the *Periplus*—a chronicle that was posted in one of the many temples consecrated to the god Baal after his return home. Sailing through the Straits of Gibraltar, he turned south, establishing cities along the West African coast as he went, including present-day Agadir, Morocco. Eventually, he came to a land filled with crocodiles and hippopotami and many groups of people, some herders, some "wild men," some friendly, some not. The interpreters he had brought from Morocco could not understand the languages spoken here. He sailed by what is now Senegal, Gambia, and Sierra Leone. He passed a great mountain from which a fire reached "to heaven," and from which, night and day, "streams of fire flowed into the sea." This is, almost certainly, the Mount Cameroon volcano just east of the delta of the River Niger. He may have gone almost as far as the Congo before returning.

In the last of eighteen short paragraphs in his *Periplus*, Hanno describes finding, just before turning back, an island in an African lake,

> full of wild men. By far the majority of them were women with hairy bodies. The interpreters called them "gorillas."

The males escaped by climbing precipices and hurling stones. But the females were not so lucky.

> We captured three women . . . who bit and scratched . . . and did not want to follow. So we killed them and flayed them and took their skins to Carthage.

Modern scholars take these beseiged and mutilated beings to be either what we today call gorillas, or chimpanzees. One of the details, the throwing of stones by the males, suggests to us that they were chimps. The *Periplus* is the earliest firm historical account we have of a first contact between apes and humans.[3]

————

The ancient Mayan authors of the *Popol Vuh* considered monkeys to be the product of the last botched experiment conducted by the gods before they finally got it right and managed to create us. The gods meant well, but they were fallible, imperfect artisans. Humans are hard to make. Many peoples in Africa, Central and South America, and the Indian subcontinent thought of apes and monkeys as beings with some deep connection to humans—aspirant humans, perhaps, or failed humans, demoted for some grave transgression against divine law, or voluntary exiles from the self-discipline demanded by civilization.

In ancient Greece and Rome the similarity of apes or monkeys with humans was well-known—indeed, it was stressed by Aristotle* and Galen. But this led to no speculations about common ancestry. The gods who had made humans were also in the habit of changing themselves into animals to rape or seduce young women: Like the centaurs and the Minotaur, the offspring of these unions were chimeras, part beast, part human. Still, no ape chimeras are prominent in the myths of Greece and Rome.

In India and ancient Egypt, though, there were monkey-headed gods, and in the latter large numbers of mummified baboons—indicating that they were cherished if not worshipped. A monkey apotheosis would have been unthinkable in the post-classical West—in part because the Judaeo-Christian-Islamic religion came of age where nonhuman primates were rare or absent, but mainly because the worship of animals (for example, the Golden Calf of the Israelites) was singled out as an abomination: They were pedaling away from ani-

---

* "[An ape's] face resembles that of a man in many respects . . . [I]t has similar nostrils and ears, and teeth like those of man, both front teeth and molars . . . [I]t has hands and fingers and nails like man, only that all these parts are somewhat more beast-like in appearance. Its feet are exceptional . . . like large hands . . . [T]he internal organs are found on dissection to correspond to those of man."[4]

mism as fast as they could. Apes were not widely available for examination in Europe until about the sixteenth century; the so-called Barbary ape of North Africa and Gibraltar—which is what Aristotle and Galen apparently described—is actually a monkey, a macaque.

Without exposure to the beasts most like men, it was difficult to draw the connection between beasts and men. It was easier by far to imagine a separate creation of each species, with the less vivid similarities between us and other animals (the suckling of the young, say, or five toes on each foot) understood as some trademark idiosyncracy of the Creator. The ape was as far below man, it was asserted, as man was below God. So, when, after the Crusades, and especially beginning in the seventeenth century, the West came to know monkeys and apes better, it was with a sense of embarrassment, shame, a nervous snigger—perhaps to disguise the shock of recognition at the family resemblance.

The Darwinian idea that monkeys and apes are our closest relatives brought the discomfort to the conscious level. You can still see the unease today in the conventional associations with the word "ape": to copy slavishly, to be outsized and brutal. To "go ape" is to revert, to become wild, untamed. When we handle something idly, in an exploratory way, we're "monkeying around." To "make a monkey" out of someone is to humiliate him. A "little monkey" is a mischievous or playful child. A "monkeyshine" is a prank. To "go bananas" is to lose control—reflecting the fact that monkeys and apes, who indeed love bananas, are not subject to the same social restraints that we are. In Christian Europe in the Middle Ages and early Renaissance, monkeys and apes were emblematic of extreme ugliness, of a doomed craving for the status of humans, of ill-gotten wealth, of a vengeful disposition, of lust and foolishness and sloth.[5] They were accessories—because of their susceptibility to temptation—in the "Fall of Man." For their sins, it was widely held, apes and monkeys deserved to be subjugated by humans. We seem to have weighed these beings down with a heavy burden of symbols, metaphors, allegories, and projections of our own fears about ourselves.

———

Before the outside world knew anything of his long effort to understand evolution, Darwin wrote telegraphically in his 1838 "M" notebook: "Origin of man now proved . . . He who understands baboon would do more towards metaphysics than [the philosopher John] Locke."[6] But what does it mean to understand a baboon?

One of the earliest scientific studies of the chimpanzee in its natural African habitat was made by Thomas N. Savage, a Boston physician. Writing in early Victorian times, he concluded:

> They exhibit a remarkable degree of intelligence in their habits, and, on the part of the mother, much affection for their young . . . [But] they are very filthy in their habits . . . It is a tradition with the natives generally here, that they were once members of their own tribe: that for their depraved habits they were expelled from all human society, and, that through an obstinate indulgence of their vile propensities, they have degenerated into their present state and organisation.[7]

Something was bothering Thomas N. Savage, M.D. "Filthy," "depraved," "vile," and "degenerate" are terms of abuse, not scientific description. What was Savage's problem? Sex. Chimpanzees have an obsessive, unself-conscious preoccupation with sex that seems to have been more than Savage could bear. Their zesty promiscuity may include dozens of seemingly indiscriminate heterosexual copulations a day, routine close mutual genital inspections, and what at first looks very much like rampant male homosexuality. This was a time when proper young ladies were abjured not to inquire too closely into the stamens and pistils—"the private parts"—of flowers; the renowned critic John Ruskin would later harumph, "With these obscene processes and prurient apparitions, the gentle and happy scholar of flowers has nothing to do."[8] How was a proper Bostonian physician to describe what he had witnessed among the chimpanzees?

And if he did describe it, even obliquely, did he not run a certain risk—that his readers would conclude he approved what he was chronicling? Or more than "approved." What had drawn him to chimpanzees in the first place? Why did he insist on writing about them? Were there no worthier matters deserving of his attention? Perhaps, he felt obliged to ensure that even a casual reader would note the great distance separating Thomas Savage from the subjects of his study.*

---

* Savage also wrote the first systematic account of gorillas in the wild, and was responsible for the modern use of the ancient North African word "gorilla." He took pains to

William Congreve was the leading playwright of the English comedy of manners around the turn of the eighteenth century. The monarchy had been restored after a bloody struggle with the Puritan religious schismatics who gave their name to rigidity on sexual morality. Each age is repelled by the excesses of the last, so this was a time of moral permissiveness, at least among the dominant elite. Their sigh of relief was almost audible. But Congreve was not their apologist. His ironical and satirical wit was directed at the pretensions, affectations, hypocrisies, and cynicisms of his age—but especially at the prevailing sexual mores. Here, for example, are three fragments of ruling-class dialogue from his *The Way of the World:*

[O]ne makes lovers as fast as one pleases, and they live as long as one pleases, and they die as soon as one pleases; and then, if one pleases, one makes more.

You should have just so much disgust for your husband as may be sufficient to make you relish your lover.

I say that a man may as soon make a friend by his wit, or a fortune by his honesty, as win a woman with plain dealing and sincerity.[9]

Bearing in mind Congreve's role as daring social critic of sexual manners, now consider this excerpt from a 1695 letter he wrote to the critic John Dennis:

I can never care for seeing things that force me to entertain low thoughts of my Nature. I don't know how it is with others, but I confess freely to you, I could never look long upon a monkey without very Mortifying Reflections; tho I never heard any thing to the Contrary, why that Creature is not Originally of a Distinct Species.[10]

Somehow, the sexual imbroglios of upper-class twits that he chronicled did not generate as many Mortifying Reflections as a visit to the

---

repudiate popular notions of gorillas carrying off attractive women for unspeakable purposes—the theme echoed a century later to enormous public acclaim in the motion picture *King Kong.*

zoo. Plays such as Congreve's were themselves being criticized as breaking down "the Distinctions between Man and Beast. Goats and Monkeys, if they could speak, would express their Brutality in such Language as This."[11] Monkeys were beginning to bother Europeans. And Congreve put his finger on the problem: What does it say about us if monkeys are our close relatives?

From the earliest encounters that history records between apes and men, to parents hurrying their children past the monkey cages before awkward questions are posed, we've felt an unease—and the unease has been greater the more puritanical the observer. "The body of an ape is ridiculous . . . by reason of an indecent likeness and imitation of man," wrote the cleric Edward Topsell in his 1607 work *Historie of Foure-Footed Beasts*. Charles Gore, "a man of rock-like faith" and a successor of Samuel Wilberforce as Anglican Bishop of Oxford, was a conflicted habitué of the London Zoo: "I always return an agnostic. I cannot comprehend how God can fit those curious beasts into his moral order." He once shook his finger at a chimpanzee and rebuked it aloud, in the presence of an attentive small crowd of which he was wholly unaware: "When I contemplate you, you turn me into a complete atheist, because I cannot possibly believe that there is a Divine Being that could create anything so monstrous."[12] If, say, ducks or rabbits with a penchant for sexual excess were under review, people would not have been nearly so bothered. But it's impossible to look at a monkey or ape without ruefully recognizing something of ourselves.

Simians have facial expressions, social organization, a system of mutually understood calls, and a style of intelligence that's familiar. They have opposable thumbs and five fingers on each hand which they use as we do. Some walk upright on two legs, at least occasionally. They are awfully, uncomfortably, like us. Might their mores suggest alternative sexual arrangements that might be erosive of the social fabric? * And other ruminations about human affairs might be roused by close attention to monkeys and apes—on the prevalence of coercion and violence, for example, or about public sanctions on sexual intimidation, rape, and incest. These are weighty and sensitive

---

* The soldiers of Alexander the Great—not otherwise known for their prudishness—are said, in their India campaign, to have put monkeys to death for their "lasciviousness."[13]

matters. The behavior of monkeys and apes—particularly the ones that look most like us—is an awkward business. Better to put it aside, better to ignore it, better to study something else. Many people would rather not know.

———

Carl Linnaeus, the eighteenth-century biologist, founded the science of taxonomy—the goal of which is to classify every organism on Earth.[14] He set himself the task of recording the similarities and differences of all the plants and animals then known, and arranging them all into a web—or, better, a tree—of relatedness. It was he who introduced many elements of the now-standard classification scheme: species, genus, family, order, class, phylum, and kingdom, moving from less to more inclusive categories. Each of these categories is called a "taxon" (plural, "taxa"). So we humans, for example, are of the animal kingdom, the vertebrate phylum, the class of mammals, the order of primates, the family of Hominidae, the genus *Homo*, and the species *Homo sapiens*. In other words, we're animals, not plants or fungi or bacteria; we have backbones, so we're not invertebrates such as worms or clams; we have breasts to supply milk to the young, so we're not reptiles or birds; we're primates, not rats or gazelles or raccoons; and we're Hominidae, not orangutans or vervet monkeys or lemurs. We are of the genus *Homo*, in which taxon there is but one species (although once there were others—maybe many others). This is how we classify ourselves today. And it's almost the same as what Linnaeus proposed.

After accruing vast experience with his new discipline of taxonomy, classifying thousands of beasts and vegetables, Linnaeus contemplated the status of an animal of special interest—himself. Then he reconsidered. By his standard criteria, Linnaeus would have placed human beings and chimpanzees in the same genus.* His scientific integrity urged him to do so. But he well understood what an abomination, how scandalous such a step would have been judged by the Swedish Lutheran Church—indeed, by every religious establishment of which he knew. So Linnaeus trimmed his sails, made a social com-

---

* Jean-Jacques Rousseau, in 1753, had gone further and classified chimps and men as members of the same *species*, the power of speech being at the beginning, in his view, not "natural to man."[15] Congreve had toyed with something similar.

promise, and placed us in a genus by ourselves—although he outraged many by declaring us, with the apes and monkeys, a member of the same order.

It's hard to fault him. Like Copernicus, Galileo, and Descartes, he was about as brave as his age would allow. Many naturalists placed humans in a separate order; by Darwin's time this would become the conventional wisdom. Many clerics (and some naturalists) placed us in a separate kingdom. The evidence may not have warranted it, but isolating humans in their own genus, their separate first-class compartment, was a popular step, reassuring to human vanity. In 1788, in a reflective and undefensive mood, Linnaeus wrote:

> I demand of you, and of the whole world, that you show me a generic character . . . by which to distinguish between Man and Ape. I myself most assuredly know of none. I wish somebody would indicate one to me. But, if I had called man an ape, or vice versa, I would have fallen under the ban of all the ecclesiastics. It may be that as a naturalist I ought to have done so.[16]

One of the scientific names for the common chimpanzee then was *Pan satyrus*. Pan was an ancient Greek deity, part man, part goat, associated with lust and fertility. A satyr was a closely associated chimera—at first represented as a man with a horse's tail and ears and an erect penis. Clearly the rampant sexuality of chimps was the defining characteristic in this early naming of the species. The modern classification is *Pan troglodytes*, troglodytes being mythical creatures who live in caves and beneath the Earth—a much less appropriate designation, since chimps reside exclusively on (and slightly above) the Earth. (The Barbary apes of North Africa do sometimes live in caves; the only other primates known routinely to have lived in caves are humans.) Linnaeus had mentioned a *Homo troglodytes*, but it's unclear whether he had ape or human in mind. Or something inbetween.

A systematic comparison of the anatomies of apes and humans was performed during the opening salvos of the Darwinian Revolution by T. H. Huxley. He described his research program in these words, notable among other respects for their extraterrestrial perspective:

> [L]et us endeavour for a moment to disconnect our thinking selves from the mask of humanity; let us imagine ourselves scientific Satur-

nians, if you will, fairly acquainted with such animals as now inhabit the Earth, and employed in discussing the relations they bear to a new and singular "erect and featherless biped," which some enterprising traveller, overcoming the difficulties of space and gravitation, has brought from that distant planet for our inspection, well preserved, may be, in a cask of rum. We should all, at once, agree upon placing him among the mammalian vertebrates; and his lower jaw, his molars, and his brain, would leave no room for doubting the systematic position of the new genus among those mammals, whose young are nourished during gestation by means of a placenta, or what are called the "placental mammals" . . .

There would remain then, but one order for comparison, that of the Apes (using that word in its broadest sense), and the question for discussion would narrow itself to this—is Man so different from any of these Apes that he must form an order by himself? Or does he differ less from them than they differ from one another, and hence must take his place in the same order with them?

Being happily free from all real, or imaginary, personal interest in the results of the inquiry thus set afoot, we should proceed to weigh the arguments on one side and on the other, with as much judicial calmness as if the question related to a new Opossum. We should endeavour to ascertain, without seeking either to magnify or diminish them, all the characters by which our new Mammal differed from the Apes; and if we found that these were of less structural value, than those which distinguish certain members of the Ape order from others universally admitted to be of the same order, we should undoubtedly place the newly discovered tellurian [terrestrial] genus with them.

I now proceed to detail the facts which seem to me to'leave us no choice but to adopt the last mentioned course.[17]

Huxley then compares the skeletal and brain anatomies of apes and humans. The "manlike apes" (chimps, gorillas, orangutans, gibbons, and the gibbon-like siamangs—the first three called "greater" and the last two "lesser" apes) all have the same number of teeth as humans; all have hands with thumbs; none has a tail; all arose in the Old World. The skeletal anatomies of chimps and humans are strikingly similar. And "the difference between the brains of the Chimpanzee and of Man," he concluded,[18] "is almost insignificant."

From these data, Huxley then drew the straightforward conclusion that contemporary apes and humans are close relatives, sharing a recent ape-like common ancestor. The conclusion scandalized Victorian England. The outraged reaction of the wife of the Anglican Bishop of Worcester was typical: "Descended from apes! My dear, let us hope that it is not true, but if it is, let us pray that it will not become generally known."[19] Here it is again: the fear that knowledge of the true nature of our ancestors might unravel the social fabric.

———

In recent years it has been possible to go much further, to the very heart of life, to the Holy of Holies, and compare, nucleotide by nucleotide, the DNA molecules of two animals. We can now quantify the kinship of different species. We are able to establish molecular pedigrees, DNA genealogies, which provide the most powerful and compelling evidence that evolution has occurred, as well as tantalizing clues on its mode and tempo. The new tools of molecular biology have yielded insights wholly unavailable to previous generations.

Every animal with a backbone has a bloodstream in which hemoglobin is the oxygen carrier. Hemoglobin is composed of four different kinds of protein chains wrapped about one another. One of them is called beta-globin. A particular region of the ACGT sequence codes for beta-globin in all these animals, but only about 5 percent of the region is occupied by the actual instructions for this protein chain. Much of the remaining 95 percent are nonsense sequences—so here mutations can accumulate without being winnowed out by selection. When the beta-globin regions of the DNA are compared across the primate order,[20] humans are found to be more closely related to chimps than to anyone else. (The human-gorilla connection comes in a close second.) A new basis for our chimp connection is uncovered: not just the bones, the organs, and the brains, but also the genes—the very instructions for making chimps and humans—are almost indistinguishable.

The DNA sequence that codes for beta-globin is roughly fifty thousand nucleotides long; that is, along a given strand of the DNA molecule, fifty thousand As, Cs, Gs, and Ts in a particular sequence describe precisely how to manufacture the beta-globin of the species in question. If the sequences of humans and chimpanzees are compared nucleotide by nucleotide, they differ by only 1.7%. Humans and gorillas differ by 1.8%, almost as little; humans and orangutans, 3.3%;

humans and gibbons, 4.3%; humans and rhesus monkeys, 7%; humans and lemurs, 22.6%. The more the sequences of two animals differ, the more remote (both in relatedness and, usually, in time) is their last common ancestor.

When ACGT sequences that are mainly active genes are examined, a 99.6% identity is found between human and chimp. At the level of the working genes, only about 0.4% of the DNA of humans is different from the DNA of chimps.[21]

Another method is first to take the DNA from a human being, unzip the double helix, and separate the two strands. Then do the same for a comparable DNA molecule of some other animal. Put the two strands together and let them link up. You've now made a "hybrid" molecule of DNA. Where the complementary sequences are closely the same, the two molecules will tightly bind to each other, forming part of a new double helix. But where the DNA molecules from the two animals differ a great deal, the bonding between the strands will be intermittent and weak, and whole sections of the double helix will be flopping loosely. Now take these hybrid DNA molecules and put them in a centrifuge; spin them up so the centrifugal forces tear the two strands apart. The more similar the ACGT sequences are—that is, the more closely related the two DNA strands are—the more difficult it will be to tear them apart. This method does not rely on selected sequences of DNA information (that coding for beta-globin, for example) but on vast amounts of hereditary material, making up whole chromosomes. The two methods—determining the ACGT sequences of selected portions of DNA, and DNA hybridization studies—give remarkable overall agreement. The evidence that humans are most closely related to the African apes is overwhelming.

On the basis of all the evidence, the closest relative of the human proves to be the chimp. The closest relative of the chimp is the human. Not orangs, but people. Us. Chimps and humans are nearer kin than are chimps and gorillas or any other kinds of ape not of the same species. Gorillas are the next closest relatives, both to chimps and humans. The more remote the kinship—when we go to monkeys or lemurs or, say, tree shrews—the less the similarity in sequence. By these standards, humans and chimps are about as closely related as horses and donkeys, and are closer relatives than mice and rats, or turkeys and chickens, or camels and llamas.[22]

"All right," you might say, "maybe chimp anatomy is almost the same as mine. Maybe the chimp's cytochrome *c* and hemoglobin are almost the same as mine. But the chimp isn't nearly as smart as I am, as well-organized, as hardworking, as loving, as moral, as devout. Maybe when the genes for these traits are discovered, bigger differences will be found." Yes. Maybe you're right. And even that 99.6% identity can be misleading. A 0.4% difference is substantial, because the DNA in any cell in either species is composed of some 4 billion ACGT nucleotides; of them conservatively 1% are in working, no-nonsense portions of the DNA and constitute the genes as such.

The number of operational ACGT nucleotide pairs that are different between humans and chimpanzees must then be roughly 0.4% times 1% times 4 billion, or 160,000. If these are the working parts of genes each 1,000 nucleotides long, each of which codes for a separate enzyme, then the number of completely different kinds of enzymes that humans have and chimps don't, or vice versa, would be somewhere around 160,000/1,000 or 160. We recall that enzymes have a powerful leverage; they preside over changes in the chemistry of the cell, which can happen very fast; one enzyme can process a multitude of molecules. A hundred enzymes, if they're the right enzymes, might make a very big difference. A hundred enzymes seems more than enough to account for Huxley's metaphorical description of the difference between apes and humans: "a hair in the balance-wheel, a little rust on a pinion, a bend in a tooth of the escapement, a something so slight that only the practised eye of the watchmaker can discover it." Some enzymes would affect estrus, some stature, some fur, some climbing and leaping abilities, some development of the mouth and larynx, some changes in posture, toes, and gait. Many of them would be for a bigger brain with a bigger cerebral cortex, and new ways of thinking beyond the reach of apes.

What's more, a hundred enzymes changed is certainly an underestimate. Probably none of the differences between chimps and humans requires entirely new enzymes to be evolved. A small number of changes, maybe only a change in a single nucleotide, is adequate to render an enzyme inoperable or to change its function. And many of the differences may not be in the genes themselves, but in the promoters and enhancers, the regulatory elements of the DNA that control when and for how long certain genes should be operational. So even a 0.4% difference could, for all we know, imply profound differences in certain characteristics.

Still, chimps are nearer relatives to us than any other animal on Earth. A typical difference between your DNA—all of it, including the untranscribed nonsense—and that of any other human being[23] is roughly 0.1% or less. By this standard, chimps differ from humans only about 20 times more than we differ from one another. That seems awfully close. We must be very careful that those "mortifying reflections" of which Congreve spoke do not make us exaggerate the differences and blind us to our kinship. If we want to understand ourselves by closely examining other beings, chimps are a good place to start.

———

Fledgling students of animal behavior are warned against anthropomorphizing. The word literally means changing into human form—attributing human attitudes and states of mind to other animals whose thoughts are not vouchsafed to us. Fairy tales, Aesop, La Fontaine, Joel Chandler Harris, and Walt Disney are among the foremost exponents of the genre. Darwin was guilty of a kind of anthropomorphizing and, even more flagrantly, so was his student George Romanes. The temptation of sentimental self-deception was considered so insidious, and the sin of anthropomorphizing so grave an error, that an influential school of American psychology arose in the first half of the twentieth century which taught that animals enjoyed *no* internal mental states, no thoughts and no feelings. Its practitioners talked about "the myth of consciousness." We must, its founder said, "make a clean break with the whole concept of consciousness." Real scientists, it was claimed, are concerned with no more than what can be observed of the actual behavior of animals. Sensory inputs go in, behavioral outputs come out, and that's that. Animals feel no pain. Animals are mechanical black boxes. Behaviorism, as it was called, was an example of the ultrapragmatic streak in American science. It had something in common with Descartes's automata, although it allowed far less room for free inquiry. It came close to deciding that humans don't have any thoughts or feelings either.

A concerted but fair-minded attack on at least the more extreme forms of behaviorism has been mounted by the biologist Donald Griffin. In the following passage, Griffin refers to "parsimony"—in science, the doctrine that in deciding between two adequate explanations, we should choose the simpler. It's also called "Occam's Razor."

> According to the strict behaviorists, it is more parsimonious to explain animal behavior without postulating that animals have any mental experiences. But mental experiences are also held by behaviorists to be identical with neurophysiological processes. Neurophysiologists have so far discovered no fundamental differences between the structure or function of neurons and synapses in men and animals. Hence, unless one denies the reality of human mental experiences, it is actually parsimonious to assume that mental experiences are as similar from species to species as are the neurophysiological processes with which they are held to be identical. This, in turn, implies qualitative evolutionary continuity (though not identity) of mental experiences among multicellular animals.
> The possibility that animals have mental experiences is often dismissed as anthropomorphic because it is held to imply that other species have the same mental experiences a man might have under comparable circumstances. But this widespread view itself contains the questionable assumption that human mental experiences are the only kind that can conceivably exist. This belief that mental experiences are a unique attribute of a single species is not only unparsimonious; it is conceited. It seems more likely than not that mental experiences, like many other characters, are widespread, at least among multicellular animals, but differ greatly in nature and complexity.
> . . . Extreme forms of behaviorism tend to become little more than irrelevant pleas of willful ignorance . . .
> Some behavioral scientists vigorously proclaim that they are not interested in animal awareness even if it does occur. Their antipathy sometimes seems to be so strong as to suggest that they really do not *want* to know about any thinking in which animals might engage.[24]

It's possible, we submit, to carry the fear of anthropomorphism too far. There are excesses worse than a surfeit of sentiment. There must be some interior state, some thoughts and feelings among the monkeys and apes, and if they are genetically our close relatives, if their behavior is so similar to ours as to be familiar, it's not unreasonable to attribute to them feelings similar to ours as well. Of course, until better communication with them is established, or until we understand much more about how their brains and hormones work, we can't be sure. But it's plausible, it's an effective teaching tool, and in this book on a few occasions we attempt to portray what it might be like inside the head of another animal.

———

By now, the reader will have at least suspected that the interior monologues of the preceding chapter—the first and third by a middle-ranking female, the second by a high-ranking male—are not intended to refer, exactly, to humans. Instead, we've tried to depict what it's like to be a chimp in chimp society. Systematic, long-term observation of chimp groups in the wild is a new field of science. We've relied chiefly on the courageous, insightful, and pioneering work of Jane Goodall at the Gombe Reservation in Tanzania, as well as studies by Toshisada Nishida and his colleagues in the Mahale Mountains, also in Tanzania, and by Frans de Waal, who investigated a troop of chimpanzees in a two-acre enclosure in the Arnhem Zoo in the Netherlands.[25] Every event dramatized in the last chapter is based on the accounts of these scientists. Their observations speak to us of a way of life that is unmistakably familiar, rich with the *Sturm und Drang* of human relations. Of course, no human has ever been inside a chimp's mind, and we cannot be sure how they think. We have taken liberties. We make no apologies for doing so, but stress that it is intended only as a way to think about the chimps.

We must be careful of circular reasoning here—foisting human mental and emotional processes on the chimps, and then triumphantly concluding at the end of our narrative how much like us they are. If we're to understand ourselves better by looking closely at chimpanzees, we'll have to give great weight to what they do and comparatively little to what we imagine is going on inside their heads. We must be careful not to deceive ourselves. The behaviorists were not wholly misguided.

We didn't mention that chimps sleep in trees and that they spend a great deal of time grooming one another. Although chimps do not seem as much transfixed by oral sex as some other primates (cunnilingus is an almost invariable part of foreplay among the orangs[26]), we used the now-popular phrase to "suck up" to someone because it seems to us, in its present English-language associations at least, to approximate some of the nuance of chimpanzee submission. (The gestural vocabulary of chimpanzee submission does include kissing the alpha's thigh.)

Many behavioral differences exist between chimps and humans, just as between chimps and gorillas or between gibbons and orangutans. But we are struck by how much the core of chimpanzee social life in the wild resembles some forms of human social organization, especially under great stress—in prisons, say, or urban and motorcycle gangs, or crime syndicates, or tyrannies and absolute monarchies. Niccolò Machiavelli, chronicling the maneuvering necessary to get ahead in the seamy politics of Renaissance Italy—and shocking his contemporaries, especially when he was honest—might have felt more or less at home in chimpanzee society. So might many dictators, whether they style themselves of the right or left persuasion. So might many followers. Beneath a thin varnish of civilization, it sometimes seems, there's a chimp struggling to bust out—to take off the absurd clothes and the restraining social conventions and let loose. But this is not the whole story.

They're a little shorter, somewhat hairier, much stronger, and a lot more sexually active than most humans are. They have brown hair and brown eyes. In their natural habitats, they may live to be forty or fifty years old—which is longer than the average in any human society before the Industrial and Medical revolutions. But their average life expectancy is much less. Unlike modern humans, females past infancy are not likely to live as long as males. They alternate between walking on two feet and on all fours, using their knuckles. Chimpanzee males tend to have short fuses. They give off a faint but characteristic odor when they're nervous or excited, revealing emotions they sometimes try to hide. Chimps are not ashamed of displaying their sexual parts. By our lights they're a lot dumber than we are, but they do use and even make tools. They apparently hold grudges, nurse resentments, and harbor thoughts of revenge. They plan future courses of action.

Family ties may be strong and lasting. Aged mothers will rush to the defense of their children, even full-grown sons. Orphaned infants are tenderly raised by older siblings. They experience prolonged grief at the loss of a loved one. They suffer from bronchitis and pneumonia, and can be infected with almost any human disease, including the AIDS virus. The elderly turn gray, get wrinkles, lose teeth and hair. Chimps get drunk. They're able to learn more words of a human language than we have of any chimp language. When they look in the mirror, they recognize themselves. They are, at least to some degree, self-aware. Infants get cranky and irritable when they're weaned. Chimps form friendships, often with comrades-in-arms who hunt together and guard their turf against intruders. They share food with relatives and friends.

When raised among humans, they have been known to masturbate to pictures of naked people. (This is probably true only of those who, through prolonged contact, have come to consider themselves human. Wild chimps would no more masturbate to erotic images of humans than vice versa.) They keep secrets. They lie. They both oppress and protect the weak. Some, despite many setbacks, persistently strive for social advancement and career opportunities. Others, less ambitious, are more or less content with their lot.

Among much other innate knowledge, they are born with an understanding about how to make a bed of leaves each night up there in the trees. They are much better climbers than we, partly because they haven't lost, as we have, the ability to grasp branches with their feet. The youngsters love to climb trees and rival one another in spectacular feats of gymnastic derring-do. But when an infant has climbed too high, its mother—socializing with her friends at the base of the tree —decisively taps the trunk and the baby obediently scampers down.

The forest is crisscrossed with a network of trails made by generations of chimps going about their daily business. Each knows the local geography at least as well as the average human city-dweller knows the neighborhood streets and shops. They almost never get lost. Here and there along the trails are trees with acoustically resonant trunks. When a party of foragers spies such a tree, many run forward and drum away—both sexes, children as well as adults. There are no strings, woodwinds, or brass yet, but the percussion section is in place.

Chimps recognize one another's individual voices, and a distinctive pant-hoot may summon an ally or relative from a considerable distance. In answering a pant-hoot from, say, an adjacent valley, they lift their heads and purse their lips as if they were on stage at La Scala. Up close, they have an uncanny ability—"uncanny" means only that we haven't been smart enough to figure it out yet—to communicate with one another, not just about such straightforward matters as sex or dominance, but about much more subtle matters, such as hidden dangers, or buried food supplies. A classic set of experiments was done by the psychologist E. W. Menzel:

> [Menzel] maintained four to six young chimpanzees in a large out-door enclosure that was also connected to a smaller holding cage. He restrained all but one animal in the holding cage, while showing this chosen "leader" the hidden location of either an amount of food or an aversive stimulus such as a stuffed snake. The leader was then returned to the holding cage, and the whole group was released. According to Menzel's reports, the variable behavior of the animals indicated that they "seemed to know approximately where the object was, and what sort of object it was, long before the leader reached the spot where it had been hidden" . . . If the goal was food, they ran ahead looking in possible hiding places; if it was a stuffed alligator or snake, they emerged from their cage showing piloerection [their hair standing on end] and staying close to their companions. If the hidden item was an alligator or snake, they became very cautious in their approach and often mobbed the area, hooting in the direction of the hidden item and hitting at it with sticks. If the hidden item was food, the animals searched the area intensively and showed little fear or distress. The behaviors occurred even if the aversive stimulus had been removed before the animals were released from the holding cage, so it was not the item itself that produced these reactions.
>
> In the food tests, one male (Rocky) began to monopolize the food supply when it was located. When Belle, a female, served as leader, she attempted to avoid indicating the location of the food cache, but Rocky could often extrapolate from her line of orientation and find the food. If Belle were shown two caches, one large and one small, she would lead Rocky to the small one and, while he was busy eating, run to the larger one which she would share with other individuals. Menzel concluded that chimpanzees could communicate the direction, amount, quality, and nature of the goal, as well as attempt to conceal at least some of this information, but precisely how chimpanzees achieve such communication is still not known.[27]

The only possibilities seem to be gestures and speech.

Chimps have hundreds of different kinds of food and crave dietary variety. They eat fruit, leaves, seeds, insects, and larger animals, sometimes dead ones. Caterpillars are delicacies, and the discovery of an infestation of caterpillars becomes a memorable gastronomic event. They're known to eat soil from cliff faces, presumably to provide mineral nutrients such as salt. Mothers will offer choice tidbits of food to their infants and will snatch unusual, possibly dangerous, foods from their mouths. In the wild, adults share food occasionally, often in response to begging. There are no set mealtimes; they snack throughout the day. As a foraging party moves on, one of its members may carry a branch still laden with berries or leaves to munch as she rambles.

When in the middle of the night, in their beds of leaves in the high branches, they are awakened by the cries of predators, they clutch each other in fright, their urine and feces raining down on the forest floor below.

They love to play, children (whose energy is prodigious) more than adults, but even adult play is common—especially when there's enough to eat and large numbers of chimps gather together. Play often involves, but is not restricted to, mock fighting.

Chimp males are protective toward females and the young. They will readily risk their own lives to protect "women and children" from attack, or to rescue a youngster in trouble. Goodall writes, "Often it seems that a male cannot resist reaching out to draw an infant into a close embrace, to pat him, or to initiate gentle play."[28] When a male is discovered *in flagrante delicto* with a female, which is often, an infant may rush up to punch the male in the mouth or jump on the back of the female, most often his mother.* In such situations the male's tolerance frequently exceeds human limits.

But in a display competition for dominance, all this good-natured equanimity vanishes, and a male who ordinarily is protective of infants may pick up a small, innocent bystander and slam it to the ground in his rage. When an unfamiliar female is discovered in their territory,

---

* A young mother will not usually come again into estrus until she weans her infant. The little one, understandably enough, may interpret weaning as rejection. The mother's renewed sexual interest in adult (and sub-adult) males probably compounds the infant's agony and resentment. Perhaps we also share the Oedipus complex with the apes.

chimps are known to seize her infant by the ankles and smash it against the rocks.[29]

Chimps tend to pick on the runt of the litter, and to displace their own anger away from higher-ranking chimps (who might do them harm) to those who are milder-tempered, younger, weaker, and female. At Gombe in 1966 there was a polio epidemic which resulted in the partial paralysis of full-fledged members of the group. Crippled by their disease, they were forced to move in odd ways, dragging limbs. Other chimps were at first afraid; then they threatened the afflicted, and then attacked them.

Because aggression is episodic and friendly relations so much more common, some early field observers were tempted by the notion that chimps in a state of nature (that is, unimprisoned) are non-violent and peace-loving. This is not the case. In hunting other animals, in working the dominance hierarchy, in hustling the females, in peevish moments, and in skirmishes with other groups of chimps (the Strangers, in our narrative), they show themselves capable of great violence.

Meat contains essential amino acids and other molecular building blocks more difficult to acquire from plants. Both sexes are ravenous for meat. On rare occasions, females will attack other females in their group and steal and eat their infants. Once the little one is in hand, there are no ill feelings directed to the mother of the tiny victim. In one case, a female approached those who were eating her baby; one of the diners responded by putting out her arms to embrace and comfort the grieving mother. Chimps are known to hunt mice, rats, small birds, a twenty-kilogram adolescent bush pig, monkeys such as baboons and colobuses, and other chimps.

A successful hunt is accompanied by enormous excitement. The spectators scream, hug, kiss, and pat one another reassuringly. Those actually involved in the kill immediately begin feeding, or attempt to carry off tasty body parts. The forest is filled with screeches, barks, pants, and hoots—which attract additional chimps, sometimes from a considerable distance. Generally males help themselves to bigger portions than females. Those of high rank are more likely to distribute the spoils, and one way or another most who are actually present at the kill gain a share. Newcomers plead for morsels. Pieces will be stolen, and the chimp whose prize has been taken will be furious, perhaps indulging in a temper tantrum. Portions of meat are taken to bed for midnight snacks.

A rat may be eaten head first. A monkey or young antelope is often killed by having its head smashed against a rock or tree trunk, or by giving it a vampirish bite in the back of the neck. Almost always the brains are eaten first. This is often the prize of the hunter who performs the actual kill. Other tasty body parts include the genitals of male victims and the fetuses of pregnant female victims. Goodall reports the final, attenuated, scream of a young bush pig as a chimp, like some ancient Aztec priest, tore out its living heart. Cooking has not yet been invented, nor flatware, nor table manners, nor squeamishness. This is a world of red blood and raw meat.

Janis Carter describes[30] a juvenile chimp and a colubus monkey, about its own size, grooming one another; but when the colubus is seized by the tail and killed by a passing adult chimp, who bashes its head against a tree, the juvenile readily enough joins in devouring its erstwhile playmate. Most of the monkey (and small mammal) victims of chimp predation are infants and juveniles, often snatched from their mothers' arms. Sometimes the mother tries to rescue the infant and is herself eaten.

In this world there is no mercy shown to food, even if it's still ambulatory. Food is for eating. Those who are moved to mercy eat less and leave fewer offspring. Clearly the chimps do not recognize monkeys, or chimps of other groups, or even members of their own group as deserving of mercy or other moral considerations. They may be heroic in defending their own young, but they do not show the least compassion for the young of other groups of species. Perhaps they consider them "animals."

Hunting is a cooperative endeavor. Cooperation is essential for making the larger kills—and also for avoiding their dangers, such as an enraged bush pig charging, tusks first, to save its young. The hunters exhibit real teamwork. One chimp may softly call to another when it has detected prey in the underbrush. They smile to one another. The victim is flushed out of its cover toward other chimps who are lying in wait. Escape routes are blocked off. Ambushes are refined. Plays are called. The chimps—so passionate after the kill—were coolly planning it all out beforehand.

———

In densely forested habitats, the territory controlled by a given chimpanzee group is only a few kilometers wide. In sparsely wooded regions, it can be as much as thirty kilometers across. These are the territories that a chimp group considers its turf, its home, its fatherland or motherland, to which something like patriotic sentiments are owed. It is not to be trespassed by strangers. It's a jungle out there. The typical day range of a chimp combat patrol is a few kilometers. So if they live in heavy forest, they can fairly readily patrol a good portion of the border in a single day. But if the vegetation and food supply are more sparse and their territory accordingly larger, it may be a few days' journey from one end to the other, and longer if they go around the perimeter.

> A patrol is typified by cautious, silent travel during which the members of the party tend to move in a compact group. There are many pauses as the chimpanzees gaze around and listen. Sometimes they climb tall trees and sit quietly for an hour or more, gazing out over the "unsafe" area of a neighboring community. They are very tense and at a sudden sound (a twig cracking in the undergrowth or the rustling of leaves) may grin and reach out to touch or embrace one another.
>
> During a patrol the males, and occasionally a female, may sniff the ground, treetrunks, or other vegetation. They may pick up and smell leaves, and pay particular attention to discarded food wadges, feces, or abandoned tools on termite heaps. If a fairly fresh sleeping nest is seen, one or more of the adult males may climb up to inspect it and then display around it so that the branches are pulled apart and it is partially or totally destroyed.
>
> Perhaps the most striking aspect of patrolling behavior is the silence of those taking part. They avoid treading on dry leaves and rustling the vegetation. On one occasion vocal silence was maintained for more than three hours . . . [When] patrolling chimpanzees return once more to familiar areas, there is often an outburst of loud calling, drumming displays, hurling of rocks, and even some chasing and mild aggression between individuals . . . Possibly this noisy and vigorous behavior serves as an outlet for the suppressed tension and social excitement engendered by journeying silently into unsafe areas.[31]

In this description by Jane Goodall of a patrol at Gombe, we are taken by the ability of the chimps to overcome their fear, to exercise self-control by inhibiting their usual noisy interchanges, but particularly by their deductive abilities. These chimps are tracking. They are weighing the evidence of branches, footprints, droppings, artifacts.

We might expect that, when food is in short supply, group differences in tracking skills help determine who lives and who dies. Not just strength and aggressiveness are being selected here, but something akin to reasoning and quick-wittedness. And stealth. When one human who lived with a troop for a long time tried to accompany a patrol as it set out, they looked at him reprovingly. He was just too clumsy. He could not, as they do, slip silently through the forest.

So the long-range combat patrol wends its way toward the borders of their turf. If it's more than a day's walk, they'll set up camp at night and continue on their way tomorrow. What happens if they encounter members of another group, Strangers from the adjacent territory? If it's just one or two intruders, they'll attempt to attack and kill them. There's much less disposition here toward threat displays and intimidation. But if two parties of roughly equal strength encounter one another, now there are a great many threat displays, rocks and sticks are thrown, trees are drummed. "Somebody hold me back, I'm gonna break his knees," you can almost hear them saying. They practice threat assessment: If the patrol senses an obviously larger number of Strangers, it is likely to beat a hasty retreat. At other times chimp patrols may penetrate enemy territory or even raid its populated core area—for many purposes, including copulating with unfamiliar females. The combination of tracking, stealth, danger, teamwork, fighting hated enemies, and the opportunity for sex with strange females is enormously attractive to the males.

The delight shown by the members of a patrol after having successfully returned from dangerous—perhaps enemy-held—territory is little different from what happens when chimps unexpectedly encounter a substantial cache of food. They screech, kiss, hug, hold hands, pat one another on the shoulders and the rump, and jump up and down. Their camaraderie is reminiscent of teammates in mutual embrace just after winning the national title. At the start of a heavy rain, male chimps often perform a spectacular dance. On coming upon a stream or waterfall they display, seize vines, swing from one tree to another, and cavort high above the water in a breathtaking acrobatic performance that may last for ten minutes or more. Perhaps they are awed by the natural beauty or entranced by the white noise. Their evident joy sheds a revealing light on the eighteenth-century doctrine [32] that humans are right to enslave other animals because we are unmatched in our capacity to be happy.

The prescription offered by Sewall Wright for a successful evolutionary response to a changing environment closely matches many aspects of chimp society. The species is divided into free-ranging groups, generally comprising between ten and one hundred individuals. They have different territorial ranges, so that if the environment alters the impact will be at least slightly different from group to group. A staple food at one end of a vast tropical forest may be a rare delicacy at the other. A blight or infestation that might result in serious malnutrition or famine for chimps in one part of the forest might have negligible consequences in another. Each territorial group is enough inbred that the gene frequencies differ systematically, group to group. And yet the pattern of inbreeding is relieved by exogamy (outbreeding). There are key sexual encounters with chimps from adjacent territories, initiated either when a patrol penetrates into alien territory or when a foreign female wanders over. These unions provide genetic communication, group to group, so that if in an adaptive crisis one group were more fit than the others, the adaptation would rapidly spread to the entire chimpanzee population through a sequence of sexual contacts—perhaps hundreds of copulations in a chain linking the remotest groups of a vast tropical forest. If there's a modest environmental crisis, the chimpanzees are ready.

If this is indeed the explanation, at least in part, of the territoriality, ethnocentrism, xenophobia, and occasional exogamy that characterize chimpanzee society, we do not imagine that individual chimps understand the reasons for their behavior. They simply can't stand the sight of strangers, find them hateful and deserving of attack— except, of course, for the chimps of the opposite sex, who are unaccountably exciting. The females occasionally run away with strange males, no matter what crimes they may earlier have committed against their land and kin. Perhaps they feel something of what Euripides makes Helen of Troy feel:

> What was there in my heart, that I forgot
> My home and land and all I loved, to run away
> With a strange man? . . .
> > Ah, husband still, how shall thy hand be bent
> To slay me? Nay, if Right be come at last,
> What shalt thou bring but comfort for pains past,

And harbour for a woman storm-driven:
A woman borne away by violent men . . .[33]

Mothers know who their sons are and so can preferentially resist their (very rare) sexual advances. But fathers are not sure who their daughters are, and vice versa. Thus, when a female comes of age in a small group, there's a significant chance of an incestuous union, further inbreeding, more infant mortality, and fewer of their genetic sequences passed on to future generations. So around the time of first ovulation, a female often feels an inexplicable urge to visit the neighboring territory. This can be a dangerous undertaking, as she probably understands full well. The compulsion, then, must be strong, which in turn underscores the evolutionary importance of her mission. Combine this not uncommon itch to wander at first ovulation with the rarity of brother-sister and, especially, mother-son unions and it's clear that a high-priority, well-functioning incest taboo is operating among the chimps.

There's one aspect of chimpanzee territoriality not shared by other apes—all of whom are divided into territorial, xenophobic groups, with a little exogamy thrown in: Unlike encounters within the group, where bluff and intimidation play major roles and only rarely does anyone get seriously hurt, when two chimp groups interact there can be real violence. No main force combat has ever been observed. They prefer guerrilla tactics. One group will pick off the members of the other in ones and twos until there's no longer a viable force left to defend the adjacent territory. Chimpanzee groups are constantly skirmishing to see if it's possible to annex more turf. If the penalty for failure in combat is death for the males and alien sexual bondage for the females, the males soon find themselves caught up in a powerful selection for military skills. Genes for those skills must have been racing through the tropical forests, by exogamous mating, until nearly all chimps had them. If they didn't, they died.

Moreover, the skills that make you good on patrol and good in skirmishes also make you good in the hunt. If your combat skills are honed, you can also supply your friends, loved ones, and concubines —to say nothing of yourself—with more of that delicious red meat. Except for the part about the good eating, being a male chimp is a little like being in the army.

Chapter 16

LIVES OF THE APES

I hear the apes howl sadly
In dark mountains.
The blue river
Flows swiftly through the night.

MENG HAU-RAN
(Tang Dynasty, early 730s),
"Written for Old Friends in
Yang-jou City While Spending the
Night on the Tung-lu River"[1]

The alpha male is sitting bolt upright, jaw set, staring confidently into middle distance. The hair on his head, shoulders, and back is standing on end, which gives him an even more imposing aspect. Before him crouches a subordinate, in a bow so deep that his gaze must be fixed on the few tufts of grass directly before him. If these were humans, this posture would be recognized as much more than deference. This is abject submission. This is abasement. This is groveling. The alpha's feet may, in fact, be kissed. The supplicant could be a vanquished provincial chieftain at the foot of the Chinese or Ottoman Emperor, or a tenth-century Catholic priest before the Bishop of Rome, or an awed ambassador of a tributary people in the presence of Pharaoh.[2]

Calm and assured, the alpha male does not scowl at his nearly prostrate subordinate. Instead, he reaches out and touches him on the shoulder or head. The lower-ranking male slowly rises, reassured. Alpha ambles off, touching, patting, hugging, occasionally kissing those he encounters. Many reach out their arms and beg for contact, however brief. Almost all—from highest to lowest rank—are visibly buoyed by this king's touch. Anxiety is relieved, perhaps even minor illnesses cured, by the laying-on of hands.

Regal touching, one after the other, in a sea of outstretched hands seems familiar enough to us—reminiscent, say, of the President striding down the central aisle of the House of Representatives just before the State of the Union address, especially when he's riding high in the polls. The future King Edward VIII on his world tour, Senator Robert Kennedy in his presidential campaign, and countless other political leaders have returned home black and blue from the grasp of their enthusiastic followers.

The alpha male will intervene to prevent conflict, especially between hotheaded young males pumped up on testosterone, or when aggression is directed at infants or juveniles. Sometimes a withering glance will suffice. Sometimes the alpha will charge the pair and force them apart. Generally, he approaches with a swagger, arms akimbo. It's hard not to see here the rudiments of government administration of justice. As in all primate leadership positions, an alpha male must accept certain obligations. In return for deference and respect, for sexual and dining privileges, he must render services to the community, both practical and symbolic. He adopts an impressive demeanor, even something approaching pomp, in part because his subordinates demand it of him. They crave reassurance. They are natural followers. They have an irresistible need to be led.

Beyond the reaching out of hands there are many styles of submission, of which the most common is, in the scientific literature, demurely called "presenting." What is it that's being presented? The subordinate animal, male or female—but here we're discussing males in the dominance hierarchy—wishing to pay its respects to the alpha male crouches down and elevates its anogenital region toward the leader, moving its tail out of the way. It sometimes gives a little bump and grind. It may whimper and, grinning over its shoulder, approach the alpha, raised rump first. The subordinate's need to pay respect in this manner is so great that it may even present to an alpha who's fast asleep.

The alpha (if awake) moves forward, grasps the submissive animal from behind, closely embraces it, and not infrequently makes a few pelvic thrusts. Since this is the invariable posture and motion of chimp copulation, there can be no mistaking the symbolic significance of the exchange: The subordinate animal asks please to be fucked, and the dominant animal, perhaps a little reluctantly, complies.

In most cases these actions are only symbolic. There is no intromission and no orgasm. They fake it. You wish to pay respect to a high-ranking male, but Nature has not equipped you with appropriate spoken language. Still, there are many postures and gestures in your everyday life that have a meaning readily apprehended by everyone. If females must comply with nearly every proferred sexual invitation, the sex act itself is a vivid, powerful, and unambiguous symbol of submission. Indeed, presenting is a mark of deference and respect among all the apes and monkeys, and among many other mammals as well.

The anger of a high-ranking male is fearsome. His arousal is obvious to any bystander, because all the hair on his body is standing on end. He may charge, intimidate, and tear branches from trees. If you're not prepared to meet him in single combat, you might want to appease him, to keep him happy. You closely monitor the slightest raising of a single one of his hairs. Not only are you perpetually compliant ("I'm yours whenever you want me"), but just for your own comfort you need frequent reassurance that he's not angry with you. When he *is* angry, he exaggerates his size and ferocity and displays the weapons that he will bring to bear if the adversary does not submit. He uses his displays to keep more junior males in line, and they use theirs to advance within the hierarchy. Displays may serve as a response to a challenge, or just as a general reminder to the community at large that here's someone not to be trifled with. Of course, it's not all bluff; if it were, it wouldn't work. There must be a credible threat of violence. A kind of menace maintenance is required. If push comes to shove there may be serious fighting. But much more often the display has a ritual and ceremonial character. (Almost always the alpha wins, and if, on occasion, he loses, that doesn't usually mean that the hierarchy has been inverted; for that to happen, a consistent pattern of defeat is needed.)

The lesson being communicated is deterrence, pure and simple: "Cross me and you'll have to deal with this stature, these muscles, these teeth (note my canines), this rage." Chimpanzee strategy is encapsulated in the earliest comprehensive account we have of human military affairs, the sixth-century B.C. work *The Art of War,* by Sun Tzu: "The supreme act of war is to subdue the enemy without fighting."[3] Deterrence is *old.* And so is its prerequisite, imagination.

So law and order are maintained, and the status of the leadership preserved through the threat (and, if necessary, the reality) of violence; but also through patronage delivered to constituents, and through the widespread craving to have a hero to admire, who can tell you what to do—especially when there's a threat from outside the group. Violence and intimidation alone would not suffice—although there may be those who enjoy being chastised and bullied, who perhaps look on it as a form of affection.

Male chimps are obsessively motivated to work their way up the dominance ladder. This involves courage, fighting ability, often size, and always real skill in ward-heeler politics. The higher his rank, the fewer the attacks on him by other males and the more gratifying instances of deference and submission. But the higher his rank the more he will be obliged to take pains to reassure subordinates. The dominance hierarchy makes for a stable community not only because the high-ranking males break up fights among their subordinates, but also because the very existence of hierarchy, along with the genetic tradition of compliance, inhibits conflict. One powerful motivation to be high-ranking is that the top echelons often have preferential sexual access to ovulating females. As in all mammals, this behavior is mediated by testosterone and related steroid hormones. Leaving more offspring is what natural selection is about. For this reason alone, hierarchy makes evolutionary sense.

The alpha male, merely by virtue of his exalted status, stimulates the formation of cabals to depose him. A lower-ranking male may challenge the alpha by bluff, intimidation, or real combat, as a step towards reversing their relative status. Especially under crowded conditions, females can play a central role in encouraging and helping to implement coups d'état. But the alpha male is often prepared single-handedly to take on coalitions of two, three, or four opponents.

Alphas enforce authority; betas and others sometimes challenge it —not on abstract philosophical grounds, but as a means to selfish ends. We might guess that both warring inclinations are built into us too, a different balance in different people, with much depending on the social environment. The roots of tyranny and freedom trace back to long before recorded history, and are etched in our genes.

Over a period of years in a typical small chimpanzee group, half a dozen different males may become alpha in succession—because of death or illness of the dominant male, or because of challenges from below. On the other hand, an alpha male maintaining his status for a decade is not unknown either. Perhaps coincidentally, these terms of office are roughly those typical of human governments—ranging, respectively, from Italy, say, to France. Political assassination—that is, dominance combat in which the loser dies—is rare.

In combat, males are more likely to hit, kick, stomp, drag, and wrestle. Or throw stones and beat with clubs, if any are handy. Females are more likely to pull hair and scratch, and to grapple and roll. For all their baring of teeth, males rarely bite anyone in the group, because their canines can do terrible damage. They may flash the razors and switch-blade knives, but they hardly ever draw blood. Females with much less prominent canines have fewer inhibitions. Any given fight is likely to stimulate other fights among unrelated or even nonaligned parties. One combatant may poignantly appeal for aid from passersby, who may, in any case, be attacked for no apparent reason. Any conflict seems to raise the testosterone level in all the male bystanders. Everyone's hair stands on end. Perhaps long-standing resentments flare. General mayhem often results.

Chimps will place their fingers between the teeth of a high-ranking male and derive reassurance when the fingers are returned intact. At times of rising group tension, male chimps may touch or heft each other's testicles, as the ancient Hebrews and Romans are said to have done upon concluding a treaty, or testifying before a tribunal. Indeed the root of "testify" and "testimony" is the Latin word, *testis*. The significance of the gesture, less common now that men wear pants, is not only transcultural, but trans-species.

——

From infancy, chimps are groomed, chiefly by their mothers. They in turn clutch their mothers' fur from the moment of their birth. The infant revels in the physical contact, deriving deep and long-term psychological benefits from it. Even if their physical needs are attended to, monkeys and apes that, as infants, don't receive something like hugging and grooming, grow up to be socially, emotionally, and sexually incompetent. As the infant matures, grooming behavior is slowly transferred to others. Most adults have many grooming partners. In a grooming pair, one partner mainly does, the other is mainly done to. But even the alpha will play either role. One individual will sit serenely while the other combs through its hair, rubs all its parts, and occasionally finds a parasite (a louse or a tick—maybe getting ripped on butyric acid), which it promptly eats. Sometimes the chimps hold hands the whole time. Jittery full-grown males will return to their mothers to be groomed and reassured. Males who become irritable with one another often hastily repair to mutual grooming to calm each other down. It may have been selected for long ago, as an improvement in chimp hygiene and public health, but grooming has now become a centrally important social activity, probably lowering testosterone and adrenaline titers.

The closest human counterpart may be the back rub or the body massage, which have been raised to art forms in cultures as diverse as modern Japan and Sweden, Ottoman Turkey and Republican Rome —where, in characteristic human fashion, a specialized tool, the strigil, was employed to rub the back. Gentlemen in Restoration England idled away the hours by collectively combing their wigs. Where body lice are a problem, human parents carefully and routinely go through their children's hair. The emotional power of being groomed by the alpha male is perhaps akin to the laying-on of hands by shamans, healing ministers, chiropractors, charismatic surgeons, and kings.

Despite the importance of the male dominance hierarchy, it is by no means the only important chimp social structure, as the grooming pairs indicate. A mother and her children, or two grown siblings, have special, lifelong, mutually supportive bonds. A high-ranking son may be to a mother's social advantage. There are also long-term relationships between unrelated individuals of the same sex that might certainly be called friendships. Largely outside the male hierarchy, there's an intricate set of female bonds that often depend on the number and status of relatives and friends. These extrahierarchical alliances provide important means of mitigating or reordering a dominance hierarchy: If the alpha male is undefeated in one-on-one confrontation, an alliance of two or three lower-ranking males with supporting females may conceivably put him to flight. High-ranking males are known to establish alliances with promising younger males, perhaps co-opting them to prevent future putsches. Occasionally females will step in to defuse a tense encounter.

Alliances are made and broken. Loyalties shift. There is courage and devotion, perfidy and betrayal. No dedication to liberty and equality is evident in chimpanzee politics, but machinery is purring to soften the more hard-hearted tyrannies: The focus is on the balance of power. Frans de Waal writes:

> The law of the jungle does not apply to chimpanzees. Their network of coalitions limits the rights of the strongest; *everybody* pulls strings.[4]

In this complex, fluid social life great benefits accrue to those skilled in discerning the interests, hopes, fears, and feelings of others. The alliance strategy is opportunistic. Today's allies may be tomorrow's adversaries and vice versa. The only constant is ambition and fixity of purpose. Lord Palmerston, the nineteenth-century British Prime Minister—who described his nation's foreign policy as no permanent national alliances, only permanent national interests—would have been right at home among the chimps.

Males have special reasons to avoid permanent rivalries. In the hunt and in patrols into enemy territory, they rely on one another. Mistrust would endanger their effectiveness. They need alliances to work their way up the promotion ladder or to maintain themselves in power. So, while males are much more aggressive than females, they are also much more highly motivated toward reconciliation.

When Calhoun crowded his rats together he found a wholesale change in their behavior, almost as if their collective strategy was now to kill off enough of themselves and to lower the birth rate enough that the population in the next generation would be reduced to manageable numbers. Given all the chimp propensities that we've chronicled (and the fact, described in the next chapter, that baboons can go into a murderous, annihilating group frenzy when packed together), you might expect that chimps behave badly when overcrowded, as in zoos. In close confines a male chimp cannot escape from an attack, cannot lead a female into the bushes away from the controlling gaze of the alpha male, cannot enjoy the excitement of the hunt or the patrol or contact with females from adjacent territories. You might expect frustration levels to rise, and hierarchical encounters now to involve less bluff and more real combat. If you're not ready for a fight to the death, you'd better, you might think, find some way to mollify, appease, show deference, pay your respects, perform services, be useful—and genuflect at every step so the alpha harbors no possible misgivings about whether you know your place.

Surprisingly, just the opposite is true. In zoo after zoo, males—and especially high-ranking males—exhibit a degree of measured restraint under crowded conditions that would be unthinkable if they were free. Imprisoned chimps are much more likely to share their food. Captivity somehow brings forth a more democratic spirit. When jammed together, chimps make an extra effort to get the social machinery humming. In this remarkable transformation it is the females who are the peacemakers. When, after a fight, two males are studiously ignoring one another as if they were too proud to apologize or make up—it is often a female who jollies them along and gets them interacting. She clears blocked channels of communication.

At the Arnhem colony in the Netherlands, every adult female was found to play a therapeutic role in communication and mediation among the petulant, rank-conscious, grudge-holding males. When real fights were about to break out and the males began to arm themselves with rocks, the females gently removed the weapons, prying their fingers open. If the males rearmed themselves, the females disarmed them again. In the resolution of disputes and the avoidance of conflict,* females led the way.[5]

So, it turns out that indeed chimps are not rats: Under crowded conditions they make extraordinary efforts to be more friendly, to be slower to anger, to mediate disputes, to be polite—and the female role in calming the testosterone-besotted males is crucial. This is an important and encouraging lesson about the dangers of extrapolating behavior from one species to another, especially when they are not very closely related. Since humans are much more like chimps than like rats, we can't help wondering what would happen if women played a role in world politics proportionate to their numbers. (We're not talking about those occasional women Prime Ministers who have risen to the top by besting the men at their own games, but about proportional representation of women at all levels of government.)

———

Students of the chimpanzee call it "courtship." It's a set of ritualized gestures by which the male signals to the female his sexual intentions. But in ordinary usage courtship is a word describing a patient human attempt, over long periods of time, and often with great gentleness and subtlety, to build trust and to create the foundations for a long-term relationship. The male chimpanzee's courtship communication is much briefer and more to the point, much closer to "Let's fuck." He may swagger, shake a branch, rustle some leaves, fix her with his stare, and reach out an arm toward her. His hair will be erect. And not just his hair. An erect penis—bright red, contrasting vividly with his black scrotum—is an invariable part of chimpanzee "courtship,"

* Among the males. Among their own gender, females may carry grudges for years, and refuse to be reconciled.

which you might think is a good thing because most of the other symbolic desiderata of courtship are barely distinguishable from those used in intimidating other males. In chimpish, "Let's fuck" sounds almost exactly like "I'm gonna kill you." The significance of this similarity has not been lost on the females. They comply. A typical female rejection rate to an unrelated male's sexual overture is about 3%.

In chimpanzee etiquette, the correct response to the male courtship display is to crouch down on the ground and lift your behind invitingly. If the social niceties should elude you at first, the male will shortly set you straight. Recalcitrant females are attacked. All males in the group expect sexual access to all females, subject to necessary exclusions enforced by jealous, higher-ranking males. (Adolescent females are available for copulation even to infant males, who are sometimes ardent lovers.) Again, a significant exception is mothers and sons; although the son may give it a try, the mother tends to resist vigorously.

It's natural for us to think of the instant submission and compliance of these female apes as exacted under threat of bodily harm, as rape pure and simple, even if the female is not bitten or bruised. But this cannot be the whole story, because female primates raised alone will, on going into first estrus, present themselves readily to many passing males, to humans, and, occasionally, even to furniture. Not just some degree of compliance is hardwired and built in, but so is real sexual enthusiasm. As in the hamsters-in-motorcycle-jackets experiment, the females, if given a chance, often show a marked preference for the higher-ranking males: The Big Guy, he's all right. Perhaps also the males present themselves to those of higher rank not so much as a humiliating means of social advancement but because they genuinely enjoy submission.

As with most animals, the chimp male enters the female's vagina from behind. Often the male is in a crouching or seated position, with his hands on her waist or buttocks as she positions herself on him. To a human observer their faces are strangely expressionless. Much has been made about the difference between chimp and human sexual practices—almost certainly in an attempt to deny the closeness of the kinship. But the favorite ancient Roman sexual practice was chimplike, the male seated on a small stool and the female, often her back to his front, settling herself down on him. The style of our hunter-gatherer ancestors (if we may judge from contemporary examples) is also more like the chimps: They are often recumbent on their sides, the male embracing the female from behind. As a fashionable human sexual practice, perhaps the "missionary position" is not much older than missionaries—although, as we'll see later, there's one other animal that adopted it long before they did.

By human standards chimp sexual life is a perpetual open-air orgy —compulsive, unending, and always with the male grasping the female from behind. The average copulation rate is one or two an hour. Every hour. For each mature chimp. In estrus, of course, it's more. When the females are ovulating and able to be impregnated, their vulvas and allied nether parts swell extravagantly and turn bright pink.* In estrus, they're walking sexual advertisements, and are then far more alluring. Because estrous periods are to some degree synchronized, there are times when a chimpanzee group is a sea of bobbing, compliant, soliciting swollen red rumps. Olfactory cues also signal their sexual availability. In marginal cases a passing male, unable to determine just by looking if she's ovulating, may simply insert his finger into her vulva and take a sniff.

Chimpanzee sex isn't a long and drawn-out business. Maybe eight or nine thrusts, each taking less than a second, and they're done. The males have, by human standards, impressive recovery rates, including documented sequences of many ejaculations at five-minute intervals. Females in estrus are especially attractive in the early morning, probably because of the long and stressful celibacy imposed on the males by the necessity of having to sleep at night. As a kind of community property for the males, she may be taken every ten minutes by one male after another through mid-morning, by which time they may tire a little.

Occasionally a heroic or foolish female will refuse the male despite his transfixing stare, threatening gestures, and other signs of arousal. When he makes his approach she may scream and run away from him. Generally she doesn't get far. When some hesitation is discerned, young males will ostentatiously search for a rock, or actually find one and make as if to throw it at her. This serves almost always as a

---

* That this might have something to do with sex was first proposed, in the face of considerable Victorian skepticism and unease, by the ever-insightful Charles Darwin.[6]

convincing argument. One of the earliest studies of chimp sexual behavior suggested that female compliance occurs "by reason of the dominance or impulsiveness of the male and the desire of the female to avoid risk of physical injury by obeying his command."[7]

Despite their apparently unrestrained sexual behavior, chimps get jealous. A male who rejected the solicitation of a female in estrus, but instead copulated with her daughter, was slapped in the face by the outraged mother. Cruising migrant females from the next territory are threatened or attacked by the local females—especially if the visitors go so far as to groom with one of the resident males. The male may also blaze with sexual jealousy over a particular female's behavior—but, almost without exception, only when she is vividly pink and swollen and able to conceive. High-ranking males will then chase away aroused lower-ranking males. Although it's unlikely he's thought this out, his motive, it seems very clear, is to monopolize her around the time of ovulation so that no one but he can father her children.* As far as he's concerned, the rest of the time she can do as she pleases.

Possessiveness is hard to maintain, though, at the core of the territory where the chimp population density is high. Even the most vigilant and high-ranking males will be distracted—by hunting, say, or challenges from lower ranks, or insufficient deference, or by grooming, or by the necessity of adjudicating disputes. And during such an intervention—it may last only a few minutes—other males, patiently awaiting their chance, pounce on the off-limits female, especially if she's in estrus. Kleptogamy is on their minds. In zoos a female will, as soon as the alpha male is removed from her cage, present herself to lower-ranking males, even if this requires adroit positioning so the act can be performed through the bars of two adjacent cages. Both in the wild and in captivity, when the cuckolded male discovers what has happened, he attacks the female. Perhaps he knows that she was all too willing. Besides, it's much safer than attacking a rival male.

Even when the alpha is present, a subordinate male may catch the eye of a female who strikes his fancy and then gaze pointedly toward

* Similar behavior is known among other social animals—in gorillas, for example, where the alpha permits a female to mate with lower-ranking males, but only if she's pregnant. Among wolves, only the alpha male and the alpha female breed, but the female mates with other members of the pack when she's not in heat.[8]

some nearby bushes. Nonchalantly, he then ambles off, often followed after a discreet interval by the female. Sometimes their infidelity is observed. Motivated by jealousy or by the wish to ingratiate himself to the leader, the informer rushes up to the alpha in great excitement, takes his arm, points, and leads him to the treacherous couple. At other times the female may inadvertently reveal what is going on by uttering a high-pitched scream at the moment of her orgasm. After being discovered in this way more than once, females do not usually abandon the risky practice of clandestine rendezvous; instead, they learn to suppress the scream, converting it into a kind of husky pant.

Frans de Waal reports that, following a long grooming session between a high-ranking and a low-ranking male,

> a subordinate male may invite the female and enjoy a copulation without interference by the others. These interactions give the impression that males obtain "permission" for an undisturbed mating by paying a price in grooming currency . . . Perhaps sexual bargaining represents one of the oldest forms of tit for tat, one in which a tolerant atmosphere is created through appeasing behavior.[9]

To achieve reliable sexual monopoly during her estrus, the ardent male must usher the female away from the multitude. Scientists who study chimps call this "consortship," and distinguish it from "courtship." The proposition is put to the female as follows: He takes a few steps away and looks at her over his shoulder. If she does not instantly follow, he shakes a nearby branch. If this provides insufficient inducement, he will chase her and, if need be, attack her. More often she goes quietly, especially if he's high-ranking. Then, off somewhere alone in the forest, he has her to himself. It is a distant intimation of monogamy.

Consortship typically lasts for weeks, and is not without its perils. The happy couple may be attacked by predators or patrols from the neighboring territory; and the male's status in the dominance hierarchy may be undergoing active review during his absence. Jane Goodall reports a few cases in which the young female's mother invites herself along on the consortship; "as far as the male is concerned," she is a "most unwelcome chaperone." Here, where conception is most likely, the incest taboo is particularly vivid—no case is known of a male chimp ever inviting his own mother or sister to be his consort.

Why do the females put up with all this? Certainly males are larger and stronger than females and can and will hurt them, if that's what's needed to get their way. But this is only in one-on-one interactions. Why don't females band together to defend themselves against a sexually predatory male? If two or three aren't enough, six or eight would be. This is known, but rare, in the wild. (It is the custom among the chimps in the Tai National Forest in the Ivory Coast.) But it's more common when they're in closer quarters, as in the Arnhem colony in the Netherlands. Here the social conventions are different. If a male solicits a female and she's uninterested, she so indicates, and that, usually, is that. If he makes himself obnoxious, he may be attacked by one or more other females. It is astonishing that so striking a characteristic of chimpanzee life in the wild as male sexual oppression of females can to such an extent be reversed merely because they're all crowded together in a minimum security prison. We've already seen how, under these conditions, restraint, coalition building, and peacemaking by females come to the fore. Societies in which females have something approaching equality are also societies that benefit from their political skills.

In a state of freedom—where it's possible to avoid your rivals by taking your sweetheart on a little trip into the country, and where you can escape a bully by running away—the circumspection required in crowded conditions is relaxed. Here testosterone is at full throttle and gentlemanly behavior is uncommon. The primate expert Sarah Blaffer Hrdy[10] speculates that, among wild chimpanzees, female compliance to male sexual demands is the single mother's desperate strategy for safeguarding her children. The males, Hrdy proposes, nursing their resentment at any rejection, might attack the children of an unresponsive mother (perhaps at a later time), or at least not protect them against attack by others.* In the brutal world of the chimpanzee, she suggests, the female does what the males ask in order to bribe

---

* This is not just an unpleasant circumstance of chimpanzee life; it occurs among gorillas, baboons, and many other apes and monkeys. Over a fifteen-year-long study of gorillas near the Virunga volcano in Rwanda, more than a third of all infant mortality was directly due to killing by gorilla males. Infanticide for them is a way of life.[11]

them, so they will not kill (and, who knows, if they're in a good mood might even help save) her children.* If Hrdy is right, perhaps the males are not oblivious of the bargain struck. Do they threaten the children *in order* to make the mothers come around? Do they attack children at random as a cautionary lesson for any mothers toying with noncompliance? Have chimp males organized a protection racket, with the females and the young as their victims?

Let's leave aside the possibility of conscious extortion, and think for just another moment about Hrdy's speculation. The females don't provide food for the males. They don't seem to be any better at grooming than the males. Perhaps the only commodity—certainly the most valuable commodity—they can offer to protect their children is their bodies. So they make the best of a desperate situation. Now a male is less likely to attack and more likely to protect her baby. But when circumstances change, when aggression is inhibited because of crowding, the females can finally say "No"—without having their heads handed to them for it.

Again, we must not imagine that chimps think all this through. They must have some other, more immediate reinforcement of their behavior. Hrdy raises the question of the selective advantage of orgasms, especially multiple orgasms, among female apes and humans. In a monogamous couple, what evolutionary benefit does it confer? she asks, and argues that none is apparent. But if instead we imagine the female copulating with many males in order that none of them harm her offspring, then, Hrdy conjectures, the orgasm—reinforcing successive matings with many partners—plays a vital role.

To what extent female sexual compliance is coerced by the males and to what extent it is entered into voluntarily and exuberantly is still not clear.

———

* Something similar is observed in other, quite different non-monogamous species—for example, hedge sparrows. The alpha male works hard to prevent copulation by betas, but only in the females' fertile period. However the female, even in the fertile period, may dart away on occasion for surreptitious matings with the betas. Only in this case will a beta help feed her chicks. Again the females are using male preoccupation with sex to induce them to help her little ones.[12]

Nucleic acids compete, individual organisms compete, social groups compete, perhaps species compete. But there is also competition on a very different level: Sperm cells compete. In a single human ejaculation there are some 200 million sperm cells, the fittest among them with tails lashing, racing against each other, speeding along at an average clip of five inches per hour, each striving—or so it seems—to be first to reach the egg. A surprising number, though, from normal, fertile males have deformed heads, multiple heads or tails, kinked tails, or are just motionless, dead in the water. Some swim straight, others in convoluted paths that may turn back on themselves. The egg may actually choose among sperm cells. Chemically, it cries out to them, egging them on. Sperm cells are equipped with a sophisticated array of odor receptors, some oddly similar to those in the human nose. When the sperms obediently arrive in the vicinity of the calling egg, they don't seem to have sense enough to stop swimming and thrashing, and molecules on the egg's surface may cast out a kind of fishing line, hook the sperm, and reel it in. The fertilized egg then promptly establishes a barrier that turns away all future sperm cells who may come blundering in. These modern findings are rather different from the conventional view of the passive egg waiting to be claimed by the champion sperm.[13]

But there is, in an ordinary impregnation, something like one success and 200 million failures. So conception, while controlled to a significant degree by the egg, is still in part the result of a competition among sperm cells for speed, range, trajectory, and target recognition, at least.*

Odds anywhere approaching 200-million-to-1 in every conception, continued once a generation through geological ages, imply an extremely strong selection of sperm. Leaner, more streamlined sperm cells with more swiftly lashing flagellas that can swim straight and that have superior chemical sensors will probably arrive first; but that has very little to do with the characteristics, once grown up, of the individual so conceived. Getting to the egg first with genes for

---

* A sperm that carries the smaller Y chromosome—the one that makes a male—weighs slightly less than one that carries the bigger X chromosome that makes a female; if lighter sperms travel faster, this may be why slightly more males are conceived than females.

boorishness, say, or stupidity, seems a dubious evolutionary benefit. A great deal of effort would appear to be squandered in natural selection among the sperm cells.[14] But then it seems odd that so many sperm cells are dysfunctional. We do not understand why this should be.

Many other factors affect which sperm succeeds: Who's conceived must depend on the progress of the egg into the fallopian tubes, the precise moment of ejaculation, the position of the parents, their rhythm of motion, subtle distractions or encouragements, cyclical hormonal and metabolic variables, and so on. At the heart of reproduction and evolution, again we find a surprisingly strong random component.

The monkeys and apes are preeminent among animals where many males mate, one after the other, with the same female. They can hardly contain themselves, jumping up and down with excitement, awaiting their turn. In chimpanzees, as we've noted, there may be dozens of copulations in quick succession with an ovulating female. So the act itself cannot be prolonged or rich in nuance. Several pelvic thrusts, roughly one a second, and it's over. For an average male there's a copulation maybe once an hour, all the livelong day. For females in estrus it's much more than that.

In ten or twenty minutes many males may have copulated with the same female. So consider the sperm cells of these various male chimps, racing against one another. Essentially, they set out from the same starting line. The probability of insemination by a given male is proportional to the number of sperm cells delivered, other things being equal; and thus the chimps with the largest number of sperm cells per ejaculation, the chimps able to copulate the most times in succession before exhaustion sets in, have an advantage. Having more sperm cells requires larger testicles. The very large testicles of male chimps amount to about a third of a percent of their entire body weight—twenty or more times the endowment, relatively speaking, of primates who are monogamous or who live in breeding units of one male and several females. In general it is found that males have considerably larger testicles for their body size in species where many males mate with each female. Not only is there selection for testicular volume, but also for an interest in copulation. This may be one of the routes—there are many mutually reinforcing trajectories, as we've described—to the highly sexual social proclivities of our primate order. Because men, compared to male chimps, have such relatively small testicles, we might guess that promiscuous societies were uncommon in the immediate human past. But a few million years ago, say, our ancestors may have been substantially more indiscriminate sexually and substantially better endowed.

———

A mother and her adult daughter who have been foraging separately for a few hours may merely look at each other and give a few grunts when they meet; but if they have been separated for a week or more, they are likely to fling their arms around each other with grunts or little screams of excitement, then settle down for a session of social grooming. [15]

Chimpanzee females and their young have deep bonds of affection, while the adolescent and adult males seem more often mesmerized by rank and sex. The young revel in rough-and-tumble play together. Infants whimper and scream if they find themselves out of sight of their mothers. Youngsters will come to the aid of their mother if she is being attacked, and vice versa. Siblings may show each other special, affectionate consideration throughout their lives, and take care of the young during childhood if—as is common—the mother has died before the children are grown. Occasionally chimps of either sex will endanger themselves to help others, even those who are not close relatives. Male bonding on a hunt or patrol is palpable. Clearly there are opportunities—especially when the testosterone titers are low—for civil, affectionate, even altruistic behavior in chimpanzee society.

Adult males, despite the dominance hierarchy, spend considerable time alone. After the birth of their first baby or two, most females spend their entire lives with others. So females are both required to develop more refined social skills and have more opportunity to do so. As is usual among monkeys and apes—with rare exceptions—only one child is born at a time. Except when they're in estrus, their time is spent mainly with the children. This is key for the next generation: As we've mentioned, apes and monkeys that are not regularly cared for, nursed, held, fondled, and groomed by an adult tend to become socially awkward, sexually inept, and disastrous as parents when they grow up.

Females are not born knowing how to be competent mothers; they must be taught by example. The investment of time required of the mother is substantial: The young are not weaned until they're five or six years old, and enter puberty around age ten. For much of the time until weaning they're unable to care for themselves. They're very good, though, in clutching their mother's hair as they ride upside-down on her belly and chest. So long as they allow the infant to nurse whenever it wants, perhaps several times an hour, chimp mothers are usually infertile, and unattractive to males. This is called "lactational anestrus." Without the males constantly hassling them for sex, they're able to spend much more time with the kids.

Chimp mothers use corporal punishment only very rarely. Infants learn the conventional modes of threat and coercion by closely observing older male role models. Infant males soon attempt to intimidate females. This may take some effort; females, especially high-status females, may not take kindly to being bullied by some young whippersnapper. The upstart's mother may help him in his efforts at intimidation. But before reaching adulthood nearly every male has obtained submission from nearly every female. Nursing male infants—including those still years away from weaning—routinely and successfully copulate with adult females. Adolescent males emulate adult males carefully (aping every nuance of their intimidation displays, for example), wish to be their apprentices and acolytes, are simultaneously nervous and submissive and hopeful in their presence. They're looking for heroes to worship. It even happens that an adolescent who's been cruelly attacked by an adult male will leave his mother and follow the aggressor everywhere, submission signals flashing, longing for acceptance at some future and glorious time.

———

From a human perspective chimpanzee social life has many nightmarish flourishes. And yet, despite its excesses, it's hauntingly familiar. Many spontaneous groupings of men are oriented around hierarchy, combat, blood sports, and loveless sex. The combination of dominant males, submissive females, differential but scheming sub-ordinates, a driving hunger for "respect" up and down the hierarchy, the exchange of current favors for future loyalty, barely submerged violence, protection rackets, and the systematic sexual exploitation of all available adult females, has some marked points of similarity with the lifestyles and ambiance of absolute monarchs, dictators, big-city bosses, bureaucrats of all nations, gangs, organized crime, and the actual lives of many of the figures in history adjudged "great."

The horrors of everyday life among the chimpanzees recall similar events in our history. We find humans behaving like chimps at their worst in endless succession in the daily press, in modern popular fiction, in the chronicles of the most ancient civilizations, in the sa-cred books of many religions, and in the tragedies of Euripides and Shakespeare. A summary of human nature based on the plays of Shakespeare would define "man," wrote Hippolyte Taine, as

> a nervous machine, governed by a mood, disposed to hallucinations, transported by unbridled passions, essentially unreasoning . . . and led at random, by the most determinate and complex circumstances, to pain, crime, madness, and death.[16]

We're not descended from chimps (or vice versa); so there's no necessary reason why any particular chimp trait need be shared by humans. But they're so closely related to us that we might reasonably guess that we share many of their hereditary predispositions—perhaps more effectively inhibited or redirected, but smoldering in us never-theless. We're constrained by the rules that, through society, we im-pose on ourselves. But relax the rules, even hypothetically, and we can see what's been churning and fermenting inside us all along. Beneath the elegant varnish of law and civilization, of language and sensibility—remarkable accomplishments, to be sure—just how dif-ferent from chimpanzees are we?

For example, consider the crime of rape. Many men find depictions of rape arousing—especially if the woman is portrayed as enjoying it despite her initial resistance. Most American high school and college students (of both sexes) believe that a man is justified in forcing a woman to have sex—at least when the woman behaves provoca-tively.[17] More than a third of American college men acknowledge some propensity to commit rape if they were guaranteed they could get away with it.[18] The percentage goes up if some euphemism such as "force" appears in the question instead of "rape." The actual risk of an American woman being raped in her lifetime is at least one chance in seven; almost two-thirds of the victims have been raped when they were minors.[19] Perhaps men in other nations are less fascinated with rape than Americans are; perhaps mature men, with lower testosterone titers, are less comfortable with rape than adolescents are.[20] But it would be hard to argue that there's no biological predisposition for men to rape.

While a range of causal factors have been proposed, most rapists turn out to be not slavering psychopaths, but ordinary men given the opportunity and acting on impulse,[21] sometimes repeatedly and compulsively. Some students of the subject see rape as a biological strategy (entered into without his conscious understanding) to propagate the rapist's genes;[22] others see it as a means for men (again largely unconsciously) to maintain through intimidation and violence their domination over women.[23] The two explanations do not seem mutually exclusive; and both seem to be operative in chimp society. Also, a significant minority of women are aroused by fantasies of rape, and, in one study, women who have been raped by an acquaintance seem disturbingly more likely to continue dating their assailants than those who were subjected only to attempted rape by an acquaintance.[24] This is at least reminiscent of the compliance pattern of female chimps.

Over a set of hereditary predispositions human society lays down a kind of stencil that permits some to be fully expressed, some partially, and some hardly at all. In cultures where women have roughly comparable political power with men rape is rare or absent.[25] However strong any genetic propensity toward rape might be, social parity appears to be a highly effective antidote. Depending on the structure of the society, many different brews of human proclivities can be elicited.

———

Chimpanzee society has an identifiable set of rules that most of its members live by: They submit to those of higher rank. Females defer to males. They cherish their parents. They care for their young. They have a kind of patriotism, and defend the group against outsiders.

They share food. They abhor incest. But they have, so far as is known, no lawgivers. There are no stone tablets, no sacred books in which a code of conduct is laid out. Nevertheless, there is something like a code of ethics and morals operating among them—one that many human societies would find recognizable and, as far as it goes, congenial.

Chapter 17

ADMONISHING THE CONQUEROR

Perhaps no order of mammals presents us with
so extraordinary a series of gradations as this
[step by step, from humans to apes to monkeys
to lemurs]—leading us insensibly from the
crown and summit of the animal creation down
to creatures, from which there is but a step, as
it seems, to the lowest, smallest, and least
intelligent of the placental Mammalia. It is as if
nature herself had foreseen the arrogance
of man, and with Roman severity had provided
that his intellect, by its very triumphs, should
call into prominence the slaves, admonishing
the conqueror that he is but dust.

T. H. HUXLEY
Evidence as to Man's Place in Nature[1]

The Archbishop of York is Primate of England. The Archbishop of Armagh is Primate of Ireland. The Archbishop of Warsaw is Primate of Poland. The Pope is Primate of Italy. The Archbishop of Canterbury is Primate of the planet, at least as far as his Anglican communicants are concerned. These ancient titles come from the medieval Latin word *primus*, which in turn derives from older Latin words meaning "principal" and "first." The ecclesiastic use was straightforward: A primate of a region was the chief ("first") of all its bishops. In recent centuries the title has devolved often to little more than an honorific. Other titles have taken precedence. But "Prime Minister" and "President" and "Premier" come from similar linguistic roots, all meaning "first."

When Linnaeus was drawing up the family tree of life on Earth he was, as we've noted, afraid to include humans among the apes. But despite widespread opposition, it was impossible to deny some deep connections of monkeys, apes, and humans.* So all were classified into the order (for him, one taxon higher than genus) that he called primates. Scientists who study non-human primates—of course, they're all primates themselves—are called primatologists.

This other meaning of "primate" also derives from the Latin for "first." It's hard to see by what standard a squirrel monkey, say, could be considered "first" among the lifeforms of Earth. But if a case is made that humans are "first," then the tarsiers, bushbabies, mandrills, marmosets, sifakas, aye-ayes, mouse lemurs, pottos, lorises, spider monkeys, titis, and all the rest are dragged in along with us. We're "first." They're our close relatives. So they, in some sense, must be "first" also—an undemonstrated and suspect conclusion in a biologi-

---

* Apes are bigger and smarter than monkeys, and lack tails. The apes are the chimpanzees, gorillas, gibbons, siamangs, and orangutans. The siamangs are about as closely related to gibbons as chimps are to humans.

cal world that runs from virus to great whale. Perhaps, instead, the argument goes the other way, and the humble status of most members of the primate tribe casts doubts on the lofty title we have appropriated to ourselves. It would make things so much easier for our self-esteem if those other primates weren't—anatomically, physiologically, genetically, and in their individual and social behavior—so much like us.

Surely there is at least a hint in the word "primate," not just of self-congratulation, but of the idea, fully realized in the practices of our own time, that we humans arrogate command and control of all life on Earth into our own hands. Not *primus inter pares*, first among equals, but just plain *primus*. We've found it convenient, even reassuring, to believe that life on Earth is a vast dominance hierarchy—sometimes called "The Great Chain of Being"—with us as the alphas. Sometimes we claim that it wasn't our idea, that we were commanded by a Higher Power, the most Alpha of Alphas, to take over. Naturally, we had no choice but to obey.

About two hundred species of primates are known. Conceivably, in the quickly dwindling tropical rainforest another species or two—nocturnal or elegantly camouflaged—may have so far escaped our notice. There are about as many species of primates as there are nations on Earth. And like the nations, they have their different customs and traditions, which we sample in this chapter.

———

Take the baboons—"the people who sit on their heels," as the !Kung San people of the Kalahari Desert respectfully call them. Hamadryas baboons are different from savanna baboons (from whom they diverged perhaps 300,000 years ago), and free baboons behave very differently from baboons crowded together in zoos (the latter "insolently lascivious," as an eighteenth-century naturalist described them). One telltale trait they all have in common: Sharing meat is virtually unknown among baboon males of either species, although it's fairly widespread among the chimps.

At sunrise the baboons rouse themselves from their sleeping cliffs and break up into a number of smaller groups. Each group wends its separate way over the savanna, foraging, scampering, playing, intimidating, mating—all in a day's work. But at the end of the day, all the groups converge on the same distant waterhole, and it may be a different waterhole on different days. How do the groups, out of sight of one another for most of the day, know to wind up at the same waterhole? Have the leaders negotiated the matter as the sun was rising over the sleeping cliffs?

Adult male hamadryas baboons are almost twice as large as the females. They display a leonine mane, enormous, almost fanglike canine teeth, and a ruthless character. These males were deified by the ancient Egyptians. They utter deep and prolonged grunts as they copulate. Their faces are "the color of raw beef steak—as different from the mousey grey-brown females as if they belonged to two different species."[2] As females approach sexual maturity, they are chosen by particular males and herded into harems. Squabbling among competing males over ownership of the females may have to be worked out. A high priority of the males is maintaining and improving their status in the dominance hierarchy.

Hamadryas harems characteristically comprise from one to ten females; the males are concerned to keep peace among the females and to make sure that they do not so much as glance at another male. This is a bondage with little hope of escape. A female must follow her male about for the rest of her days. She must be sexually submissive: the least reluctance and she is bitten in the neck. It is not unknown for a hamadryas female to have her skull punctured and crushed in the massive jaws of the male for a minor infraction of the behavioral code he ruthlessly enforces.[3] Conflict and tension around her are high when she's ovulating, and somewhat muted when she's pregnant or nursing the young. Unlike the chimps, you can see sexual coercion in the very posture of the baboon copulatory style: The male typically grasps the female's ankles with his prehensile feet while mating, guaranteeing that she cannot run away. Compared to hamadryas behavioral norms, chimps live in an almost feminist society.

In a quarrel among females, one will sometimes threaten her rival with her teeth and forearms and, at the same time, alluringly present her rump to the male; with this postural offer of a deal, she sometimes induces him to attack her adversary. Subordinate male savanna baboons, as well as barbary apes, may use an infant—an unrelated infant, a bystander infant, or maybe an infant he is baby-sitting—as a hostage or shield or placatory object when approaching a high-ranking male. This tends to calm the alpha down if he's in a grumpy mood.

The hamadryas male's larger size and ferocious temperament doubtless are useful when the troop is imperilled by predators, or in conflict with other groups. But, as in the rest of the animal kingdom, when there are conspicuous differences in stature between the sexes (usually, it's the males who are bigger), there's exploitation and abuse of the smaller and weaker (usually the females).* Another distinction of the hamadryas baboons is that, alone among nonhuman primates so far as we know, two groups have been observed to ally themselves in combat against a third.[4]

Among savanna baboons, where the size difference between the sexes is not so striking, there are no harems. They are great walkers; it's not unknown for a troop to cover twenty miles a day. Unlike chimps and hamadryas baboons, here it's the male who leaves the natal troop around puberty—again, probably as an evolutionary device to avoid incest, and genetically to connect semi-isolated populations. When he attempts to enter a new troop, objections are likely to be raised by the resident males. Acceptance by the group often requires the time-honored method of submission, bluff, coercion, and alliance-making in the male hierarchy. But in many cases another strategy works well: make friends with a particular female in the troop and her children. He grooms her. He baby-sits and cares for her young. No killing off the young here in order to bring her into ovulation, as with rats and lions. If all goes well, she sponsors his entry into the troop. We can imagine a certain exhilaration as, gingerly, he attempts to enter the new community, his gaffes and old enemies left behind, a clean slate before him, and success dependent almost entirely on his social skills.

The males are more flighty and tempestuous than the females. Social stability is mainly provided on the female side. Indeed, since savanna baboon males are transients, the only hope for coherent group structure lies with the females. In all things, female baboons are comparatively conservative; it is the testosterone-pumped males who take the risks.

The female dominance hierarchy is largely hereditary. Daughters

* The fact that in every human ethnic group and culture males have been on average larger than females has not escaped the notice of primatologists. It may have something to do with the penchant of men for sexism, coercion of women, rape, and harems when they can get away with it. The key question is to what extent anatomy is destiny, a point to which we will return.

of alpha females are given unusual deference, even as juveniles, and have a good chance of achieving alpha status when they grow up. Every close relative of the dominant female may outrank every other member of the troop—a royal family. Submission and dominance in the female hierarchy of savanna baboons and many other monkey species is conveyed in the time-honored idiom of presenting and mounting, the heterosexual metaphor again adapted to another purpose.

———

For reasons not fully understood but worth speculating on, much more attention—at least in public discussion and until recently—has been given to hamadryas baboons than to their savanna cousins. Sometimes the impression has been left that hamadryas behavior is representative of all nonhuman primates, or even all primates. For example, the hamadryas males, in a species in which nothing else is owned, have a clear sense of females as private property. But this is by no means true of all primates. The hamadryas baboons, it turns out, provide perhaps the most extreme example of hierarchy and brutality in the entire order of the primates. This behavior was especially marked under a set of cruel circumstances devised by humans who meant them no harm:

Living with apes or monkeys in the wild did not much appeal to primatologists until recently. More typical was an expedition back to his native South Africa by Solly Zuckerman, anatomist to the Zoological Society of London:

> On the 4th of May, 1930, I succeeded in collecting on a farm near Grahamstown in the Eastern Province twelve adult females from one troop of baboons. Four of these were non-pregnant. Five were pregnant; one had an embryo 2.5 mm. in length; another one of 16.5 mm.; the third one of 19 mm.; the fourth one of 65 mm.; and the fifth an apparently full term male foetus with a crown-rump length of 230 mm. Three were lactating, and their babies were caught alive. One infant was estimated to be four months old, and the other two were each about two months.[5]

He dutifully noted how much fresh semen there was at various depths within the reproductive tracts of his female victims; "collected," it turns out, is a euphemism for "killed." Baboons had been officially declared "vermin" in South Africa, because they're smart enough to defeat the efforts of farmers to safeguard their crops. A bounty was paid for each dead baboon. So a few baboons "collected" for science hardly mattered, compared with the wholesale slaughter being organized by the farmers. Through such studies Zuckerman "had the luck to discover from post-mortem study that ovulation in mature females occurs in the middle of the monthly sexual cycle."[6] The corresponding discovery about the human menstrual cycle was made around the same time.

He had long been interested in the standing of humans among the primates, and was dissecting baboons in South Africa while still a teenager.[7] But he was not wholly unmoved by the plight of the hunted baboons, and later quoted this early-twentieth-century account:

> Hugging her baby tight to her breast, she regarded us with a world of sadness in her eyes, and with a gasp and shudder she died. We forgot for the moment that she was but a monkey, for her actions and expression were so human, that we felt we had committed a crime. Muttering an oath, my friend turned and walked rapidly off, vowing that this was the last time he would shoot a monkey. "It isn't sport, it's downright murder," he declared, and I fervently agreed with him.[8]

If you wanted to meet a baboon—and you lived in a country where they didn't roam about in the wild—you could always go to the local zoo and see the bedraggled and deracinated inmates, lifers pent up in tiny cubicles. After World War I, some European zoos thought it would be better, as well as more "humane," if a large number of baboons could be gathered together in a partly open enclosure admitting observation by city-bound primatologists. The London Zoo was among them, and Dr. Zuckerman was playing a central role in the organization of one of these multiyear experiments:

In the spring of 1925, about one hundred hamadryas baboons were introduced into moat-bordered Monkey Hill, about 33 by 20 meters in area. So each baboon had, on average, less than 7 square meters, or some 60 square feet, indeed about the size of a small prison cell. It had been intended that this be an all-male group, but through an "accidental inclusion" six of the hundred baboons proved to be female. After a time, the oversight was rectified and the group was augmented by a further thirty females and five males. By late 1931, 64% of the males were dead, and 92% of the females:

> Of the thirty-three females that died, thirty lost their lives in fights, in which they were the prizes fought for by the males. The injuries inflicted were of all degrees of severity. Limb-bones, ribs, and even the skull, have been fractured. Wounds have sometimes penetrated the chest or abdomen, and many animals showed extensive lacerations in the ano-genital region . . . The fight in which the last of these females lost her life was so protracted and repellent—from the anthropocentric point of view—that the decision was made to remove the five surviving females from the Hill . . . The very high percentage of females killed in the London Colony suggests . . . that the social group of which they formed a part was in some way unnatural.[9]

Despite this last qualification, the hamadryas colony at the London Zoo reinforced a widespread belief in an unconstrained Darwinian struggle for existence. Even though baboons would quickly have exterminated themselves from the world if the events at Monkey Hill were characteristic of life in the wild, many people felt that they had now glimpsed Nature in the raw, a brutal Nature, red in claw and fang, a Nature from which we humans are insulated and protected by our civilized institutions and sensibilities. And Zuckerman's vivid descriptions of the unrestrained sex lives of the baboons—he was one of the first to stress that baboon social organization may be determined largely by sexual considerations—increased the contempt that many humans felt toward the other primates.

What had gone wrong on Monkey Hill? First, almost all of the baboons introduced into the "colony" were unknown to one another. There was no long-term mutual habituation, no prior establishment of dominance hierarchies, no common understanding in these harem-obsessed males of who was to have many females and who none at all. No kinship-based female dominance hierarchy had been established. Unlike the situation in the wild, there were many more males than females. Finally, these baboons were crowded together to a degree rarely experienced in their natural state.

Because of their powerful jaws and spectacular canines, baboon males within a troop hardly ever fight among themselves in earnest, although corporal punishment is visited on the females for the slightest infractions. But in the London Zoo, dominance hierarchies had to be established, dedicated attempts were made to steal females, escape from a formidable attacker was cut off by the moat, and the calming influence of many sexually compliant females was almost entirely wanting. The result was carnage. In all six and a half years, only one infant survived. When the males would fight over them, the adult females would listlessly wait, as if "paralysed." The battered, lacerated, punctured females would be sexually used by a quick succession of males.

But the females were not mere passive instruments:

[W]hen her overlord's back was turned she quickly presented to the bachelor attached to her party, who mounted for a moment. The overlord then slightly turned his head, whereupon the female rushed to him, her body low to the ground, presenting and squealing, and threatening her seducer with grimaces and with quick thrusts of her hands on the rocks. This behaviour immediately stimulated an attack by the overlord . . . Closely pursued, the bachelor fled. On another occasion the same female was left alone for forty seconds while her overlord chased a bachelor around Monkey Hill. In that space of time she was mounted and penetrated by two males to whom she had presented. Both of these immediately made off after their contact with the female, who again responded to the return of her male in the manner described above.[10]

When females were killed, the males would continue to drag them around, one male after another, to fight over them, and to copulate with their corpses. When the keepers, grimly watching this necrophilial tableau unfold, felt it necessary—for "anthropocentric" reasons—to enter the compound and remove the dead body, the males, in concert, would violently object and resist. Zuckerman, writing back then in the 1920s, used and may have coined the phrase "a sexual object"[11] in describing the lot of the female baboon.

We've seen in Calhoun's experiments with rats that—even when there's plenty of food, even when there are as many males as females —severe crowding induces violent and other modes of behavior that many would describe as aberrant and maladaptive. We've also seen in the Arnhem chimp colony how, under similar circumstances, new modes of behavior come to the fore to inhibit violence. From the baboons in the London Zoo we learn that if you take a species given to sexual violence in the best of conditions, provide a small number of sexual prizes to be fought over, arrange to have no pre-existing social order in which the animals know where they fit, and now crowd everybody together with no hope of escape, mayhem is the likely outcome. Monkey Hill reveals a deadly intersection of sex, hierarchy, violence, and crowding that may or may not apply to other primates.*

In Nature, as Zuckerman recognized, hamadryas baboons live much more peaceably. Dominant males are surrounded by a small corona of females, their offspring, and a few affiliated "bachelor" males. These harems wander over the landscape in bands, collecting food. Hundreds of baboons, a kind of gathering of the tribes, camp out each night near one another on sleeping cliffs. Fights to the death for possession of the females (or for any other reason) hardly ever happen. Everyone knows his, and especially her, place. The females are of course routinely abused, bitten on average once a day, but not so deeply as to draw blood. They are certainly not all killed off because they might be interested in other males, as happened in the London Zoo.

Hamadryas baboons in *very* small groups behave very differently: A bachelor baboon male watches a couple—on their first date—in an adjacent cage. Days go by, and he is forced to observe their deepening sexual relationship while he sits alone. When he's then introduced into their cage, he makes no effort to attack the male or to lure the female away. He respects their relationship. He looks away when they have sex. He is a model of rectitude and circumspection, even if he's of larger stature than either.[12]

Unsurprisingly, there are ways of arranging a primate society so its structure collapses and almost everybody dies. Shall we think of primates who find themselves in such circumstances as criminals? Are

---

* Something similar happened when a number of fugitive Englishmen, without a well-established dominance hierarchy (the alpha male and his close followers had been put overboard in a small boat), along with a few Polynesian women settled tiny Pitcairn Island in 1790, after the mutiny on H.M.S. *Bounty*.

they accountable for their actions? Do they have free will? Or shall we attribute the bulk of the responsibility for what happens to those through whose miscalculation the social environment was established? For a society to be successful, it must be consonant with the nature and character of the individuals who must live in it. If those contriving social structures overlook who these individuals are, or sentimentalize their nature, or are incompetent social engineers, disaster can result.

Zuckerman has consistently argued that almost nothing about human nature or evolution can be learned by studying monkeys and apes—quite the opposite of many students of animal behavior who believe that understanding primates might provide a direct route to understanding humans: "[M]y unbending critical attitude to attempts to explain human behavior by analogies from the animal world must have been acquired at a very early age."[13] He describes Konrad Lorenz, Desmond Morris, and Robert Ardrey—who popularized, with at least some excesses, the idea that we have something to learn about ourselves from studying other animals—as "three writers who are equally adept at devising superficial analogies."[14]

As "Prosector" of the London Zoo—the officer in charge of animal autopsies—Zuckerman later submitted the manuscript of a book, entitled *The Social Life of Monkeys and Apes*, for approval to his superior in the zoo's dominance hierarchy. It was promptly rejected on grounds of displaying an undecorous explicitness on matters sexual (for example, "The overlord's attention is caught by the perineal region of one of his females, usually when her sexual skin is swollen. He bends his head forward, his hand reaches out, his lips and tongue move and, having thus stimulated the sexual response in the female, he mounts and copulates"[15]). Zuckerman submitted the book for publication anyway. In his autobiography, *From Apes to Warlords*, published forty-six years later—amidst much vivid detail of those years—he makes only the most tangential reference to the events at Monkey Hill.

At the start of World War II Zuckerman studied the consequences of aerial bombardment on civilian populations—his anatomical knowledge could there be put to good use. He soon moved on to analyzing the effectiveness of aerial bombing in the accomplishment of strategic goals, where his skeptical proclivities came in handy: The

RAF's Bomber Command (and the U.S. Army Air Corps), he found, had consistently exaggerated the potential of massive aerial bombardment to lessen the enemy's will to fight and to shorten the war.

After the war Zuckerman headed the London Zoo, and through a few turns in his career wound up to be the principal scientific adviser to the British Ministry of Defence, where his expertise in understanding dominance hierarchies may have been germane. Created a Life Peer, Lord Zuckerman worked for many years to slow the nuclear arms race.

————

Baboons as a whole represent only one small corner in the vast arena of primate behavior. We could just as easily have focused

> on any of a number of lemur species, species in which females rather routinely dominate males. We could have decided to make an example of the shy and nocturnal owl monkey . . . where males and females cooperate in child care with the male playing the major role in carrying and protecting the infant, or we could have focused on the gentle South American monkeys known as "muriqui" . . . who specialize in *avoiding* aggressive interactions, or any of a host of other primate species in which we now know that females play an active role in social organization.[16]

Consider the gibbon. Its preternaturally long arms permit it to make great balletic leaps through the canopy of the forest—sometimes ten meters or more from branch to branch—that put champion human gymnasts to shame. Gibbons are, apparently without exception, monogamous. They marry for life. They produce haunting songs heard a kilometer or more away. Adult males often sing long solos in the darkness just before sunrise. Bachelors sing longer than old married males, and at a different time of day. Wives prefer duets with their husbands. Widows bear their grief in silence and sing no more.

Gibbons are also territorial and their matins serve to keep intruders away. A nuclear family, typically parents and two children, tends to control a small turf. Defense of the home territory is accomplished not so much by throwing stones or raining blows as by singing anthems. Perhaps there are cadences, timbres, frequencies, and amplitudes that other gibbons, contemplating a little poaching, find especially impressive and daunting. At least sometimes, an aging father will confer responsibility for territorial defense on his adolescent son, passing the patriotic torch on to the younger generation. In other equally poignant instances, adolescents are banished from the home territory by the parents, perhaps to avoid the temptations of incest. Adult males and females behave pretty much alike, and have nearly equal social status. Primatologists describe the females as "codominant," and the partners in a marriage as "relaxed" and "tolerant."[17]

Gibbon life seems downright operatic. It's easy to conjure up feverish love solos, duets sung in praise of marital felicity, and ritual intimidation chants cast into the forest night: "We're here, we're tough, we sing good songs. Better leave our turf alone." Perhaps there are gibbon Verdis singing power-transfer arias, rich with pathos, soulful lamentations on the passing of glory and of time.

Or consider the bonobo. This is a reclusive species or subspecies of chimpanzee that lives in a single group in Central Africa, south of the Zaire River.[18] Bonobos have certain traits that render them conventionally ineligible for the local zoo, which may be one reason that they're not nearly so well known as the common chimp we've described in the preceding chapters. Bonobos, given the Linnaean name *Pan paniscus*, are also called pygmy chimpanzees; they're smaller and more slender and their faces protrude less than the usual variety, *Pan troglodytes*, which we'll here and there continue to describe simply as chimpanzees.* Bonobos often stand up and walk on two legs. (They have a kind of webbing of skin between their second and third toes.) They stride with their shoulders squared and do not slouch as much as chimps do. "When bonobos stand upright," writes de Waal, "they look as if they had walked straight out of an artist's impression of prehistoric man."[19]

Unlike chimp females, among whom estrus is advertised and is a time of pronounced sexual receptivity, bonobo females display genital swellings about half the time; and they're nearly always attractive to the adult males. We recall that common chimps, *Pan troglodytes*, like almost all animals, have sex with the male entering the female's vagina from behind, his front against her back. But in bonobos, about a quarter of the time, the matings are face-to-face. This is the position

---

* Those who study chimps and bonobos, so the joke goes, are called panthropologists.

the females seem to prefer, probably because their clitorises are large and positioned far forward compared to chimps. Bonobos indicate their mutual attraction by prolonged gazing into one another's eyes, a practice which precedes almost all their matings, and which is unknown among common chimps. The initiation of sexual activity among the bonobos is mutual, unlike the chimps, where it is peremptory and nearly always by the males. While in general, especially in larger social contexts, male bonobos dominate females, this is not always the case, especially when they're alone together. At night, in the forest canopy, a male and a female will sometimes snuggle up together in the same nest of leaves. Adult chimps never do.

The sexual activity of common chimps, which by human standards seems obsessive to the point of mania, is almost puritanical by bonobo standards. The average number of penile thrusts in an average copulation—a measure of sexual intensity that primatologists are drawn to, in part because it can be quantified—is around forty-five for bonobos, compared to less than ten for chimps. The number of copulations per hour is 2½ times greater for bonobos than for chimps—although these observations are for bonobos in captivity, where they may have more time on their hands or more need for mutual comfort than when they are free. Less than a year after giving birth, bonobo females are ready to resume their lives of sexual abandon; it takes three to six years for chimp females.[20]

Bonobos use sexual stimulation in everyday life for many purposes besides mere satisfaction of the erotic impulse—for quieting infants (a practice said once to have flourished also among Chinese grandmothers), as a means of resolving conflict among adults of the same sex, as barter for food, and as a generic, all-purpose approach to social bonding and community organization. Less than a third of the sexual contacts among bonobos involve adults of opposite sexes. Males will rub rumps together or engage in oral sex in ways unheard of among the more prudish chimps; females will rub their genitalia together, and sometimes prefer it to heterosexual contacts. Females characteristically engage in genital rubbing just before they're about to compete for food or for attractive males; it seems to be a way of reducing tension. In times of stress, a bonobo male will spread his legs and present his penis to his adversary in a friendly gesture.

Despite these differences in nuance, bonobos are still chimpanzees. There's a male dominance hierarchy, although not nearly as pronounced as among common chimps; dominant males have preferential access to females, although males do not always dominate females; there are submissive gestures and greetings; the size of groups is about the same as with chimps, a few dozen; adolescent females wander over to adjacent groups; the males preferentially hunt animal prey, although apparently not in hunting parties; males are proportionately larger than the females by about the same ratio as among chimps; and encounters between groups sometimes become violent—although groups may also on encountering one another, behave very peaceably and laid-back. Infanticide and all other killing of bonobo by bonobo are, so far, unknown. Their standard initial response on meeting unfamiliar humans, as we ourselves experienced, is a very chimp-like, and adequately intimidating, charging display.

Grooming is most frequent between males and females and least common between males and males, the reverse of chimp practice. The grin serves not mainly as a gesture of submission, but performs a range of functions similar to those of the human smile. Male bonding is much weaker than in chimp society, and the social position of females much stronger. Certain mothers and sons associate closely until the son becomes an adult; among chimps the relationship tends more often to be broken off when the young male reaches adolescence. Social skills for resolving conflicts are much more highly developed among the bonobos than among the chimps, and dominant individuals are much more generous in making peace with their adversaries.

If we feel a certain revulsion at having hamadryas baboons as relatives, we may take some comfort from our connection with the gibbons and the bonobos. Indeed, we're far more closely related to the apes than to the monkeys. Chimps and bonobos are certainly members of the same genus and, according to some taxonomic classifications, even the same species. Given that, it's startling how different they are from one another. Perhaps many of the distinctions between the two—ranging from the frequency, increased variety, and social utility of sex to the relatively higher status of females—are due to the evolution in the bonobos of a new step: abandoning the monthly badge of ovulation, graduating from estrus. Perhaps when ovulation is not evident at a glance or a sniff, females can be viewed as more than sexual property.

The primates are so rich in potential that even a small change in anatomy or physiology may provide an aperture to a universe never dreamt of in the rude sleeping pallets made each night in the low branches of the once-vast tropical forests.

SOME SKETCHES FROM LIFE

Monkeys:

Monkeys are liable to many of the same non-contagious diseases as we are . . . Medicines produced the same effect on them as on us. Many kinds of monkeys have a strong taste for tea, coffee, and spirituous liquors: they will also, as I have myself seen, smoke tobacco with pleasure. Brehm asserts that the natives of north-eastern Africa catch the wild baboons by exposing vessels with strong beer, by which they are made drunk. He has seen some of these animals, which he kept in confinement, in this state; and he gives a laughable account of their behaviour and strange grimaces. On the following morning they were very cross and dismal; they held their aching heads with both hands, and wore a most pitiable expression: when beer or wine was offered them, they turned away with disgust, but relished the juice of lemons. An American monkey, an Ateles, after getting drunk on brandy, would never touch it again, and thus was wiser than many men. These trifling facts prove how similar the nerves of taste must be in monkeys and man, and how similarly their whole nervous system is affected.[21]

Eastern Mountain Gorillas:

When two animals meet on a narrow trail the subordinate gives the right-of-way; subordinates also yield their sitting place if approached by superiors. Sometimes the dominant animal intimidates the subordinate by starting at it. At most it snaps its mouth or taps the body of the other animal with the back of its hand.[22]

Monkeys:

[P]hallic threatening, derived from a sexual domination gesture (mounting), . . . has been described among many monkey species in both the Old World and the New. Among guenons and baboons, a few males always sit with their back to the group keeping guard, displaying their strikingly colored penis and their sometimes similarly strikingly colored testicles. If a stranger to the group approaches too closely, the guards actually have an erection; so-called "rage copulations" also take place.[23]

Squirrel monkeys:

The displaying monkey vocalizes, spreads one thigh, and directs the fully erect penis toward the head or chest of the other animal. The display is seen in its most dramatic form when a new male is introduced into an established colony of squirrel monkeys . . . Within seconds all males begin to display to the strange monkey, and if the new male does not remain quiet with its head bowed, it will be viciously attacked.[24]

Brown capuchin monkeys:

An estrous female will shadow the dominant male for days. At frequent intervals she approaches him closely, grimaces at him while giving a distinctive vocalization, pushes him on the rump, and shakes branches at him. When she is ready to copulate, she charges him, he runs away, she follows, and when he stops running, they mate.[25]

Orangutans:

At midcycle, a female orangutan will seek out the dominant male in her vicinity. At other times during her cycle, young males and subordinate males will sometimes cluster around her and it appears that she is being forced to mate with them. She resists, she screams, she fights, but they mate with her anyway. It is either a good act, or it's the equivalent to rape. Primatologists try not to use that term. People tend to get upset.[26]

Lemurs:

In *Lemur catta*, the incidence of aggression within groups is high, particularly between males. Aggression takes the form of-chasing, cuffing, scent marking, and, in males, stink fighting . . . Acts of submission include retreat or cowering as a dominant animal approaches, and low-ranking males habitually walk with lowered head and tail carriage, lagging behind the group and generally avoiding other animals. Females are much less frequently aggressive than males, and the female dominance hierarchy is less easy to detect, although the few agonistic encounters observed suggest that it is stable. Yet, "at any time . . . a female may casually supplant any male or irritably cuff him over the nose and take a tamarind pod from his hand."[27]

Monkeys:

In most monkeys with multimale groups, tolerant or cooperative relationships among males are rare or unknown. Male-male grooming, for example, is virtually nonexistent in rhesus monkeys . . . [I]f grooming ever occurs, it is given entirely by subordinates to dominant males . . . , unlike the more reciprocal system in chimpanzees. As another example, Watanabe . . . studied alliance formation among Japanese macaques. Out of 905 cases only 4 alliances were between adult males. Relationships between males in these groups are thus primarily competitive.[28]

Stumptail monkeys:

The two newcomer adult females . . . were thus repeatedly mounted as well as bullied by the three subadult males and the higher ranking juvenile male throughout their stay. This forced mounting might be considered as rape, in the sense that the female was obviously unreceptive and unwilling. She kept crouching while the male forcibly lifted her hindquarters, shook and even bit her, and ignored her screams and dismount signals.[29]

Stumptail monkeys:

At the very moment that the round-mouthed expression appeared on the female's face and the hoarse vocalizations were uttered, the equipment registered a sudden acceleration of her heart rate, from 186 to 210 beats per minute, and intense uterine contractions.

Actually, this experiment concerned reassurance behavior. The female's partners were other females . . . [It] can be demonstrated that the sexual posture that stumptails often adopt during reconciliation is accompanied by physiological signs of orgasm. This is not to say that sexual climax is achieved during every reconciliation. . . . [Nature] has provided stumptails with a built-in incentive for making up with their enemies.[30]

Colobine monkeys:
   [I]nfants are often passed around to other females from soon after birth. This pattern may continue for the first few months of life. In particular contrast to some macaques and baboons, every colobine infant has free access to every other infant, and females of all ranks have free access to all infants. Swapping of infants may be one of the roots of the [comparatively] nonaggressive colobine society . . .
   A very interesting feature of colobine intertroop encounters is the fact that they have readily available means of avoiding such contact. As arboreal animals occupying upper story vegetation which provides a relatively unobstructed view of surroundings, and as possessors of loud, sonorous vocalizations, colobine groups could rather easily avoid contact. Nevertheless, contact is frequent. Colobines maintain troop separation by one or a combination of the following: variable movement patterns, the male whoop vocalization, and male vigilance behavior.
   . . . Excitement is high during this stage, which includes tremendous leaping and running through the tree tops, as is evidenced by frequent defecation and urination. Another indication of high excitement and/or tension is the fact that males may have penile erections . . .
   The most common dominant signals include grinning, staring, biting air, slapping the ground, lunging, chasing, bobbing the head, and mounting another animal. Submissive gestures include presenting the hindquarters, looking away, running away, turning one's back to another animal, and being mounted . . . The higher the animal's position in the dominance hierarchy, the wider the personal space it controls which a less dominant animal may not enter without first clarifying its intent.[31]

Monkeys:
   [A]s long as the infant monkey should be riding on its mother, whether it is injured or even dead, its mother will continue to

carry it. If she stops carrying it, an adult male is likely to go to her and to bark at her and in this sense make it clear to her that she should continue carrying the infant. We had one case in our small colony at Berkeley where a mother carried her dead infant for two days and dropped it, and then the dominant adult male of the troop picked the infant up and carried it for two more days before discarding it.[32]

Vervet monkeys:

In 1967, T. T. Struhsaker reported that East African vervet monkeys gave different-sounding alarm calls to at least three different predators: leopards, eagles, and snakes. Each alarm elicited a different, apparently adaptive response from other vervets nearby. Struhsaker's observations were important because they suggested that nonhuman primates might in some cases use different sounds to designate different objects or types of danger in the external world . . .

Seyfarth, Cheney, and Marler . . . began by tape-recording alarm calls given by vervets in actual encounters with leopards, eagles, and snakes. Then they played tape-recordings of alarm calls in the absence of predators and filmed the monkeys' responses.

[W]hile adult vervet monkeys restrict their eagle alarm calls to a small number of genuine avian predators, infants give alarm calls to many different species, some of which present no danger. Eagle alarms given by infants, however, are not entirely random and are restricted to objects flying in the air . . . From a very early age, therefore, infants seem predisposed to divide external stimuli into different classes of danger. This general predisposition is then sharpened with experience, as infants learn which of the many birds they encounter daily pose a threat to them . . .

[But] . . . experiments offer no proof that primates in the wild recognize the relationship between a vocalization and its referent.[33]

Squirrel monkeys:

The Gothic variety of the male squirrel monkey provides a most graphic example. He signals 1) his aim to dominate another male, 2) his intention to assault him, *and* 3) his amorous ideas about a female—all three—by shoving his erect phallus into the face of the other monkey while grinding his teeth. The courtship display is identical to the aggressive display. Ethologists have found this crossed-wire phenomenon in numerous reptilian and lower forms.[34]

Hamadryas baboons:
[Y]oung males . . . present in situations which provoke fear. They employ sexual approach in obtaining access to each other and to entice a fellow for play. They masturbate and mount each other. They mount and are mounted by adult males and by adult females, their heterosexual activities not provoking aggressive responses from the overlords. They engage in manual, oral and olfactory ano-genital examination with animals of their own age and with adults of both sexes. They frequently end a sexual act by biting the animal with whom they have been in contact. This end to sexual activity, which is not usually seen in the behaviour of adults, often appears to be playful.[35]

Baboons:
Sir Andrew Smith, a zoologist whose scrupulous accuracy was known to many persons, told me the following story of which he was himself an eye-witness; at the Cape of Good Hope an officer had often plagued a certain baboon, and the animal, seeing him approaching one Sunday for parade, poured water into a hole and hastily made some thick mud, which he skilfully dashed over the officer as he passed by, to the amusement of many bystanders. For long afterwards the baboon rejoiced and triumphed whenever he saw his victim.[36]

Baboons:
In Abyssinia, Brehm encountered a great troop of baboons who were crossing a valley: some had already ascended the opposite mountain, and some were still in the valley: the latter were attacked by the dogs, but the old males immediately hurried down from the rocks, and with mouths widely opened, roared so fearfully, that the dogs quickly drew back. They were again encouraged to attack; but by this time all the baboons had reascended

the heights, excepting a young one, about six months old, who, loudly calling for aid, climbed on a block of rock, and was surrounded. Now one of the largest males, a true hero, came down again from the mountain, slowly went to the young one, coaxed him, and triumphantly led him away—the dogs being too much astonished to make an attack.[37]

Titis and other small monkeys:

Hidden among the tangled branches and vines of the Neotropical forests are the most paternal of primate fathers. The monogamously mated males of the small titi (*Callicebus* sp.) and night monkeys and the tiny Callimiconidae and Callitrichidae are unique in the intensity and duration of their relations with infants . . . Males of these species share all parental duties except nursing, and although the extent of participation is quite variable within species, they are generally the major caretakers of infants . . .

Males in these species are often strongly attracted to infants. Immediately after birth, they have been observed trying to sniff, touch, or hold the still-bloody newborn, and they sometimes even lick off the covering birth fluids . . . Within hours of birth, males carry infants on their backs, groom them, and protect them . . . Large portions of a male's day are devoted to infant care, and the most devoted fathers return their infants to the mother only to suckle . . .

Males also permit infants to take food from their hands and mouths . . . The food items shared are those that infants have difficulty obtaining or processing themselves, such as large mobile insects or hard-shelled fruit . . .

Fiercely protective, males will defend infants against any real or imagined threat. In captivity, tiny lion tamarin males have flung themselves against intruders as intimidating as woolly monkeys, macaques, and humans.[38]

Chapter 18

THE ARCHIMEDES OF THE MACAQUES

Some ascribe this to his natural genius; while others think that incredible effort and toil produced these, to all appearances, easy and unlabored results. No amount of investigation of yours would succeed in attaining the proof, and yet, once seen, you immediately believe you would have discovered it—by so smooth and so rapid a path he leads you to the conclusion . . . Such was Archimedes.

PLUTARCH
"Marcellus," in *The Lives of the Noble Grecians and Romans*[1]

We humans have not evolved from any of the two hundred other primate species alive today; rather, we and they have evolved together from a succession of common ancestors. As we reconstruct the primate family tree, we discover who our closest relatives are. The behavior of the primates varies so widely, even between species in the same genus, that it really does make a difference for our view of ourselves which ones are our nearest relatives.

The answer, as we've already described, seems to be that the chimps are our closest kin, sharing some 99.6% of our active genes. We know from DNA sequencing, as you would of course suspect, that bonobos and ordinary chimps are a lot more like each other than either of them are like us.[2] But 99.6% is very close. We must share many characteristics of both. (Indeed, there must be behavioral traits that we share with our most *distant* primate cousins.)

By using molecular and anatomical evidence, together with the record in the rocks, the entire family tree of primates can be drawn, at least approximately, and a timeline placed upon it. The evidence from the bones and from the molecules are not in perfect accord, although they are beginning to converge; in this book, we have given weight to gene sequencing and DNA hybridization data. According to the molecular evidence, gorillas branched off from the evolutionary line leading to us about 8 million years ago; the still unidentified, now-extinct common ancestor of humans and chimpanzees separated from the gorillas maybe a million years later. Very quickly thereafter, the lines to chimps and humans began evolving toward their separate destinies.[3] On a planet that's been inhabited a thousand times longer, that's pretty recently, as recent as the last two weeks in the life of a fifty-year-old human. This doesn't mean that humans and chimps *themselves* began 6 million years ago; only that our common twig in the evolutionary tree branched out then.

———

To understand a little more about our primate nature and its development, let's cast our minds back toward the end of the Mesozoic Age, around 100 million years ago. That would be about a year ago in the life of a middle-aged person. There were mammals even then; but they were not easy to find. The daytime was ruled by the dinosaurs; among them, some of the most fearsome killing machines ever to evolve on land. Our mammalian ancestors, it is thought, were timid, weak, and small; they were in fact typically the size of mice. Like all reptiles and amphibians today, some of the dinosaurs may (this is still a controversial point) have been cold-blooded; if so, in the chill of the evening, especially in winter, they closed up shop—especially the smaller ones that both preyed on mouse-sized mammals and were more vulnerable to the cold. But the mammals themselves were warm-blooded, and so could stay out all night.

Imagine a moonlit darkness in which their adversaries lay senseless, strewn across the landscape in stupors of sleep. This was the chance for our ancestors to scamper about their humble business—catching grubs, nibbling leaves, mating, caring for the young. But to function well in the dark, they had to be very good at using senses other than sight; and in that epoch the mammalian brain evolved along with elaborate machinery for enhanced hearing and smell, their hedge against whatever dinosaurs hunted at night..

Asleep in burrows during the day, our ancestors perhaps tossed fitfully, dreaming daymares filled with row after row of needle-like teeth and nimble, hair-raising scampers to safety. They may have been frightened all their lives, their hearts in their throats at every daylit step, longing for nightfall.

Sixty-five million years ago, a bolt from the blue—the impact of a small world—seems to have cataclysmically altered the planetary environment, wiping out the dinosaurs and permitting the mammals, wholly insignificant until then, to flourish and diversify. We do not know if there were primates so early, or if some other mammal quickly evolved into the first primate. We do know from fossil evidence that tiny monkey-like beings, weighing perhaps a few ounces, with teeth about a millimeter long, lived in what is today Algeria just after the extinction of the dinosaurs.[4] By 50 million years ago (six months ago in the life of our fifty-year-old) there were arboreal primates living in subtropical Wyoming.[5] The canine teeth of the males were twice as long as the females'. If we can judge by what this difference means in contemporary monkeys, the males bullied the females, established dominance hierarchies, competed with each other, and probably maintained harems. All that's been with us since the beginning of the primate order.

The first primates are judged to have been much more like early mammals (with longer snout, eyes to the sides of the head, and claws) than are modern monkeys, apes, and humans. The so-called "lower" primates, or prosimians—lemurs or lorises, say—may be something like the earliest primates. You can see that they're nocturnal at a glance: Their eyes are appealingly large for their faces, the larger aperture being an adaptation for night vision in a world illuminated only by the moon and the stars.

They probably communicated in part by spraying scents from specialized glands.* They had brains—large for their body size—to think with, stereoscopic vision to see with, and hands to manipulate the environment. Typical primate dominance hierarchy rituals had probably already appeared, including both sexes presenting their rears as a gesture of submission to the dominant male.

The early evolution of primates was marked by a profound transformation of creatures of the night into habitués of daylight; a corresponding suppression of the sense of smell[6] and elaboration of vision; developing facial muscles so moods could be communicated by expressions; a still more powerful bond between mother and child; a longer period of infantile dependence; and an improving ability of the newer, higher brain centers of the cerebral cortex to moderate aggression and other behavior patterns emanating from the older, lower layers. All this in turn led to major changes in primate society: The less aggression, the more a true communal life is possible; the longer the childhood, the more parents can teach their young. Alliances and support groups, reconciliation, reassurance, forgiveness, remember-

---

* Ring-tailed lemur males will smear a pheromone they generate onto their tails and then wave these prominent, black-and-white-banded appendages at each other, wafting the smell into the air. This is mainly a competition for females: Apparently the most aromatic lemur tends to win the most attractive female. In one lemur species, all adult males may have their tail waves answered in the same evening, because all adult females come into heat together, by the light of the silvery (and full) moon.

ing the past behavior of specific individuals, and planning future actions swiftly evolved. Our ancestors were by now well along a path toward greater alertness, intelligence, communications skills, love.

After the extinction of the dinosaurs, mammals moved out into the daylight. For a while, they must have felt safe and free. But the growing, multiplying, and diversifying mammals eventually became too good a meal to pass up. They began to eat each other. And new predators evolved, including birds of prey. The day shift became increasingly dangerous. For example, in a study of modern South American harpy eagles, 39% of the "prey items" returned to the nest turn out to be body parts of monkeys.[7] In daylight you have to be on your toes. Mutual defense—scanning the skies, say, and air raid sirens when an eagle is spied—becomes vital.

Foraging baboons, faced with predators, typically respond by closing ranks and moving faster.[8] Certain collective behavior that we readily describe as military constitutes an adaptive response of very ancient standing to the threat of predation. Competent predators can force the potential prey to evolve rapidly—toward binocular vision, arboreal acrobatics, mutual support, quickly disinhibited combat skills, intelligence, and general military virtues.

Monkeys are born with an ability to recognize the significance of various facial expressions—although just how to respond to such expressions depends on experience and training. There are single brain neurons that are preferentially triggered when the monkey sees the eyes or mouth or fur of another monkey. There is even a kind of brain cell specifically responsive to a crouching or bowing posture. Facial expressions and body posture have a meaning in the primates that's hardwired, and not merely a matter of social convention. The male rhesus monkey's come-hither look is to thrust out his chin and pucker his lips; if you're a rhesus monkey (of either sex), it's important, even early in your career, to know what this means.

One of the uses to which the evolving primate brain has been put is the storing up of grudges. Monkeys generally make up—often by ceremonially mounting each other—within minutes after a fight. Chimp males, with females frequently in a peacemaking role, may take hours or days. But among themselves the females are less forgiving; they may hold grudges for the rest of their lives. Humans of both sexes can take anywhere from moments to millennia. Even among monkeys, a smoldering resentment against an individual is often broadened to encompass his or her relatives. Among the many new social forms invented by the primates are feuds and vendettas, sometimes extending over many generations—intimations of the beginnings of history.

As in most mammals, primate aggression, dominance, territoriality, and the sex drive are mediated by testosterone circulating in the blood, and generated mainly by the testicles. Almost certainly this was true of the earliest primates, and long before. The more testosterone and other androgens the developing fetal brain receives, the more of these masculine characteristics the animal will exhibit when he grows up. The lower the testosterone levels in a male, the more subdued will be these proclivities and the more likely that he will present himself for mounting by other males. But the testosterone levels also respond to the mantle of leadership. When presented with females in estrus and no high-ranking males around, the testosterone level of lower-ranking males soars. Within certain limits, primates rise to the occasion. The office makes the monkey.

Males of many primate species (although, on average, not humans) show a marked preference for female sexual partners who have already produced offspring; younger females may have to make special efforts to be alluring.[9] We have described the vigilance with which chimp alpha males guard their females, but only during ovulation. Nevertheless, sex has evolved in primates into something much more than simply the means for the replication and recombination of DNA sequences. Year-round, virtually compulsive sex with many partners —described by human observers as "promiscuous," "depraved," "perverse," and "indiscriminate"—is there for a reason. It serves as a mechanism of socialization. This is clearest among the bonobos. Despite sexual jealousy, it holds the group together. It provides bonds of affection, common goals, means of identification with others, and a gentling of dangerous aggression. The essence of primate living arrangements is a gregarious communal life, which partakes of many recognizable aspects of human culture and society. One of the chief motivations for this communal life is sex.

Adult role models are essential among animals in which childhood learning plays so central a role. Dominance hierarchies soften violence (but not aggression) within the group. Cooperation is important in any hunt, critical in hunting large animals, and sometimes essential in evading predators. In a survey of thirty primate species in the wild, the probability that any given individual will be eaten by year's end is found to be one chance in sixteen.[10] Evading predators must be very high on the primate agenda—and communal life provides early warning and collective defense.

Vervet monkeys have ventured a little out of the comparative safety of the forest and into the open savanna, where there is less cover for them, and more danger. By playing recordings of their calls back to them, they reveal that they have specific, readily understood alarm cries that elicit specific actions—for a python or black mamba (whereupon all stand on tiptoes and peer anxiously about them in the grass), for a Martial eagle (whereupon all look up into the sky and dive into deep foliage), and for a leopard (whereupon all quickly scramble up into the trees). Different predators elicit different cries and different evasive behavior. The responses are in part learned. Infants frantically sound the eagle alarm even when a non-raptor is spied flying overhead, and sometimes in response to a falling leaf. Gradually, they get better at making these distinctions. They learn from experience and from others. They have a range of other grunts, some of which scientists think they understand; vervets leave at least a superficial impression of conversing with each other. Gregariousness, by several different routes, spurs social intelligence, which seems to be, of all the species of life on Earth, most highly evolved in the primates.

The vervet fear of snakes is shared by baboons, chimps, and many other primates. You expose wild rhesus monkeys to snakes and objects that look like snakes and they jump out of their skins. Do the same experiment with laboratory-raised rhesus monkeys who have never seen a snake and, although some of them are afraid, you find that they're much less distraught. In one experiment the wild chimps' snake phobia became almost manageable when every time the chimp saw a snake it also was offered a banana.[11] So is the fear of snakes not hereditary, but somehow taught by mothers to their babies? Or is there an inborn fear that's softened in laboratory monkeys because they become habituated to harmless, snake-like objects—hoses, for example? Which is it: heredity or environment? Is knowledge of what a snake looks like, and that snakes mean primates no good, encoded in the DNA? Or are baby primates just watching adults closely and copying what they do?

Almost certainly the answer is a mix between the two. There seems to be an inborn snake-aversion program in the brains of primates. But this is not a closed program, inaccessible to new information from the outside world. Instead it's an open program that can be modified by experience—for example, "I've seen a lot of snakes in my time that don't do me much harm, so I'll be a little more relaxed around them," or, "Every time I see a snake, a banana miraculously appears; snakes have their good points too." Most primate programs are open, adaptive, malleable, adjustable to new circumstances—and therefore necessarily partaking of ambivalence, complexity, inconsistency.

In a typical modern chronology,[12] the line that would lead to us split off from Old World Monkeys about 25 million years (m.y.) ago; from the gibbons, 18 m.y. ago; from orangutans around 14 m.y. ago; from gorillas some 8 m.y. ago; and from the chimps approximately 6 m.y. ago. Bonobos and common chimps went their separate ways only about 3 m.y. ago. Our genus, *Homo*, is 2 million years old. Our species, *Homo sapiens*, is maybe 100,000 to 200,000 years old—the equivalent of the last day in the life of that fifty-year-old.

Committed to a communal social life, under intense selection pressure from predators, with brains evolving rapidly and education of the young effectively institutionalized, the primates have been developing new forms of intelligence. Their curiosity, experimental bent, and intellectual quickness are partly responsible for their success.

———

Here is an account, by a Japanese primatologist, of a remarkable set of events that transpired in a colony of macaques isolated on a small island called Koshima. Initially, in 1952, there were only twenty of the monkeys; the number almost trebled over the following decade. The natural food supply on Koshima was inadequate, so the monkeys had to be provisioned—with sweet potatoes and wheat dumped on the shore by the primatologists who were observing them.

As anyone knows who's ever been to a picnic at the beach, sand sticks to food and makes it unpleasantly gritty. In September 1953 a one-and-a-half-year-old female named Imo figured out that she could rinse the sand off her sweet potatoes by dunking them in a nearby brook.

After Imo, the next individual to learn potato washing was Imo's playmate, who did so in October. Imo's mother and another male peer began to wash in January 1954. In subsequent years (1955 and 1956), three of Imo's lineage (younger brother, elder sister, and niece) and four animals from other lineages (two were a year younger and two were a year older than Imo) started to do so. Thus, with the exception of her mother, all the individuals that learned potato washing quickly were either peers or young close relatives of Imo . . .

After 1959, features of information transfer changed. Sweet potato washing was no longer a new mode of behavior: when infants were born, they found most of their mothers and elders washing potatoes and learned this behavior from them as they learned the group's usual food repertoire. Infants are taken to the edge of the water during the period when they are dependent on mothers' milk. While their mothers wash potatoes, infants watch carefully and put into their mouths pieces of potatoes that mothers drop in the water. Most of the infants acquire potato washing around 1 to 2.5 years old . . .

[I]n the second period (1959–present, the period of "precultural propagation"), acquisition of potato washing occurred independent of sex and age. During the second period, virtually all individuals . . . acquired this habit through their mothers or playmates when they were infants or juveniles.

But there was still the problem of sandy wheat—until Imo's second epiphany:

In 1956, when Imo was 4 years old, she took a handful of mixed wheat and sand to the brook. When it was dropped on the water, the sand sank and the floating wheat could be skimmed off the water's surface, now clean again. This "placer-mining" technique* was also adopted by some of the other monkeys, and soon more and more animals learned it . . .

Compared to potato washing, placer mining was quite slow to propagate . . .

Placer mining appears to require more understanding of complex relations between objects and may be particularly difficult to learn because a monkey must "discard" his food first, while in potato washing he can keep the potato from the beginning to the end.[13]

---

* Used in panning for gold.

Imo was a primate genius, an Archimedes or an Edison among the macaques. Her inventions spread slowly; macaque society, like traditional human societies, is very conservative. Perhaps the fact that she came from a high-ranking family in a species given to hereditary matriarchy aided acceptance. As is usually true, adult males were the slowest to catch on, obstinate to the last; a female invented the process, other females copied her, and then it was taken up by youngsters of both sexes. Eventually, infants learned it at their mother's knee. The reluctance of the adult males must tell us something. They are fiercely competitive and hierarchy-ridden. They are not much given to friendships or even to alliances. Perhaps they felt impending humiliation—if they were to imitate Imo, they would be following her lead, becoming in some sense subservient to her, and thereby losing dominance status. They would rather eat sand.

No other group of macaques anywhere in the world is known to have made such inventions. By 1962, it is true, macaques on other islands and the mainland, recently provisioned with potatoes, began washing their food before eating it. But it is unclear whether this was due to independent invention or to cultural diffusion: In 1960, for example, Jugo—a macaque who had become adept at washing potatoes—swam from Koshima to a nearby island where he stayed for four years and may have trained the resident macaques.[14] Perhaps there were other macaque Archimedes; perhaps not. Imo is the only one we know for certain.

It took a generation for these two obviously useful inventions to become widely accepted.[15] The conservative, near immobility of popular prejudice, the reluctance to adopt a new practice even if its advantages are clear, is a tendency not restricted to Japanese macaques.[16] Perhaps the stolidity of the adult males is partly a matter of learning abilities declining with age. Human teenagers seem so much more adept than their parents at, say, operating a personal computer or programming a videocassette recorder. But this doesn't explain why adult *female* macaques learned so much more readily than their male counterparts.

We can see how such inventions made in different, nearly isolated, groups can lead to cultural differentiation even in monkeys. A much more innovative species of primate, in which various groups are in occasional contact, conflict, or competition, might, we would guess, devise spectacular new forms of culture and technology.

———

An early Algerian myth held that long ago apes could talk, but were rendered mute for their transgressions by the gods. There are many similar stories in Africa and elsewhere.[17] In another widespread African story, apes *can* talk, but prudently refuse to do so—because talking apes, their intelligence in this way made manifest, will be put to work by humans. Their silence is proof of their intelligence. Occasionally the indigenous people would introduce a visiting explorer to a chimp with many remarkable skills and tell him that it could also speak. But, at least while the explorer was there, none ever did.

Lucy was a chimpanzee celebrity. She was one of the first of the apes to learn to use a human language. The mouth and throat of the chimp are not configured for speech as ours are. In the 1960s, the psychologists Beatrice and Robert Gardner wondered whether chimps might be intellectually capable of language but prevented from speaking by the limitations of their anatomy. Chimps have phenomenal dexterity. So the Gardners decided to teach a chimp named Washoe a gestural language, Ameslan, the American sign language used by hearing-impaired humans. Here each gesture can represent a word, rather than a syllable or a sound, and in this respect Ameslan is more like Chinese ideograms than the Greek, Latin, Arabic, or Hebrew alphabets.

Young female chimps proved to be adept pupils. Some of them eventually acquired vocabularies of hundreds of words. Julian Huxley —T. H. Huxley's grandson, and a leading evolutionary biologist—had argued that "plenty of animals can express the fact that they are hungry, but none except man can ask for an egg or a banana."[18] Now there were chimps eagerly requesting bananas, oranges, chocolate candies, and much else, each represented by a different sign or symbol. Their communications were often clear, unambiguous, and apparently in context, as has been attested to by delighted audiences of hearing-impaired people watching films of signing chimps. They were able, it is said, to use their signs in a fairly consistent elementary grammar, and to invent from the words they knew phrases that they had never before encountered. Chimps were found to generalize a word such as "more" into new contexts—such as "more go" and "more fruit."[19] A swan evoked the spontaneous neologism, in independent and widespread use among humans, "water bird."

Lucy was one of the first. It was she who signed "candy drink" after first tasting a watermelon, and "cry hurt food" after her first experience with a radish. She became, it is said, able to distinguish the meaning of "Lucy tickle Roger" from "Roger tickle Lucy." Tickling is close to grooming. When idly turning the pages of a magazine, Lucy made the sign for "cat" when she turned to a picture of a tiger, and "drink" when she came upon a wine advertisement. Lucy had a human foster mother; she was, after all, only a few years old during the whole of her laboratory experience with language, and young chimps especially crave emotional support. One day, when her foster mother, Jane Temerlin, left the laboratory, Lucy gazed after her and signed, "Cry me. Me cry."

Ameslan-literate apes have often been spied signing to themselves when they thought no one else was present. Perhaps this was just wordplay, trying to get the new skill down pat. Or perhaps it was an experiment to see if they could conjure "fruit," say, out of the air with no humans present, just by producing the right words. It had worked well enough when humans were around.

To what extent Lucy and her fellows understood the gestural language they were using, and to what extent they were merely memorizing sequences of signs whose true meaning they failed to grasp, is a subject of scientific debate. To what extent young humans learning their first language do the one or the other is also subject to debate.

Perhaps only the hits were recorded and not the misses; that is, maybe Lucy and other chimps judged Ameslan-literate generated a wide range of signs more or less at random which, when they made contextual sense, were written up by the human observers and discussed at scientific meetings, but which, when irrelevant or unintelligible, were ignored. This is the anecdotal fallacy* that haunts this branch of science. But the anecdotes are plentiful and striking.

One of the most thoroughgoing examinations of the linguistic and grammatical abilities of apes was done by the psychologist Herbert Terrace and his colleagues, who recorded on videotape nearly twenty thousand signing attempts generated by a male chimp named Nim.[20] He mastered over one hundred different gestural signs. Nim would

---

* Also called the fallacy of the enumeration of favorable circumstances. No dishonest intent is implied; it is merely one of those failures of logic that humans are prey to. We tend not to be dispassionate observers.

regularly sign "Play me" or "Nim eat" in context and with apparent understanding. But there was no evidence, Terrace concluded, that Nim put more than two signs together in any consistent manner appropriate to the context. The average length of his sentences was less than two words long. His longest recorded sentence was "Give orange me give eat orange me eat orange give me eat orange give me you." It does seem a little frantic, but oranges are tasty, chimps are not known for their patience, and anyone who has spent time with a small excited child will recognize the syntax. Note that four of the words are non-redundant ("give me orange you"), and that no words irrelevant to this urgent request are included among the sixteen. Emphasis through repetition is common in human languages. But the simplicity of chimp sentences has rendered their use of language unimpressive in the minds of many psychologists and linguists. Nim was also belittled for interrupting his trainers' signing with his own, for being too imitative (repeating remarks of his trainer), and for not inventing grammatical rules such as the subject-predicate sequence.

This work has in turn been criticized. Chimps require close emotional ties for social tasks and, one would think, especially for something as difficult as language; instead, Nim had sixty different trainers over a four-year period. There is a tension between a loving, one-on-one environment that might be needed to teach language skills and the emotionally sterile protocols needed so that scientific results are, with high reliability, uncontaminated by the enthusiasm of the experimenters. It has frequently been found that apes sign most creatively in spontaneous circumstances in their everyday life, and not in experimental sessions. Too, there was great emphasis on drill in the Nim experiments, the very opposite of spontaneity. The complaint about Nim interrupting the signing of his trainer has itself been belittled, because Ameslan speakers may sign simultaneously without stepping on each other's lines, an advantage of signing over speech. Delayed imitation is just what human children do when they first learn language. For all these reasons, just how much grammatical dexterity apes have is still an open question.[21]

But clearly chimps can use something like the rudiments of language with much greater facility than had been thought possible before the experiments of the Gardners. They can unambiguously associate certain signs with certain people, animals, or objects—unsurprising when there are *monkeys* with different alarm cries and evasion strategies for different species of predators. Chimps have mastered an elementary vocabulary of a few hundred words, comparable to what a normal human two-year-old can do. Chimps who have some knowledge of these signs and who are raised together have been known spontaneously to sign to one another. There is at least one case in which a young chimp, uninstructed by any human, is said to have learned dozens of signs from another chimp knowledgeable in Ameslan.[22]

"We may consider it proven," said the psychologist William James, "that the most elementary single difference between the human mind and that of brutes lies in this deficiency on the brute's part to associate ideas by similarity." He held this to be a more fundamental cause of human uniqueness than reason, language, and laughter—all of which, he taught, emerge from recognizing similarities among ideas.[23]

Some chimps were taught a common symbol to describe any one of three foods, and another to describe any one of three tools. Then they were taught the individual names of other foods and other tools and asked to put them in the proper categories—not the new foodstuffs or tools themselves, but the arbitrary *names* of the new foodstuffs and tools. They did exceptionally well.[24] How is this possible, unless chimps reason, form abstract ideas, and "associate ideas by similarity"? Another domesticated chimp, Viki Hayes, was given two piles of pictures, one of humans, the other of nonhumans, and then handed a stack of additional pictures and invited to categorize. Her performance was perfect, with one small exception: She placed the picture of herself among the humans.

The psychologist Sue Savage-Rumbaugh[25] and her colleagues devised a keyboard with 256 lexigrams on its two sides. Each lexigram stands for something of interest to a chimp—"tickle," "chase," "juice," "ball," "bug," "blueberry," "banana," "outdoors," "videotape," and so on. The lexigrams do not depict their referent; rather, they show geometrical or abstract figures that only by arbitrary convention are connected with what they stand for. The scientists tried to teach this lexigraphic language to an adult bonobo, but she was an indifferent student. Her six-month-old son, Kanzi, often accompanied his mother to these training sessions and was mainly ignored by the scientists. Two years later, having observed the laboratory routine in depth but never having been trained (for example, by being given a banana for typing the banana lexigram), Kanzi demonstrated that he was learning what they were trying to teach his mother. (His interest eventually became hard to miss: He would leap on her hand, her head, or the keyboard at the moment she was about to select a lexigram.) The focus of the study switched to him.

By age four he had mastered the board, and would routinely use lexigrams to request, confirm, imitate, choose an alternative, express an emotion, or just comment. He would indicate a future course of action and then do it. In combining two action lexigrams, he would predict (or better, reveal) the impending sequence of events; if he typed "chase, tickle" he would chase and then tickle the experimenter or another chimp, and only very rarely tickle before chasing. Kanzi typed "hide peanut," and then did just that. It seems hard to deny that Kanzi has a mental image of his intended future actions, and in appropriate sequence. As time went on, he developed other grammatical rules, especially putting the action before the object, rather than vice versa ("bite tomato," rather than "tomato bite"). Inventing grammar is much more impressive than merely being taught it.

Still, after some years about 90% of Kanzi's utterances were only a single symbol;* rarely did they comprise more than two symbols. This is the same pattern found for Nim. Perhaps we are coming up against some fundamental limitations in chimp capacity for language.

Kanzi has shown, again via an accidental discovery, that he can understand hundreds of words of *spoken* English. Place earphones on his head, situate yourself in another room, make a request of him through a microphone, and the video camera reveals him doing what he is asked. Done this way, no gestural cues can be unconsciously communicated from human to ape. Typical of over 600 novel requests, perfectly complied with, were "Put the backpack in the car," "Do you see the rock? . . . Can you put it in the hat?" "Take the mushrooms outdoors," "Knife the orange," "Eat the tomato," and "I want Kanzi to grab Rose." Even some of Kanzi's errors are not so bad. Asked "Can you put the rubber band on your foot?" he promptly put it on his head.[26] His performance was comparable to that of a 2½-

---

* One expert reviewer likens this sentence to saying that "90% of the materials dug from a gold mine are not gold ore."

year-old human who was tested in the same set of experiments. Other bonobos are also found to understand spoken English.

Kanzi loves to play ball. Hide a ball in one of seven designated sites in the laboratory's fifty-five-acre forest, tell him by lexigram or spoken word where the ball is, and Kanzi with high accuracy makes for the site, searches, and finds the ball.[27] In this case there is a reward for understanding spoken English. But in most cases Kanzi receives no reward except the approval of humans and perhaps some gratifying sense of the power of communication. The motives of a young child learning language may not be very different.

In a different laboratory, a chimp named Sarah was able to recognize that red characterized an apple more than green did (she had not been exposed to the Granny Smith variety), and a square with a stem was a better representation of an apple than a square without a stem. She was also able to associate the *words* for each of these properties of an apple with the *word* for apple—and these words were not in Ameslan, but in a symbolic language of plastic tokens she had been taught, the tokens not resembling the objects in question.[28] ("Apple," for example, was represented by a small blue triangle.) How is this possible, unless chimps are able to abstract and categorize?

Other experiments have shown chimps capable of reasoning by analogy and by transitive inference, described by the discoverers of this aspect of chimp thought as " 'A r B, B r C, therefore A r C,' where r is some transitive relation, such as greater than."[29] (There may, for all we know, be critics who do not even understand the preceding sentence but who deny that chimps reason.) Still other experiments have been interpreted as showing that chimpanzees impute states of mind to others, or, as the psychologists David Premack and G. Woodruff put it, that chimps have "a theory of mind."[30]

Where chimps are linguistically deficient, at least so far, is in grammar and syntax. They are bereft of subordinate clauses, articles and prepositions, tenses, conjugation of verbs, and the like—as are small humans first learning language. The absence of such grammatical machinery prevents the lucid expression of even fairly simple ideas; misunderstandings tend to accumulate. Compounded by small vocabularies, it's a little like a middle-aged American, relying on barely remembered high-school French, attempting to be understood in rural Provence. A better analogy might be the "pidgin" languages that emerge at the interface between two or more fully realized but very different human languages; despite their linguistic facility, the speakers revert to something like chimpish. Oddly, no one has made a serious and systematic effort to teach apes grammar and syntax,[31] so we can't be sure it's beyond their reach. "Until then," writes a modern linguist, "one cannot entirely close off the possibility, unlikely as it may be, that apes could acquire language in its fullest sense."[32]

Savage-Rumbaugh and her co-workers toy with the possibility that chimps and bonobos exhibit impressive facilities to learn something of human languages because they have their own languages, vocal or gestural, that we have not yet deciphered.[33] In announcing the location of prey, or predators, or a hostile patrol, rudimentary language would be strongly favored by natural selection. Long before humans and chimps went their separate ways, considerable aptitudes for thought, invention, and language were probably percolating in our primate ancestors.

But partly because of Terrace's work, and partly because of the perceived difficulty of doing clean, controlled, non-anecdotal experiments on so emotional a being as a chimpanzee, financial support for many of these studies has nearly disappeared. In one case, the colony where apes had been taught Ameslan had fallen on hard times. Years had passed. Support was drying up. No one seemed interested in conversing with the chimps anymore. The grounds had become weedy and overgrown. The inmates were about to be shipped to laboratories for medical experimentation. Before the end, they were visited by two people who had known them in the old days. "What do you want?" the visitors asked in Ameslan. "Key," two chimps are said to have signed back from behind bars, one after the other. "Key." They wanted out. They wanted to escape. Their request was not granted.[34]

———

When chimps approach sexual maturity, their behavior changes. Both sexes are then much stronger than humans and given to occasional, unpredictable bouts of rambunctiousness and violence. So as the chimps get older, almost inevitably the experimenters find themselves driven to use steel cages, collars, leashes, and electric cattle prods. The chimps must feel, bit by bit, betrayed by the humans and less inclined to cooperate in their strange language games. Accordingly, back in the days when the research was generously supported, it was thought prudent to terminate experiments on teaching chimps language—requiring, as it does, close face-to-face daily contact—when they begin to mature. As a result, we do not know what the linguistic abilities of an adult chimpanzee might be. Lucy, like some aging child actor, was forced into retirement just a little after puberty. The laboratory in which she had demonstrated her accomplishments in sign language was closed.

Jane Goodall, who had by then spent a decade and a half living with chimps in the wild, was astonished on meeting Lucy:

> Lucy, having grown up as a human child, was like a changeling, her essential chimpanzeeness overlaid by the various human behaviours she had acquired over the years. No longer purely chimp yet eons away from humanity, she was man-made, some other kind of being. I watched, amazed, as she opened the refrigerator and various cupboards, found bottles and a glass, then poured herself a gin and tonic. She took the drink to the TV, turned the set on, flipped from one channel to another then, as though in disgust, turned it off again. She selected a glossy magazine from the table and, still carrying her drink, settled in a comfortable chair. Occasionally, as she leafed through the magazine she identified [in Ameslan] something she saw . . .[35]

In the second half of her life, Lucy lived with other chimpanzees on a small island in Gambia. Her adjustment to Africa was slow and difficult, and she became

> an emaciated, hairless wreck . . . She had been born and raised in the United States, and in pampered upper-middle-class circumstances . . . Lucy, the fastidious, toilet-trained chimpanzee princess . . . slept on a mattress, sipped soda, developed schoolgirl crushes, and would sit in the living room during the afternoon and leaf through magazines.[36]

But after a year or two in Gambia, thanks to the loving care of Janis Carter, she began to adjust. She had regular contact with humans and was often the first chimp to greet visitors to the island. She was used to humans. Her relations with other chimps were more strained. She had missed out on the rollicking childhood of a chimp in the wild.

In 1987 Lucy's skeleton was discovered. The most likely reconstruction of events is that humans came to the island, killed Lucy, probably by shooting her, and skinned her body. Her hands and feet, the very organs that had made her famous, were missing.[37] Those responsible have never been found.[38]

ON IMPERMANENCE

In the life of a man, his time is but a moment, his being an incessant flux, his senses a dim rushlight, his body a prey of worms, his soul an unquiet eddy, his fortune dark, and his fame doubtful. In short, all that is of the body is as coursing waters, all that is of the soul as dreams and vapours; life a warfare, a brief sojourning in an alien land; and after repute, oblivion. Where, then, can man find the power to guide and guard his steps? In one thing and one alone: the love of knowledge.

MARCUS AURELIUS, *Meditations*[39]

Chapter 19

WHAT IS HUMAN?

Having proved mens & brutes bodies on one
type: almost superfluous to consider minds.

CHARLES DARWIN

Notebooks on Transmutation of Species[1]

We humans are the dominant species on the planet, a status affirmed by several standards—our ubiquity, our subjugation (politely called domestication) of many animals, our expropriation of much of the primary photosynthetic productivity of the planet, our alteration of the environment at the Earth's surface. Why us? Of all the promising lifeforms—implacable killers, professional escape artists, prolific replicators, nearly invisible beings that no macroscopic predator can find—why did one primate species, naked, puny, and vulnerable, manage to subordinate all the rest and to make this world, and others, its domain?

Why are we so different? Or are we? Unambiguous definitions of humans—definitions that include almost all members of our species, but no one else—can be produced from anatomy or from DNA base sequencing. But they fail the purpose. They explain nothing that we recognize as fundamental about ourselves. Perhaps sometime in the future we will discover that unique sequences of As, Cs, Gs, and Ts encode for particular sequences of amino acids that constitute particular proteins that catalyze particular chemical reactions that motivate particular behavior that we might agree is characteristically human. But so far no such sequence has been found.

If, then, we can discern no clear-cut distinction in our chemistry (or anatomy) that explains our dominant role, the only ready alternative is to survey our behavior. It seems plausible that the sum of our everyday activities would be sufficiently defining, but a surprisingly large number of those activities can be performed by apes. For example, here's a description of the accomplishments of Consul, the first chimp acquired, in 1893, by the zoo in Manchester, England:

> [He was] able to put on his own coat and hat, seat himself in his own carriage for a drive, sit at table with company, use his knife and fork with propriety, pass his plate for a fresh supply of food, use his ser-

viette [napkin], wash his hands after meals, put coals on the fire, ring the bell for the maid, go into the kitchen for a romp with the girls, walk into his hotel, shake hands with his friends, kiss the barmaid, smoke his pipe, and mix his own drinks.[2]

True, Consul's deportment may be dismissed as mere mimicry; but that may also be said of those of us who marvel at his abilities.

Is there *anything* we do that's uniquely human—that all or almost all of us, of every culture, throughout history, do and that no other animal does? You might think something along these lines would be easy to find, but the subject is redolent with self-deception. We have too much of a stake in the answer to be unbiased.

Philosophers of marauding high-technology civilizations have often argued that humans deserve a category distinct from and above all the other animals.* It is not enough that humans have a different assortment of the qualities evident in the other animals—more of some traits, fewer of others. A radical difference in kind, not some fuzzy-edged difference in degree, is needed, longed for, sought.

Most of the philosophers adjudged great in the history of Western thought held that humans are fundamentally different from the other animals. Plato, Aristotle, Marcus Aurelius, Epictetus, Augustine, Aquinas, Descartes, Spinoza, Pascal, Locke, Leibniz, Rousseau, Kant, and Hegel were all proponents "of the view that man differs radically in kind from [all] other things"; except for Rousseau, they all held the essential human distinction to be our "reason, intellect, thought, or understanding."[3] Almost all of them believed that our distinction arises from something made neither of matter nor of energy that resides within the bodies of humans, but of no one else on Earth. No scientific evidence for such a "something" has ever been produced. Only a few of the great Western philosophers—David Hume, for instance—argued, as Darwin did, that the differences between our species and others were only of degree.

Many famous scientists, while fully accepting evolution, have parted company with Darwin on this question. For example, Theo-

---

* Many of them would not have included the word "other," and even today there are those who bristle at being called—even by scientists speaking generically and without affect—"animals."

dosius Dobzhansky: "*Homo sapiens* is not only the sole tool-making and the sole political animal, he is also the sole ethical animal."[4] Or George Gaylord Simpson: "[M]an is an entirely new kind of animal . . . [T]he essence of his unique nature lies precisely in those characteristics that are not shared with any other animal,"[5] especially self-awareness, culture, speech, and morality. The difference between humans and non-human animals according to a number of contemporary philosophers[6] goes like this:

> Precisely because they are incapable of conceptual thought, animals . . . are not only (1) incapable of sentence-making that includes statements about the past and future, (2) unable to fabricate tools for remote future use, (3) devoid of a cumulative cultural inheritance that constitutes a long historical tradition, but they are also (4) incapable of any behavior that is not rooted in the perceptually apprehended present situation.

Apart from quibbles about how long is long in (3), every one of these confident assertions now appears false, on the basis of the sort of evidence we have presented or are about to present in this book. Even if we ourselves are not personally scandalized by the notion of other animals as close relatives, even if our age has accommodated to the idea, the passionate resistance of so many of us, in so many epochs and cultures, and by so many distinguished scholars, must say something important about us. What can we learn about ourselves from an apparent error so widespread, propagated by so many leading philosophers and scientists, both ancient and modern, and with such assurance and self-satisfaction?

One of several possible answers: A sharp distinction between humans and "animals" is essential if we are to bend them to our will, make them work for us, wear them, eat them—without any disquieting tinges of guilt or regret. With untroubled consciences, we can render whole species extinct—for our perceived short-term benefit, or even through simple carelessness. Their loss is of little import: Those beings, we tell ourselves, are not like us. An unbridgeable gap has thus a practical role to play beyond the mere stroking of human egos.[7] Darwin's formulation of this answer was: "Animals whom we have made our slaves, we do not like to consider our equals."[8]

———

We now proceed, in Darwin's footsteps,[9] to examine some of the multitude of proffered definitions of ourselves, explanations of who we are. We will try to see whether they make sense, especially in the light of what we know about the other beings that share the Earth with us.

One of the earliest attempts at an unambiguous characterization of humanity was Plato's: Man is a featherless biped. When news of this advance in the art of definition reached the philosopher Diogenes, so the story goes, he introduced a plucked chicken into the weighty deliberations of Plato's celebrated Academy, asking the assembled scholars to salute "Plato's man." This is of course unfair, because chickens are ordinarily born with feathers, just as they are ordinarily born with two feet. How we mutilate them afterwards does not change their fundamental nature. But the academicians took Diogenes' challenge seriously and added another qualification: Humans were redefined as featherless bipeds with broad flat nails.

Surely this does not get us very far to the essence of human nature. The Platonic definition might suggest, though, a necessary if not a sufficient condition, because standing on two legs is essential for freeing the hands, hands are the key to technology, and many people think our technology defines us. Still, raccoons and prairie dogs have hands and no technology, and bonobos walk upright a good part of their lives. We will address chimpanzee technology shortly.

———

In his classic justification of free enterprise capitalism, Adam Smith asserts that "the propensity to truck, barter, and exchange one thing for another . . . is common to all men, and to be found in no other race of animals."[10] Is this true? Private property was proposed as the central difference between humans and the other animals by Martin Luther in the sixteenth century, and by Pope Leo XIII in the nineteenth.[11] Is *this* true?

Chimpanzees are fond of trade, and understand the idea very well: food for sex, a back rub for sex, betrayal of the leader for sex, spare my baby's life for sex, virtually anything for sex. Bonobos take these exchanges to a new level. But their interest in barter is by no means restricted to sex:

[Chimpanzees] are famous for their tradesmanship. Experimental studies indicate that the ability comes without any specific training. Every zookeeper who happens to leave his broom in the baboon cage knows there is no way he can get it back without entering the cage. With chimpanzees it is simpler. Show them an apple, point or nod at the broom, and they understand the deal, handing the object back through the bars.[12]

With regard to females at least, chimp males have a well-developed sense of private property (raised to institutional status among the hamadryas baboons), and a rudimentary sense of private property attaches to food and to some tools.

The Wealth of Nations was published in 1776, well before any serious study had been made of the lives of the apes, even in captivity. However, Smith's argument about the uniqueness of trade among humans is embedded in a deeper misreading of the animal world:

In almost every other race of animals, each individual, when it is grown up to maturity, is entirely independent and in its natural state has occasion for the assistance of no other living creature. But man has almost constant occasion for the help of his brethren, and it is in vain for him to expect it from their benevolence only. He will be more likely to prevail if he can interest their self-love in his favor and show them that it is for their own advantage to do for him what he requires of them.[13]

But the gregariousness of the primates is one of their hallmarks. Mutual aid in working both sides of the predator/prey relationship and in conflict with other groups of the same species is widespread, not just among the primates, but among most mammals and birds.

While selfishness, exploitation, and trade are commonplace in chimpanzee society, we cannot use this fact along with our kinship with chimps to justify laissez faire economics. Nor can we use it to discredit free market societies on the grounds of their being ape-like.* Cooperation, friendship, and altruism are also chimp traits, but this is

---

* On July 14, 1858, Friedrich Engels wrote in a letter to Karl Marx: "Nothing discredits modern bourgeois development so much as the fact that it has not yet succeeded in getting beyond economic forms of the animal world."

not an argument for some competing socialist economic doctrine. Recall the macaques who would rather go hungry than administer an electric shock to other, not closely related macaques—going so far as to reject even substantial material incentives. Is this a rebuke to advocates of capitalism? At least as far back as Aesop, animal behavior has been used to buttress this or that economic theory. Even in our ideological debates, we make the other animals work for us.

———

"Man is a social animal," wrote Aristotle, or, as it is sometimes translated, "Man is a political animal." This was meant to be characteristic of humans, but not defining; again, a necessary but not a sufficient condition. The subtle and volatile factionalism of chimp and bonobo societies shows how far off the mark this is as a distinction of humanity. The social insects—ants, bees, termites—have much better organized and much more stable social structures than humans. Particular aspects of human social behavior fare no better, although a great many such definitions have been proposed: For example, humans tenderly cherish their young, but so do most other mammals and birds.

"Courage is the peculiar excellence of man," Tacitus recorded the Roman aristocrat Claudius Civilis as saying.[14] Even if the heroic exploits of mother birds shamming a broken wing, or of elephants and chimps saving their young from predators or rushing water, or of the beta hind staring the wolf in the eye so her companions can escape— even if such examples were unknown in the time of this Claudius, didn't he know about dogs? He was put in chains and brought before Nero. History does not record how much of the "peculiar excellence" was available to him in his hour of need.

Another ancient definition of humans, tracing back to Aristotle, is a "rational animal."[15] This is the distinction pointed to by many of the key figures in Western philosophy. But the categorizing chimps, reasoning by analogy and transitive inference, the conversing bonobos, and the culturally innovative macaques remind us that other animals reason also; not as well as the great Western philosophers, to be sure —but the philosophers believed not in a difference of degree, but in a radical difference in kind.

"[M]an differs from irrational creatures in this, that he is master of his actions," was a tenet of St. Thomas Aquinas in his *Summa Theologica*. But are we "masters" of our actions always and in all circumstances? Do other animals never exhibit "mastery"? In giving, as was his practice, selected pros and cons for the propositions discussed, Aquinas—debating "whether choice is to be found in irrational animals?"—mentions a case where a stag at a crossroads seemed to choose one path by excluding the alternatives. This is rejected as evidence of choice because "choice properly belongs to the will, and not to the sensitive appetite which is all that irrational animals have. Therefore, irrational animals are not able to choose." He also held that "irrational animals" could not command, "since they are devoid of reason." All this may have satisfied generations of philosophers, and established a tradition that influenced Descartes, but is it not clear that Aquinas—consider his starting point of "irrational animals"—was begging the question, assuming what he was trying to prove?[16]

"Actions directed towards a goal do not occur in any other animals at all," in a like vein wrote Jakob von Uexküll, a once influential expert on animal behavior.[17] But we need only think of the chimp holding a club behind his back and searching for his rival, or collecting stones to throw at an enemy, or the female prying his fingers open and removing the stones, to realize how much in error such statements are.

For the philosopher John Dewey, what distinguishes us is memory:

> With the animals, an experience perishes as it happens, and each new doing or suffering stands alone. But man lives in a world where each occurrence is charged with echoes and reminiscences of what has gone before, where each event is a reminder of other things.[18]

This claim is manifestly untrue for many animals, and chimps above all live in a world "charged with echoes and reminiscences." The cat experiencing a hot stove avoids the stove thereafter; elephants and deer soon grow wary of hunters; dogs who have been beaten cower when the rolled-up newspaper is raised; even worms, even one-celled protozoa can be taught to run a simple maze. The dominance hierarchy is a frozen memory of past coercion. How oblivious of the real life of nonhuman animals is Dewey's attempt to define us!

Many human sexual practices have been thought to be defining.

Maybe it's kissing: "Only mankind kisses. Only mankind has the reason, the logic, the happy faculty of being able to appreciate the charm, the beauty, the extreme pleasure, the joy, the passionate fulfilment of the kiss!" rhapsodizes a small book on the subject.[19] But chimps routinely and exuberantly kiss.

Maybe what's special about us is our reproductive posture: "It seems plausible to consider that face-to-face copulation is basic to our species."[20] But face-to-face copulation is common among the bonobos.

Concealed ovulation and female orgasm[21] have been thought unique to humans, but bonobos do not garishly advertise their ovulations, and female chimps, bonobos, stumptail monkeys and, probably, many other primate females have orgasms—as determined in part by equipping them with physiological sensors before they mate, in the style of an experiment by Masters and Johnson.

Maybe it's our mode of sexual coercion: "That rape . . . is an exclusively human character seems to be beyond serious doubt," opined a scientist writing on primates in 1928.[22] But rape is known among orangutans and stumptails, violent sexual coercion is a commonplace among baboons and chimps, and the doubt is serious indeed.

Maybe it's the elaboration and duration of our sexual foreplay; in this at least some humans may lead the other primates.[23] But this is learned behavior, as the prevalence of premature ejaculation, especially among adolescent boys, and the self-taught ability of many men to postpone ejaculation make clear. In the integration of sexual acts into everyday social life humans are probably down toward the bottom of the primate list. Most human cultures demand that even socially condoned sexual behavior be carried on in private;[24] we can see something of the sort in chimp consortship, and in clandestine encounters out of sight of the dominant males.

Maybe our distinction is the traditional and striking gender-specific division of labor: The men hunt and fight; the women gather and nurture.[25] But this cannot be a defining characteristic, because chimps have a similar division of labor: Patrols, group defense, and throwing missiles are all mainly male responsibilities; caring for the young and using tools to crack open nuts are mainly female responsibilities. Also, women's and men's jobs are in our time becoming increasingly indistinguishable.

Our long childhood, the years between birth and puberty, is essential for our education, but it is not as long as an elephant's; and the progressively earlier arrival of sexual maturity in the human life cycle over the last few centuries is whittling down our childhood so that it is now only a little longer than the chimpanzees' (who sexually mature around age ten). Play is so central to our growing up that it was once suggested[26] to call our species *Homo ludens* ("the man who plays"). But play can be seen throughout the mammalian class, especially when maturity is long delayed.

The Roman philosopher Epictetus, a former slave, held the distinguishing characteristic of humans to be personal hygiene.[27] He must have known about birds, cats, and wolves but argued that "when . . . we see any other animal cleaning itself, we are accustomed to speak of the act with surprise, and to add that the animal is acting like a man." But he then complains that many men are "dirty," "stinking," and "foul" and do not share this "distinguishing" characteristic. Such a man is advised to "go into a desert . . . and smell yourself."

Humans have been called the only animal that laughs. However, chimps smile and laugh a lot.[28] The Athenian Stranger, in Plato's *Laws*,[29] says humans are "afflicted with the inclination to weep more than any other animal." But this inclination varies widely from culture to culture, and whimpering and crying is a fact of daily life among the chimps, children and adults alike.[30]

Humans—who enslave, castrate, experiment on, and fillet other animals—have had an understandable penchant for pretending that animals do not feel pain. On whether we should grant some modicum of rights to other animals, the philosopher Jeremy Bentham stressed that the question was not how smart they are, but how much torment they can feel. Darwin was haunted by this issue:

> In the agony of death a dog has been known to caress his master, and every one has heard of the dog suffering under vivisection, who licked the hand of the operator; this man, unless the operation was fully justified by an increase of our knowledge, or unless he had a heart of stone, must have felt remorse to the last hour of his life.[31]

From all criteria available to us—the recognizable agony in the cries of wounded animals, for example, including those who usually utter hardly a sound*—this question seems moot. The limbic system in the

---

* For example, water buffalo in Southeast Asia, which are routinely castrated by crushing their testicles between two rocks.[32]

human brain, known to be responsible for much of the richness of our emotional life, is prominent throughout the mammals. The same drugs that alleviate suffering in humans mitigate the cries and other signs of pain in many other animals. It is unseemly of us, who often behave so unfeelingly toward other animals, to contend that only humans can suffer.

Murder, cannibalism, infanticide, territoriality, and guerilla warfare are not unique to humans, as described in preceding chapters. Ants have slaves and domesticated animals and main force warfare.

"The use of punishment in the attempt to train their young in anything other than avoidance," writes Toshisada Nishida, "seems exclusively limited to humans . . . No nonprimate mammals are known to teach by discouragement."[33] But his exception of the nonhuman primates says much. Also, many animals coerce and punish the young as part of the educational process, aiding smooth entrance into the dominance hierarchy. It's a little like hazing and initiation rites in our species.

Humans have institutionalized marriage and advocated monogamy, at least as an ideal; but gibbons, wolves, and many species of birds practice monogamy and mate for life. The courtship dances of animals are surely a kind of marriage ceremony. The following characteristics are described as typical of human marriage:

> There is some degree of mutual obligation between wife and husband. There is a right of sexual access (often but not invariably exclusive). There is an expectation that the relationship will persist through pregnancy, lactation, and childrearing. And there is some sort of legitimization of the status of the couple's children.[34]

But all of this is known in other animals, for example among the gibbons, plus primogeniture.

The nineteenth-century philosopher and theologian Ludwig Feuerbach—known for his influence on Karl Marx—proposed that the distinction of humans is recognition of ourselves as a species.[35] But many animals readily distinguish members of their own species from members of all others—for example, through olfactory cues. And humans are notable for demonizing members of their own species, declaring them less than human, to disinhibit sanctions on murder—especially during wartime.

Humans are sometimes said to be better at making class distinctions than other primates are,[36] but primate dominance hierarchies, some of them hereditary, seem to embrace a fineness of social discrimination that in some respects exceeds even our own.

We conclude that none of these sexual and social traits seem to work as defining characteristics of the human species. The behavior of other animals, especially the chimps and bonobos, renders such pretensions specious. They are just too much like us.

——

Knowledge and behavior patterns that are not hardwired into our genetic material, but rather are learned and passed on within a given group from generation to generation, are called culture. Could culture be the defining mark of humanity?

"Culture," says a major article in *The Encyclopaedia Britannica*,

> is due to an ability possessed by man alone. The question of whether the difference between the mind of man and that of the lower animals is one of kind or of degree has been debated for many years, and even today [1978] reputable scientists can be found on both sides of this issue. But no one who holds the view that the difference is one of degree has adduced any evidence to show that non-human animals are capable, to any degree whatever, of a kind of behaviour that all human beings exhibit.

The author then gives three examples of behavior that he thinks characterizes humans, and concludes, "There is no reason or evidence that will lead one to believe that any animal other than man can have or be brought to any appreciation or comprehension whatever of such meanings and acts."[37]

And what are these three examples? One is "defining and prohibiting incest." But this prohibition, at least for the father-daughter and mother-son varieties, is, as we've described, prevalent, indeed nearly invariable, among the primates—who have elaborate conventions to guarantee high levels of outbreeding. The taboo applies to many other animals as well. In studying Kenyan birds known as bee-eaters, the biologist Stephen Emlen carefully noted the identity and behavior of each bird; in eleven years of work he was unable to find even a single case of incest, either between siblings or between parent and off-spring. (The other two examples given in the *Britannica* article are "classifying one's relatives and distinguishing one class from another," which chimps do well enough—at least for mother-child and sibling kinship—and "remembering the sabbath to keep it holy," which is an institution unknown in many *human* cultures.)

Despite the common description of the incest prohibition as a taboo —that is, learned—it seems to be, to a considerable degree, innate. It serves as a hereditary ethical proscription, evolved for good genetic reasons, and reinforced by the conventions and rules of society (although, for all that, functioning imperfectly—very imperfectly in civilized society).

Clearly chimps have at least the rudiments of culture. In different forests, they must deal with different local geographies and ecologies. They remember over weeks—maybe over years—termite mounds, drumming trees, or, in one account, the site of a noteworthy combat. Such matters are common knowledge. Each group, with its own terrain and its own sequence of historical events, has its own miniature culture. Mutually isolated groups of chimps have different conventions in fishing for termites or driver ants, in using leaves as sponges for soaking up drinking water, in how they hold on to each other during grooming, in some aspects of the gestural language of courtship, and in hunting protocols.[38] And thanks to Imo, the macaque genius who figured out how to separate the wheat from the sand, we even have some insight into the emergence and spread of new discoveries and new cultural institutions among the primates.

The celebrated philosopher Henri Bergson—an exponent of the "revolt against reason" and best known for the idea that some immaterial "vital impulse" permeates life and makes evolution go—wrote that "man . . . is alone in realizing that he is subject to illness."[39] But chimps have a vast pharmacopoeia all around them, and a kind of folk or herbal medicine. For example, for chimps both at Gombe and at Mahale, leaves of a plant called *Aspilia* are a kind of dietary staple, preferentially eaten in the early morning. Despite the wrinkled noses of those partaking (the taste is bitter), it's consumed by both sexes, all ages, the healthy as well as the sick. But there's something odd about it: The chimps eat these leaves regularly, but consume very few of them at any one time—so their nutritional value is in doubt. In the rainy season, though, when apes are plagued by intestinal worms and other illnesses, ingestion increases dramatically. Analysis of *Aspilia* leaves reveals the presence of a powerful antibiotic and an agent that kills nematodes. It's a good guess they're treating themselves. Among other examples, a chimp sick with an intestinal disorder ingested large amounts of the shoots of a plant, different from *Aspilia* and not ordinarily a part of its diet, which also proved to be rich in natural antibiotics.[40]

How is "chimpanzee ethnomedicine" possible? Could it be based on some kind of hereditary information: You feel sick and suddenly you have a craving for a leaf whose shape or aroma is implanted in your brain from the beginning—like the goslings who are said to be born with a hereditary fear of the silhouette of a hawk? Or, more probably, is this *cultural* information passed on—by emulation or instruction—from generation to generation, and subject to rapid change if the available medicinal plants change, or if new diseases arise, or if new ethnomedical discoveries are made? Except that there are apparently no professional herbalists or medical specialists among the apes, chimpanzee folk medicine does not seem so different from human folk medicine. There's a common complaint for which everyone knows what medicine to take. It's something you learn as you grow up. Why the medicine works is a mystery to them—as it still is, in many cases, to us.

Some scholars have imagined that sexual repression was the first and inaugurating facet of human culture.[41] Unrestrained expression of sexual desire—especially among young men and women—will destroy the framework of society, it is suggested, so early human cultures must have placed severe restraints on sexual activity, and encouraged guilt, modesty, hard work, cold showers, and clothing. However, there are many human cultures, often in the tropics, with frameworks apparently uncompromised by the fact that adults go around unselfconsciously stark naked—or perhaps with a thin vine or cotton belt that conceals no sexual parts. In South America, Yanomamo women are wholly unclothed, except for such a belt; the men tie their foreskins to their belts (although they are embarrassed should the penis slip free).[42] In New Guinea and elsewhere, men cover up by wearing gourd sheaths that immodestly exaggerate their proportions. Before the Europeans arrived, the aboriginal peoples of Australia, even those in chilly climates, wore no clothes at all. In ancient Greece, Egypt, and Crete, adult nakedness was common, at least for slaves and athletes (although women spectators were excluded from the Olympic games on the grounds that it would be immodest for them to watch male athletes competing in the nude). Nudist camps seem to be models of decorum. Restraints on the permissible can be much less severe than the more repressive cultures ever imagine—as Captain James Cook's crews discovered in Tahiti.

Victorian sexual attitudes are clearly not characteristic of our species. Moreover, sexual jealousy is a common cause of domestic violence among monkeys and apes; despite their more relaxed sexual standards, they have inhibitions in place. All primate societies, humans and everyone else, set limits on acceptable practice. Sexual repression and associated feelings of shame cannot be the hallmark of our species.

Another aspect of cultural life sometimes thought to be uniquely human is art, dance, and music. But given pencils or paints, chimps with considerable drive and deliberation make art that, though exclusively nonrepresentational as far as we can see, is thought presentable in some circles.[43] Male bower birds decorate their nests guided by an aesthetic that resonates with ours; they regularly replace picked flowers, feathers, and fruit that are no longer fresh; their art evolves through the summer. Gibbons fling themselves balletically through the high forests, and chimps can be counted on to rock and roll at waterfalls and in rainstorms. Chimps delight in resonant drumming, and gibbons in song. Although we like to think it has reached its greatest elaboration in us, culture is not restricted to humans, or even[44] to the primate order.

Here is a 1932 joint assessment of primate and human culture by Solly Zuckerman:

> At the one extreme there is the monkey or ape with its harem, frugivorous [fruit-eating], without any vestige of cultural processes. At the other extreme is man, usually monogamous, omnivorous, whose every activity is culturally conditioned. Socially there are no obvious comparisons between man and ape. [45]

Put aside the facts that chimps eat meat, that most monkeys and apes have no harems, and—a fact known even in 1932—that in many cultures humans are not "usually" monogamous; and compare Zuckerman's assessment with Toshisada Nishida's, in a much later overview of twenty-five years of research on chimpanzees in the Mahale mountains:

> [T]he following social behavioral patterns are known to be present in both the chimpanzee and our own species: strong tendency to avoid incest, long-lasting mother-offspring relationship, male philopatry [males remaining in the group they are born into], strong antagonism among groups, cooperation among males, development of reciprocal altruism, triadic awareness [for example, sexual triangles], alliance fickleness strategy, revenge system, sex difference in political behavior . . . [46]

Much of this may be genetically, as well as culturally, determined, but "socially" there *do* seem to be some "obvious comparisons" between man and ape.

———

Consciousness and self-awareness are, in the West, widely esteemed as the essence of being human (although the absence of self-awareness is considered a state of grace and perfection in the East); the origin of consciousness is imagined to be an unfathomable mystery, or—not so different—the consequence of the insertion of an immaterial soul into each human being, but into no other animal, at the moment of conception. Consciousness may not be so mysterious a trait, though, that supernatural intervention is needed to explain it. If its essence is a lucid awareness of the distinction between the inside of the organism and the outside, between you and everyone else, then, as we've argued, most microorganisms are to this degree conscious and aware; and then the origin of consciousness on our planet dates back more than 3 billion years. There were vast numbers of microscopic creatures then, buffeted by sea swells and ocean currents, reveling in the sunlight, each with a rudimentary consciousness—perhaps only a microconsciousness, or even a nano- or picoconsciousness. [47]

Every cell in a healthy body can make the distinction between itself and others, and those that cannot, that suffer from auto-immune diseases, quickly kill themselves or fall prey to disease microorganisms. But maybe you're thinking that a cell distinguishing itself from another cell (in your body or in the primeval sea) is not what is generally meant by consciousness or self-awareness, that even for exceptionally unreflective humans there's more to it than that. Yes. As we've said, only the most rudimentary kind of consciousness can be imagined in the early history of life on Earth. Of course, there's been substantial evolution since then. Do we know—it might be a very hard thing to know—whether any other animals have our kind of self-awareness?

This is often thought to be a key facet of our humanity, especially because of what else it makes possible:

> The attribute of self-awareness, which involves man's capacity to discriminate himself as an object in a world of objects other than himself, is . . . central to our understanding of the prerequisites of man's social and cultural mode of adjustment . . . A human social order implies a mode of existence that has meaning for the individual at the level of self-awareness. A human social order, for example, is always a moral order . . . It is man's capacity for and development of self-awareness that makes such unconscious psychological mechanisms as repression, rationalization and so on of adaptive importance for the individual.[48]

A fish, a cat, a dog, or a bird catching sight of itself in the mirror apparently understands the image only as another member of the same species. If unhabituated to mirror images, male animals may attempt to intimidate the reflection; it must be sensed as a rival male. The image intimidates back and the animal may flee. Eventually, it accommodates to the silent, odorless, and harmless image and learns to ignore it. By mirror reflection criteria, these animals don't seem very smart. It is said that human children must usually be about two years old before they grasp that their mirror image is not some other child with a talent for imitation. In recognizing what a reflection is, monkeys also are like fish, cats, dogs, birds, and human infants. They don't get it. But some apes are like us.

In 1977 the psychologist Gordon Gallup published an article entitled "Self-Recognition in Primates."[49] When chimpanzees born in the wild were confronted with a full-length mirror, at first—like other animals—they thought the reflection was someone else. But within a few days they had it figured out. Then, they'd use the mirror to preen, and to examine inaccessible parts of themselves, looking over their shoulders to view their backs, for example. Gallup then anesthetized the chimps and painted them red—in places that they could see only in the mirror. Upon regaining consciousness and resuming the pleasures of self-examination in the mirrors, they quickly discovered the red marks. Did they reach out to the ape in the glass? Instead, they groped their own bodies, touched the painted areas repeatedly, and then smelled their fingers. They trebled the time they spent each day examining their mirror images.*

Among the other great apes, Gallup found mirror self-awareness in orangs, but not in gorillas. Later, he found it in dolphins. We are conscious, he proposes, when we know that we exist, and have a mind when we monitor our own mental states. By these criteria, Gallup concludes, chimps, orangs, and dolphins are conscious and have minds.[50]

"As to what concerns fidelity, there is no animal in the world so treacherous as man," said Montaigne.[51] But male fireflies skillfully interpose their own blinks so as to make the courting message of their rivals disagreeable to the females. Some chimp females vampirishly stalk young mothers of their group, waiting for the chance to steal and eat their newborns. Many primates seek surreptitious matings when the alpha's attention is elsewhere. Few of the male alliances rippling through the dominance hierarchy persist beyond their utility. Deception in the social relations of animals, and even self-deception in animals, is an emerging and productive topic in biology; whole books are written about it.[52]

Chimps sometimes lie. They also sometimes try to outwit others who are lying. This fact surely affords us a glimpse inside their minds:

An especially telling example is the duplicity displayed by chimpanzees trying to keep the locality of cached food to themselves, and the cunning of others at beating the bluff . . . You cannot—logically cannot—tell lies unintentionally; even the idea of self-deception in-

---

* Watching themselves in the mirror wearing hats is also a wildly popular and apparently gripping experience.

volves the intentional model, one part of the self trying to put it over
on the rest. The dissembling chimpanzee appears to be acting on the
understanding of what the signs he gives will mean to others, and
hence intentionally.[53]

And yet it is not so long ago that a modern philosopher, among
many others, was saying,

> It would be senseless to attribute to an animal a memory that distin-
> guished the order of events in the past, and it would be senseless to
> attribute to it an expectation of an order of events in the future. It
> does not have the concepts of order, or any concepts at all.[54]

How could he know that?

The chimpanzee's interior monologue is doubtless not up to the
standard of the average philosopher's, but that they have some notion
of themselves, what they look like, what their needs are, their past
experiences, future expectations, and how they relate to others—
enough for the purposes of a "social order"—seems beyond doubt.

———

"Language is our Rubicon," declaimed the famous nineteenth-century linguist Max Müller, "and no brute will dare to cross it." Language permits widely dispersed humans to communicate with one
another. It allows us to sample the wisdom of the past and time-binds
the generations. It is an essential tool in helping us to sharpen our
mental acuity, to think more clearly. It is an unsurpassed aid to memory. With good reason we prize it. Long before the invention of writing, language played a major role in human success. This is the main
reason that Huxley could reassuringly conclude, "Our reverence for
the nobility of manhood will not be lessened by the knowledge that
Man is, in substance and structure, one with the brutes."[55] But does
this mean that other animals must lack language, even simple language, even the capability for language? We are struck by Müller's
military, defensive metaphor, and the possibility he seems to raise that
language is within the grasp of "brutes" and that only timidity restrains
them.

A long tradition of similar confident assertions denying language to the beasts dates from the start of the European Enlightenment, perhaps beginning with a 1649 letter by René Descartes:

> The principal argument, to my mind, which may convince us that the brutes are devoid of reason, is that . . . it has never yet been observed that any animal has arrived at such a degree of perfection as to make use of a true language; that is to say, as to be able to indicate to us by the voice, or by other signs, anything which could be referred to thought alone, rather than to a movement of mere nature; for the word is the sole sign and the only certain mark of the presence of thought hidden and wrapped up in the body; now all men, the most stupid and the most foolish, those even who are deprived of the organs of speech, make use of signs, whereas the brutes never do anything of the kind; which may be taken for the true distinction between man and brute.[56]

That chimps and bonobos can engage in a rich flow of gestural and lexigraphic signs is beyond doubt. We have glimpsed the vigorous scientific debate about their ability to use language. The nervousness of some scientists about claims of chimp language is evident in many ways—including repeatedly changing the rules after the game has begun. For instance, some scientists denied that Ameslan-signing chimps have language because of an apparent absence of negations or interrogatives. As soon as the chimps began objecting and asking questions, the critics discovered some other aspect of language that the chimps presumably did not have while humans did, and *that* now became the *sine qua non* of language.[57] To a surprising extent, scientists and philosophers have merely asserted, sometimes with extraordinary vehemence, that apes cannot use language, and then dismissed evidence to the contrary because it contradicted their assumption.[58] Darwin's view, in contrast, was that some animals have the power of language, "at least in a rude and incipient degree," and that if "certain powers, such as self-consciousness, abstraction, etc., are peculiar to man," they are "mainly the result of the continued use of a highly developed language."

There is controversy over how many meaningful and non-redundant words chimps can routinely put into a sentence. But there is no dispute that chimps (and bonobos) can manipulate hundreds of signs or ideograms taught to them by humans; and that they use these words to communicate their wishes. As we've discussed, the words can stand for objects, actions, people, other animals, or the chimp itself. There are common and proper nouns, verbs, adjectives, adverbs. Chimps and bonobos can request, and therefore are clearly thinking about, things or actions not now present—food, for example, or grooming. There is evidence that—like the Ameslan-literate Lucy and the lexigram-literate Kanzi—they can put words together in new combinations to make a novel kind of sense. Some of them invent and tend to abide by at least a few simple grammatical rules. They can label and categorize inanimate objects, animals, and people using not just the things themselves, but arbitrary words *representing* the things. They are capable of abstraction. They seem sometimes to use language and gesture to lie and deceive, and to reflect an elementary understanding of cause and effect. They can be self-reflexive, not just in action, as with their mirror images, but also in language, as when a chimp named Elizabeth was cutting an artificial apple with a knife and signed, in a special token language in which she was fluent, "Elizabeth apple cut."

They know at best only about 10% the number of words in "basic English" or other minimal vocabularies adequate for everyday human life. This difference has been exaggerated—as by one distinguished linguist who argues that a finite number of human words can be combined to generate an "infinite" number of sentences, and an "infinite" number of communicable subjects, while chimps are stuck in their finitude.[59] In fact, of course, the entire range of human words and ideas is, as for apes, resolutely finite. The laboratory linguistic accomplishments of chimps and bonobos are in addition to their own repertoire of signals—in gesture, sound, and smell—of which we understand, probably, very little. "The word," the "use of signs" which Descartes denied to "brutes," is plainly present in chimps and bonobos.

No ape has ever shown linguistic abilities approaching those of a normal child entering kindergarten. Nevertheless they seem to have a clear-cut, although elementary, ability to use language. Many of us would grant that a child of two or three who has a vocabulary and verbal dexterity comparable to that of the most accomplished chimps or bonobos—no matter how glaring their deficits in grammar and syntax—has language.[60] It has been conventional wisdom in the social sciences that culture presupposes language and language presupposes a sense of self. Whether this is true or not, chimps and bonobos evidently have, at least in a rudimentary form, all three: consciousness, language, and culture. They may be much less repressed than we are and not as bright, but they, also, can think.

Most of us have a memory like this: You're lying in your crib, having awakened from your nap. You cry for your mother, at first tentatively, but when no one comes, more emphatically. Panic mounts. Where is she? Why doesn't she come? you think, or something along those lines —although not in words, because your verbal consciousness is still almost wholly undeveloped. She enters the room smiling, she reaches in and picks you up, you hear her musical voice, you smell her perfume —and how your heart soars! These powerful emotions are preverbal—as are much of our adult anticipations, passions, forebodings, and fears. Our feelings are present before they can be parsed into neat grammatical packages, to be dealt with and subdued. In those dimly remembered feelings and associations, we may glimpse something of the consciousness and emotional lives of chimps, bonobos and our immediate prehuman ancestors.

Chapter 20

THE ANIMAL WITHIN

[T]he human brain is an imperfect instrument built up through long geological periods. Some of its levels of operation are more primitive and archaic than others. Our heads, modern man has learned, may contain weird and irrational shadows out of the subhuman past—shadows that under stress can sometimes elongate and fall darkly across the threshold of our rational lives. Man has lost the faith of the eighteenth century in the enlightening power of pure reason, for he has come to know that he is not a consistently reasoning animal. We have frightened ourselves with our own black nature and instead of thinking "We are men now, not beasts, and must live like men," we have eyed each other with wary suspicion and whispered in our hearts, "We will trust no one. Man is evil. Man is an animal. He has come from the dark wood and the caves."

LOREN EISELEY
Darwin's Century[1]

We have now brought our story—our fragmentary effort to reconstruct some of the entries in the orphan's file, to cast a little light into the shadows—to the threshold of the appearance of humans on Earth. It is time to take stock.

Many of the protective ditches, moats, and minefields painstakingly dug to separate us from the other animals have now been bridged or flanked. Those driven to preserve for us some unique, unambiguous, defining characteristic are tempted to shift the definitions once again and erect a final line of defense around our thoughts. If chimpanzee and bonobo language is limited, we cannot tell much about what they think or feel, what meaning, if any, they give to their lives. They have authored, at least so far, no autobiographies, reflective essays, confessions, self-analyses, or philosophical memoirs. If we can choose particular ideas and feelings to define ourselves, no chimp can contradict us. For example, we might point to our knowledge that all of us will someday die, or that sex is the cause of babies—matters widely understood among humans, although sometimes denied. Perhaps no ape has ever glimpsed these important truths. Perhaps some have. We do not know.[2] But standing alone on such homiletic pinnacles is a hollow victory for the human species. These occasional insights are minor matters compared to the vaunted distinctions of humanity that have crumbled into dust as we have learned more about the other animals. At so fine a level of detail, the motives of those who would define us by this or that idea seem suspect, the human chauvinism manifest.

To compare humans with other animals in regard to behavior amenable to observation is just; but unfavorable comparisons on the basis of first-person accounts emanating from within the animals themselves, their reports of their thoughts and insights, are unfair if no channel of communication into their internal lives has yet been opened. Absence of evidence is not evidence of absence. Were we better able to enter into the mind of the ape, might we not find much more there than we guess?—a point made almost three centuries ago by Henry St. John, the first Viscount Bolingbroke:

> Man is connected by his nature . . . with the whole tribe of animals, and so closely with some of them, that the distance between his intellectual faculties and their . . . appears, in many instances, small, and would probably appear still less, if we had the means of knowing their motives, as we have of observing their actions.[3]

An oft-cited difference purported to exist between human beings and other animals is religion. Only humans, it is said, have religion, and that settles the matter. But what is religion? How could we know whether animals have it? In *The Descent of Man*, Darwin cites the comment, "a dog looks on his master as a god." Ambrose Bierce[4] defined reverence as "the spiritual attitude of a man to a god and a dog to a man." The omega looks on the alpha as something like a god, and the depths of his submission and self-abasement are reached in few extant religions. It is hard to know how profoundly dogs or apes feel reverence, how tinged with awe their attitudes are toward a stern "master" or a well-established alpha, whether they have a sense of the sacred, pray for forgiveness, and otherwise seek to placate and influence forces more powerful than they. Animals raised, educated, and disciplined by much stronger and wiser parents, animals spring-loaded to fit into a dominance hierarchy, animals moreover faced with the daunting presence of human beings armed with life-and-death powers and meting out rewards and punishment—such animals may well have feelings akin to what we call religious. Many mammals and all primates satisfy these conditions.

Over the course of human history, some religions, it is true, have become much more than this—at their best transcending intimidation, hierarchy, and bureaucracy, while providing comfort for the powerless. A few, rare, religious teachers have acted as a conscience for our species, have inspired millions by the example of their lives, have helped us to break out of baboonish lockstep. But none of this contradicts the thesis that a generalized religious predisposition, ready to be put to use by the local social structure, may be a commonplace in the kingdom of the animals.

Perhaps, if we were able to peer into the mind of the ape in a state of nature, we would find—among a flurry of other feelings—a sense of satisfaction about its apeness rivaling ours about our humanity. Every species may feel something similar. It would be far more adaptive than its opposite. If anything like this is true, then we would be denied even our self-congratulatory distinction of being the only animal that makes self-congratulatory distinctions.

If we have not much peered into the hearts and minds of other species and have not even studied them carefully, we may impute to them virtues and strengths as well as vices and deficiencies that in fact they lack. Consider this bit of verse by the poet Walt Whitman:

> I think I could turn and live with animals, they're so
>   placid and self-contain'd,
> I stand and look at them long and long.
>
> They do not sweat and whine about their condition,
> They do not lie awake in the dark and weep for their sins,
> They do not make me sick discussing their duty to God,
> Not one is dissatisfied, not one is demented with the mania
>   of owning things,
> Not one kneels to another, nor to his kind that lived
>   thousands of years ago,
> Not one is respectable or unhappy over the whole earth.[5]

On the basis of the evidence presented in this book, we doubt if any of Whitman's six purported differences between other animals and humans is true—at least given a little poetic license; that is, in the spirit if not the letter of the poem. Montaigne thought[6] that when we conclude that other animals have "ambition, jealousy, envy, revenge, superstition, and despair," we are simply projecting our own "sickly qualities" onto the beasts; but this goes too far, as the lives of the chimps make clear. While many commentators have exaggerated the differences between humans and "animals" and warned of anthropomorphizing, others, like Whitman and Montaigne, have romanticized and sentimentalized the animals. Both excesses serve to deny our kinship.

———

The proximate cause of human success must have something to do with the conjoining of our intelligence and our talent for making and using tools. Surely, our globe-girdling civilization arises chiefly from these two abilities. Without them, we would be nearly defenseless. But "a little dose . . . of judgment or reason often comes into play, even in animals very low in the scale of nature," Darwin wrote in *The Origin*. Late in life, he made extensive studies of what you might think is an unpromising subject, the intelligence of earthworms. He gave them intelligence tests involving the manipulation of real and artificial leaves. They did very well. Flatworms can work their way through a simple maze to get a reward; even worms have a degree of intelligence. Galapagos woodpecker finches, studied by Darwin on the voyage of the *Beagle*, use twigs to worry wood-dwelling larvae out of branches; even birds have a rudimentary technology.

Certainly we could not have invented civilization without intelligence and technology. But it would be unfair to describe civilization as the determining characteristic of our species, or as establishing the level of intelligence and manual dexterity required for our definition, especially because the first 99 percent of the tenure of humans on Earth was spent in an uncivilized state. We were humans then, as now, but we hadn't dreamed up civilization. Yet the fossil remains of the earliest known humans and hominids—dating back not just hundreds of thousands but millions of years—are often accompanied by stone tools. We had the talents, at least in partial measure. We just hadn't gotten around to civilization yet.

The contrast between the proclivity for tools in humans and the absence of tool use in so many other animals has made it tempting to define ourselves as the tool-using or the tool-making animal—as seems to have first been suggested by one of the members of Josiah Wedgwood's and Erasmus Darwin's Lunar Society, Benjamin Franklin. On April 7, 1778, James Boswell confesses to admiring Franklin's definition. The ever-grumpy and sometimes over-literal Samuel Johnson objects: "But many a man never made a tool; and suppose a man without arms, he could not make a tool." Again, if we are to define a human being, should we use traits that, without exception, every human being possesses, or traits that may be present only potentially? And if the latter, who knows what traits lie smoldering in other animals, not yet fully elicited by circumstance or necessity?

———

Blasé, matter-of-fact, encumbered by the infant (who, face to her chest, clutches her fur), she carefully positions the hard-shelled fruit on the log and smashes it open—using a stone tool procured for the purpose. Hammer and anvil. No light bulb goes off above her head. There's no chin to fist, no hint of insight struggling to emerge, no moment of revelation, no strains from *Also Sprach Zarathustra*. It's just another routine, humdrum thing that chimps do. Only humans, who know where tools can lead, find it remarkable.

Although many chimps literally do not know enough to come in out of the rain, they're able to use tools. Not only that: they're able to *premeditate* the use of tools—to acquire a tool now for some action they intend to perform later. They go large distances to find the right kind of stone or stick, and then lug it home. They seem to have had its ultimate use in mind all the while.

"It has often been said," wrote Darwin in *The Descent of Man*, "that no animal uses any tool; but the chimpanzee in a state of nature cracks a native fruit, somewhat like a walnut, with a stone." His source was that acute but easily offended Victorian observer of chimps, Thomas Savage, M.D. Chimpanzees regularly crack open hard-shell seeds and nuts with a stone hammer against a stone or wooden anvil; and they'll carry the appropriate rocks over a good fraction of a kilometer for the purpose. At other times, wooden clubs may be used as nutcrackers. In the Tai Forest in the Ivory Coast, chimps select an appropriate club, climb a cola tree, pick the choice cola nuts, and crack them open using the branch as the anvil and the club as the hammer.[7] Female chimps are more likely to employ hammer-and-anvil technology than males, and they're better at it.*

---

* Similar examples occur in other species. The playful and intelligent sea otter regularly dives to the ocean floor, retrieves hard-shelled mussels and an appropriate stone, swims to the surface, floats on its back, and then cracks open the mussels using the stone as an anvil. Some birds drop bivalves on rocks to crack them open. Egyptian vultures and black-breasted buzzards drop stones from altitude on the large eggs of emus and ostriches in order to dine on the contents.[8] In an apocryphal story,[9] the ancient Greek playwright Aeschylus is said to have been killed when a vulture (or eagle) dropped a heavy stone (or a turtle . . . accounts differ) on his bald head, which it perhaps mistook for the egg of a flightless bird.

A chimp breaks off a long grass stalk or a reed so she may use it later, hundreds of meters away, more than an hour in the future, to lure delectable termites out of a log or termite mound. She must remove superfluous leaves and twigs, shape it, shorten it, insert it into the termite tunnel with a deft twisting motion to follow the interior contours, shake it seductively to attract termites onto it, and then with great care remove it without scraping off too many. Chimps take years to perfect their technique and routinely teach it to their young, who are avid pupils. This exactly satisfies one confident definition of "the uniqueness of man's toolmaking"—namely, "the fashioning, out of natural materials, of an implement designed to be used at a distant time and on objects not now perceptually present." [10]

How difficult is chimpanzee termite fishing? What depth of intellect and manual dexterity are required? Suppose you are dropped naked into the Gombe Preserve in Tanzania and, like it or not, discover that termites are your principal hedge against malnutrition or starvation. You know they're an excellent source of protein; you know that self-respecting humans in many parts of the world regularly eat them. You manage to put aside whatever compunctions you may feel. But catching them one at a time is not going to be worth the effort. Unless you're lucky enough to encounter them when they're swarming, you're going to have to make a tool, repeatedly insert it into their meter-high mound, introduce the tool into your mouth, and strip off the clinging termites with your teeth and lips as you withdraw the tool from your mouth. Could you do as well as a chimp?

The anthropologist Geza Teleki tried to find out. He spent months in Gombe under the tutelage of a chimp named Leakey, who was adept at the technique. Teleki wrote about his findings in a famous scientific paper called "Chimpanzee Subsistence Technology." [11] The Gombe termites mainly come out at night; before dawn they expertly wall up all the entrances to their mounds. Chimps routinely begin their termite foraging by scraping away these entrance barriers. Teleki's inquiry started there:

> Having repeatedly observed [chimpanzee] individuals approach a mound, make a rapid visual scan of the surface while standing on or beside it, and reach decisively out—with a high degree of predictive accuracy—to uncover a tunnel, I was soon impressed by the apparent

ease with which tunnels could be located. In attempting to learn the technique, I applied several experimental procedures: examining in minute detail all crack patterns, protuberances, depressions and other "topographic" features in the clay. But, after weeks of futile searching for the essential clue, I had to resort to scraping mound surfaces with a jackknife until a tunnel was inadvertently exposed. My inability to find any physical features which could serve as visual clues eventually led me to realize that chimpanzees may possess knowledge far beyond my expectations.

. . . The only hypothesis which, at this point, seems to reasonably account for the observed facts is that an adult chimpanzee may know (memorize?) the precise location of 100 or more tunnels in the most familiar mounds. Moreover, since intensive probing is restricted to a short annual season, the possibility that chimpanzees retain a mental map of core mound features during the intervening 10 months must also be considered. That chimpanzees require a prolonged learning period (i.e. 4–5 years) to gain proficiency in this technique . . . , and that some individuals are known to have the capability to retain specific information for many years, provides circumstantial support for this hypothesis.

Next, Teleki looked into a selection of raw materials for the manufacture of the termite probe:

When performed by experienced chimpanzees, the selection procedure seems deceptively simple. After a brief visual scan of the nearby vegetation, a chimpanzee will usually extend a hand and deftly tear off a twig, vine or grass stalk. Sometimes the individual must move a few paces away from the mound and fetch a suitable probe, and in some cases 2–3 objects are initially selected. These may be rapidly examined and discarded until some specification is met in one, or several may be carried to the mound for subsequent selection. Whenever it occurs, the selection is made in a swift, almost casual manner, and modification is begun if necessary. Without being aware of the nuances involved, it is easy to undervalue the proficiency needed to perform these maneuvers.

Chimpanzees presumably have the experience whereby the properties of an object can be evaluated before it is applied to the task of probing, for the rate of error in selecting probes is not high . . . When probing for termites, the specifications are in fact surprisingly stringent: if the vine or grass selected is too pliant, it will buckle and collapse (accordion-like) when inserted into a twisting tunnel; if, on the other hand, the object is too stiff or brittle, it will catch on the tunnel walls and either break or resist entry to the necessary depth . . .

Despite months of observing and aping adult chimpanzees as they selected probes with enviable ease, speed and accuracy, I was unable to achieve their level of competence. Similar ineptness can only be observed in chimpanzees below the age of about 4–5 years.

Finally, putting aside the difficulties in finding the tunnel entrances and manufacturing the tools, Teleki set himself to learning how to use a competently produced tool:

I spent many hours inserting probes, pausing for the designated interval, and pulling them out again—without getting any termites. Only after some weeks of nearly total failure . . . did I finally begin to grasp the problems involved . . .

In order to collect these subterranean termites, the probing object must first be carefully and dextrously inserted to a depth of about 8–16 cm [centimeters], with appropriate turns of the wrist so that the object navigates the twisting channel. The probe must then be gently vibrated with the fingers during the prescribed pause, for without this movement the termites may not be stimulated into biting firmly onto the probe. However, if the vibration is performed too lengthily or roughly, there is an excellent chance that the probe will be cut through by the [termites'] mandibles while still in the tunnel. When these preliminary actions have been correctly performed, the probe, presumably with dozens of termites now attached, must be extracted from the tunnel. Once again there are nuances to be observed. If the object is too rapidly or clumsily pulled out, the insects are likely to be scraped off along the sides of the tunnel, which then yields nothing but a shredded probe. The hand motions must be reasonably but not overly swift and, once started, uniformly fluid and graceful. If the tunnel is particularly tortuous (a feature which can be determined during insertion of the probe), the success of the catch can be ensured by a slow twisting of the wrist while the probe is pulled out.

It is a little daunting to discover—on the very technological grounds on which human superiority is often claimed—that after months of apprenticeship, human scientists cannot do as well as preadolescent chimps. Teleki remained generous and good-natured about his failure. In the acknowledgments at the end of the paper, among thanks to various organizations for financial and logistical support, there appears this sentence: "I am, in addition, more than grateful to the patient and tolerant Leakey, whose termite-collecting skills so outstripped mine."

The chimp style of teaching nut cracking and termite fishing to the young is relaxed—by example and not by rote. The student fiddles with the tools and tries out various approaches, rather than slavishly copying every hand movement of the instructor. Gradually the technique improves. Chimps have for this reason been criticized [12] as not really having culture. (Ironically, one group of scientists denies chimps language because—as we described earlier—they are said to be too imitative, while another group of scientists denies chimps culture because they are said to be not imitative enough.)

The learning style of the great physicist Enrico Fermi was to ask colleagues to state the problems they had recently solved, but to withhold their answers: He could understand the problem only by working it through himself. Learning by doing is—in science and technology, as in many other human activities—much more effective than learning by rote. Knowing, as the chimps do, that a problem exists and can be solved with the tools at hand is most of the battle.

Baboons in Gombe eat termites, but almost entirely during the two- or three-week period in which the insects migrate. Then the baboons can be seen gathering and slurping the insects, and leaping into the air to catch them on the wing. In less bountiful times, baboons will be shooed away from a termite mound by an arriving group of chimpanzees. Sometimes the displaced baboons sit a little distance away, morosely observing the chimps working away with their tools on the mound. When the chimps are done, they leave their modified stalks and reeds at the base of the mound. But no baboon has ever been observed trying to use an abandoned tool—even though it could extend their termite season from weeks to months. Apparently the baboons just don't have it in them. They're not smart enough. Probably their brains are too small.

As chimps are much better than baboons at collecting termites, so some preindustrial humans who routinely eat termites are much better than chimps. They dig open the termite mounds, or fumigate them, or flood them with water. One of the more elegant practices is —with the tongue on the palate, or two pieces of wood gently touched to the mound's surface—to imitate the sound of raindrops, which entices the termites out of their nest.[13] Chimps have never been observed to use these techniques.* Probably they're not smart enough. Probably their brains are too small.

What we find most interesting is the overlap. Some chimps lack even probe technology, and are no better at catching termites than baboons are. Other chimps are armed with a well-developed if rudimentary technology, many steps having to be done correctly and in the right sequence for the method to work—as good as many human cultures, although nowhere near as good as some. There are human cultures barely up to the highest chimpanzee standards of termite catching, and others only on a par with the baboons.[15] No sharp boundaries are apparent here separating baboons from chimps, or chimps from humans.

Chimps also drop branches on intruders and sop up drinking water with leaves. While they cannot be described as fastidious or obsessively hygienic, chimps are known to use leaves as toilet paper and handkerchiefs, and twigs as toothbrushes. They employ sticks for digging up roots, for investigating animals in burrows and knotholes, and —like a croupier at a gaming table—for raking in otherwise inaccessible fruit. If they were able to manufacture more complex tools, they certainly would have the intelligence and dexterity to use them: In zoos, chimps try to steal the keys from the keeper's pocket. When successful, they often manage to open the lock. Like us, they can sometimes use their intelligence to escape from bondage.

Male chimps like to throw missiles—whatever is handy, generally sticks and stones. (Like the inmates of college fraternity houses, they also occasionally throw food.) Females are much less interested in missiles. Chimps would throw stones at the visitors who gawk at them in the traditional kind of zoo—if they had stones. As it is, all they

---

* Although in the Okorobiko Mountains in Guinea, chimps use large sticks to perforate the mounds; the escaping termites are then gathered up by the handful. Other chimp societies in Guinea are ignorant of this practice, although it is also employed by chimp groups in nearby Cameroon and Gabon.[14]

have is feces. When wild chimps are presented with a fairly realistic mechanical leopard, after a reassurance frenzy of screams, hugs, and mutual mountings, they find appropriate clubs and beat the effigy to death—or at least until they knock the stuffing out. Or they'll pelt it with stones. (In the same circumstance, baboons will furiously attack the leopard, but without a thought of using clubs. Baboons just don't know about tools.)

Chimps have stunned or killed by throwing stones. The directionality of their throwing is good. Where they're deficient is in range: In tense confrontations with prey or hostile peers, thrown rocks hit their targets only a few percent of the time. Adolescent boys don't do much better under comparable conditions. But even when inaccurate, a hail of stones can be off-putting.

A distinction needs to be made between tool *using* and tool *making*. Many scientists have conceded tool use to other animals, and, following Benjamin Franklin, defined humans as the sole tool-making animal; where tools are manufactured, it is suggested, language cannot be far behind.[16] But the chimpanzee termite fishery industry makes it clear that chimps, with considerable forethought, both make and use tools. Chimps also have a rudimentary stone technology, although, as far as we know, they don't manufacture stone tools in the wild. In captivity, though, Kanzi—the linguistically talented bonobo—has, imitating human models, hit stones together to produce sharp flakes, which he then uses to cut a string so he can open a box which is filled with food. (This is a causality sequence at least five steps long.) As long as it's sharp enough to cut the string, Kanzi will generally settle for the first crude stone knife he flakes off. But the thicker the rope he must cut, the larger and sharper the knife he makes.[17]

Evidence of chimpanzee talent to combine objects purposefully to make tools has actually been with us for decades:

Between 1913 and 1917, Wolfgang Kohler conducted observations and experiments on the intelligence of chimpanzees at a field station in North Africa. In one study a male chimpanzee, Sultan, was led into a room where a banana had been tied to a string and suspended from the ceiling in a corner. A large wooden box had also been placed in the center of the room, open side up. Sultan first tried to reach the fruit by jumping, but this quickly proved futile. He then "paced restlessly up and down, suddenly stood still in front of the box, seized it, tipped it . . . straight towards the objective . . . began to climb up it . . . and springing upwards with all his force, tore down the banana." A few days later Sultan was taken into a room with a much higher ceiling, where again there was a suspended banana, as well as a wooden box and a stick. After failing to get the banana with the stick alone, Sultan sat down "with an air of fatigue . . . gazed about him, and scratched his head." He then stared at the boxes, suddenly leaped up, seized a box and a stick, pushed the box underneath the banana, reached up with the stick and knocked the fruit down. Kohler was struck with the apparently thoughtful period that preceded Sultan's solution, as well as with his sudden and directed performance. Such "insightful" behavior apparently contrasted with other forms of learning, which develop gradually and depend on reinforcement. [18]

It's not hard to imagine an especially insightful chimp or bonobo wondering if there weren't some way to make a stone flake cut better or a projectile go farther.

Since the progress of human technology is a continuum, to pick a particular milestone—the domestication of fire, say, or the invention of the bow and arrow, agriculture, canals, metallurgy, cities, books, steam, electricity, nuclear weapons, or spaceflight—as the criterion of our humanity would be not just arbitrary, but would exclude from humanity every one of our ancestors who lived before the selected invention or discovery was made. There is no *particular* technology that makes us human; at best it could only be technology in general, or a propensity for technology. But that we share with others.

Like us, nonhuman primates are not all the same. They vary in focus from individual to individual and group to group. Some, like Imo, are technological geniuses. Others, like the hierarchy-besotted macaque males, are hopelessly old-fashioned and stuck in their ways. One chimp population pounds nuts, another does not. Some probe for termites, others only for ants. Some use grass stalks and vines to coax the insects out, others sticks and twigs. Females preferentially use hammers and anvils, males preferentially throw stones. None of them, so far as we know, has ever used a stick to dig out a nutritious root or tuber, although it ought to be possible and adaptive. Some individuals find technology uncongenial or intellectually too taxing and never use it, despite the obvious advantages accruing to other members of their group who are comfortable with technology. Some large groups have no technology at all. "I'm embarrassed to say," says an observer of a community of Ugandan chimps, "that the Kibale chimpanzees appear as the country bumpkins of the chimp world." He goes on to speculate that life is too easy and food too plentiful at Kibale for the challenge of deprivation to elicit the response of technology.[19]

Chimps are smart. They carry accurate mental maps of their territory in their heads. They seem to know the seasonal availability of plant foods and will congregate in some peripheral province of their territory to harvest a small stand of ripening fruits or vegetables. They have rudimentary culture, medicine, and technology. They have a startling capacity for simple language. They can plan for the future. Think again of the sensory and cognitive skills necessary to succeed in chimpanzee social life. You must recognize dozens of faces and their expressions. You must remember what each of these individuals has done to you or for you in the past. You must understand the foibles, weaknesses, ambitions of potential allies and rivals. You must be quick on your feet. You must be very flexible. But if you have all this, there's probably a great deal else about the world that, sooner or later, you can figure out and change.

———

How thoroughly the chimps and bonobos have erased the list of purported human distinctions!—self-awareness, language, ideas and their association, reason, trade, play, choice, courage, love and altruism, laughter, concealed ovulation, kissing, face-to-face sex, female orgasm, division of labor, cannibalism, art, music, politics, and featherless bipedalism, besides tool using, tool making, and much else. Philosophers and scientists confidently offer up traits said to be uniquely human, and the apes casually knock them down—toppling the pretension that humans constitute some sort of biological aristocracy among the beings of Earth. Instead, we are more like the nouveau riche, incompletely accommodated to our recent exalted state, insecure about who we are, and trying to put as much distance as possible between us and our humble origins. It's as if our nearest relatives, by their very existence, refute all our explanations and justifications. So as counterweights to human arrogance and pride, it is good for us that there are still apes on Earth.

Much of this chimp and bonobo behavior was discovered only recently. Doubtless they have other talents that have so far eluded us. We humans are biased observers, with a vested interest in the answer. The cure for this disease is more data. But the study of primate behavior, both in the laboratory and in the wild, is by and large poorly and grudgingly funded.

If we insist on absolute rather than relative differences, we do not, so far at least, discover any distinguishing characteristic of our species. Shouldn't we expect, especially with our close relatives, that the differences will be of degree and not of kind? Isn't this the lesson of evolution? If we require that we uniquely possess tools, culture, language, trade, art, dance, music, religion, or conceptual intelligence, we will not understand who we are. If, on the other hand, we are willing to admit that what distinguishes us from the other animals is more of one propensity and less of another, then we may make some progress. Then, if we wish, we can take pride in the fuller flowering of primate aptitudes that has taken place in our species.

The more an animal weighs, the more of it there is that its brain must control, and so—within certain limits—the bigger its brain needs to be. This is true between species, although not between individual members of a given species. A species with a much bigger brain for its body weight—especially in its higher brain centers—has a good chance of being, on some level, smarter. Indeed, for comparable body weights, humans tend to have bigger brains than other primates; primates than other mammals; mammals than birds; birds than fish; and fish than reptiles.[20] There is some scatter in the data, but the correlation is clear. It corresponds pretty well to the commonly accepted (by humans, of course) rank order of animal intelligence. The earliest mammals had significantly larger brains than their reptilian contemporaries of comparable body weight; and the earliest primates were similiarly well-endowed compared to other mammals. We come from big-brained stock.

Adult humans, who weigh only a little more than adult chimps, nevertheless have brains three to four times more massive. A human infant a few months old already has a larger brain than a grown-up chimpanzee.[21] It seems very likely that we're significantly smarter than the chimps because we have a significantly larger brain—despite the comparable body weights. For a factor of three to four increase in brain weight, the brain *size* (its circumference, say) must increase by about 50%. But the human brain isn't entirely a proportional scaling up of a chimp brain. Despite what Huxley found, there *is* a little bit of brain architecture—not much, but some—that humans have and the other primates at least mainly don't. Significantly, some of it seems to be related to speech.

Some parts of the brain are proportionally much larger in humans than in other primates: The cerebral cortex in general, responsible for thinking, is proportionally much larger in humans than in chimps (or in our non-human primate ancestors); so is the cerebellum, in charge of keeping us steady on our (two) feet.[22] The frontal lobes are far more prominent in humans than in chimps; they're thought to play an important role in foreseeing the future consequences of present action, in planning ahead.*

Still, purported distinctions in brain anatomy must be treated with caution: There are many primates not yet studied with sufficient care, and there have been so many erroneous claims. For example, in humans different information is stored in, and different abilities controlled by, the two hemispheres of the cerebral cortex—a surprising finding that emerges from patients whose bundle of neural fibers connecting the two brain hemispheres has been cut.[23] This asymmetry, called "lateralization," is connected with language, and, arguably, with tool use.[24] So, of course, the conceit arose that only the brains of humans are lateralized.[25] Then songbirds were found to have their songs stored almost exclusively in only one hemisphere of their brains,[26] and lateralization was discovered in chimps that had learned language.[27] In any case, the *qualitative* differences between chimp and human brains, if any, are few and subtle.

So is that all there is to it? Give the chimps a bigger brain and the power of articulate speech, maybe take away some testosterone, cancel the ads for ovulation, burden them with some more inhibitions, give them a shave and a haircut, stand them up on their hind legs, and get them out of the trees at night? Would they then be indistinguishable from the earliest humans?

The possibility that we might be "no more than" deluxe model apes,

---

* Most of the increase in our brain size and the improvements in our brain architecture occurred very quickly—in only the last few million years. There might be some bugs still to be worked out.

that the differences between them and us might be almost wholly differences of degree and not of kind, and that the differences of kind, if they exist, might be elusive—all this was a source of profound discomfort from the earliest days in which human evolution was seriously considered. Just a few years after *The Origin of Species* was published, Huxley wrote:

> [D]esiring, as I do, to reach the wider circle of the intelligent public, it would be unworthy cowardice were I to ignore the repugnance with which the majority of my readers are likely to meet the conclusions to which the most careful and conscientious study I have been able to give to this matter, has led me.
>
> On all sides I shall hear the cry—"We are men and women, not a mere better sort of apes, a little longer in the leg, more compact in the foot, and bigger in brain than your brutal Chimpanzees and Gorillas. The power of knowledge—the conscience of good and evil— the pitiful tenderness of human affections, raise us out of all real fellowship with the brutes, however closely they may seem to approximate us."
>
> To this I can only reply that the exclamation would be most just and would have my own entire sympathy, if it were only relevant. But, it is not I who seek to base Man's dignity upon his great toe, or insinuate that we are lost if an Ape has a hippocampus minor [in its brain]. On the contrary, I have done my best to sweep away this vanity . . .
>
> We are indeed told by those who assume authority in these matters . . . that the belief in the unity of origin of man and brutes involves the brutalization and degradation of the former. But is this really so? Could not a sensible child confute, by obvious arguments, the shallow rhetoricians who would force this conclusion upon us? Is it, indeed, true, that the Poet, or the Philosopher, or the Artist whose genius is the glory of his age, is degraded from his high estate by the undoubted historical probability, not to say certainty, that he is the direct descendant of some naked and bestial savage, whose intelligence was just sufficient to make him a little more cunning than the Fox, and by so much more dangerous than the Tiger?[28]

———

Suppose you own a personal computer. It's roughly the size of a typewriter, sits on your desk, and outcomputes any hundred mathematicians. There was nothing remotely like it on Earth only a few decades ago. Building on the strengths of this model, the manufacturer now introduces a relatively minor variant with a faster and more powerful microprocessor and a few new peripherals. Surely this is not as remarkable an accomplishment as the invention of the personal computer in the first place. But the new computer, you find, can perform a range of functions the old one couldn't. It can figure certain problems out in a reasonable span of time that previously would have taken—for all intents and purposes—forever. There are whole categories of problems you can now solve that you couldn't come within hailing distance of before. But if solving these problems were somehow important for the survival of the personal computer, pretty soon there would be a large number of personal computers with the added capabilities. Perhaps our uniqueness is no more than, or only a little more than, this: an enhancement of well-established pre-existing talents for invention, forethought, language, and general intelligence, enough to cross a threshold in our capacity to understand and change the world.

Still, depending on what else they are allied with, greater reasoning skills need not—necessarily and in all circumstances—be adaptive and improve survival. "Reason more than anything else *is* man,"[29] said Aristotle. Mark Twain countered:

> I think it is open to dispute . . . [The] strongest count against [man's] intelligence is the fact that with that [historical] record back of him he blandly sets himself up as the head animal.[30]

If we imagine that we are purely, or even mainly, rational beings, we will never know ourselves.

We are too weak to destroy or seriously damage the planet, or to extinguish all life on Earth. That is far beyond our powers. But what we *can* do is to destroy our global civilization and, just possibly, sufficiently alter the environment as to render our own species, along with vast numbers of others, extinct.[31] Even at levels far short of those that can cause our extinction, our technology has given us awesome powers—our ancestors would have thought them god-like. This is merely a statement of fact. It is not a remonstrance and is not intended to define us. But it leads us back again to the question of whether we have any choice in the matter, or whether there is some deeply buried part of our nature that, despite the comparative intelligence and promise of our species, will sooner or later arrange matters for the worst.

"We are conscious of an animal in us," wrote Henry David Thoreau, "which awakens in proportion as our higher nature slumbers."[32] The idea is, in a way, obvious; it emerges from even shallow introspection. It goes back at least to Plato,[33] who described how in dreams, "when the gentler part of the soul slumbers and the control of Reason is withdrawn . . . the Wild Beast in us . . . becomes rampant." That Wild Beast, Plato goes on, "will cast off all shame and prudence at such moments and stop at nothing"—including incest, murder, and "forbidden food." The idea of the beast within is also familiar to us from Sigmund Freud, who called it the "id," Latin for "it," and from neurophysiology, starting with the work of J. Hughlings Jackson.[34] A more recent incarnation can be found in the perspective of the neurophysiologist Paul MacLean,[35] who identifies many of the control centers for sex, aggression, dominance, and territoriality in a deep-lying, ancient part of the brain called the R-complex—"R" for reptile, because we share it with the reptiles, who lack much of a cerebral cortex, the seat of consciousness.

We go to great lengths to deny our animal heritage, and not just in scientific and philosophical discourse. You can glimpse the denial in the shaving of men's faces; in clothing and other adornments; in the great lengths gone to in the preparation of meat to disguise the fact that an animal is being killed, flayed, and eaten. The common primate practice of pseudosexual mounting of males by males to express dominance is not widespread in humans, and some have taken comfort from this fact. But the most potent form of verbal abuse in English and many other languages is "Fuck you," with the pronoun "I" implicit at the beginning. The speaker is vividly asserting his claim to higher status, and his contempt for those he considers subordinate. Characteristically, humans have converted a postural image into a linguistic one with barely a change in nuance. The phrase is uttered millions of times each day, all over the planet, with hardly anyone stopping to think what it means. Often, it escapes our lips unbidden. It is satisfying to say. It serves its purpose. It is a badge of the primate order, revealing something of our nature despite all our denials and pretensions.

The danger seems so obvious. Surely there is something in us deeply seated, self-propelled, and on occasion able to evade our conscious control—something that can do harm despite what we understand to be our best intentions: "The good that I would I do not; but the evil which I would not, that I do."[36]

Sometimes, we use our "higher nature," our Reason, to awaken the Wild Beast. It's that stirring animal that terrifies us. If we acknowledge its presence, some fear, we will be sliding toward a perilous fatalism: "That's what I'm like," the criminal might plead. "I've tried to behave myself, abide by the law, be a good citizen, but there's only so much you can ask of me. I've got an animal inside. It's human nature, after all. I'm not responsible for my actions. Testosterone made me do it."[37] Such views, if widely held, could unravel the social fabric, it is feared; therefore, it is better to suppress knowledge of our "animal" natures and pretend that those who perceive and discuss such natures are undermining human self-confidence and playing with fire.

Maybe what we're afraid we'll find if we look too closely is some resolute malevolence lurking in the heart of man, some unquenchable selfishness and blood lust; that down deep we're all mindless crocodilian killing machines. It's an uncomplimentary self-image and of course, if widely held, it *would* work to undermine human self-confidence. In an age when the global environment is within our power to ruin, the notion is not cheering for our future prospects.

What is odd about this point of view—apart from the notion that criminals and sociopaths really take heart from the scientific finding that humans have evolved from other animals—is how selectively it makes contact with the data about animals and, especially, about our closest relatives, the primates. There we can also find friendship, altruism, love, fidelity, courage, intelligence, invention, curiosity, forethought, and a host of other characteristics that we humans should be glad to have in greater measure. Those who deny or decry our "animal" natures underestimate what those natures are. Isn't there much to be proud of, as well as to be ashamed of, in the lives of the monkeys and apes? Shouldn't we be glad to acknowledge a connection with Imo, Lucy, Sultan, Leakey, and Kanzi? Remember those macaques who would rather go hungry than profit from harming their fellows; might we have a more optimistic view of the human future if we were sure our ethics were up to their standards?

And if our intelligence is our distinction, and if there are at least two sides to human nature, shouldn't we be sure to use that intelligence to encourage the one side and restrain the other? When we reconfigure our social structures—and in the last few centuries we've been tinkering with them like mad—isn't it better and safer to have our best understanding of human nature firmly in mind?

Plato was afraid that when the superimposed social controls are slumbering, the wild beast within will incline us to incest "with a mother or anyone else, man, god, or brute," and other crimes. But monkeys and apes and other "wild beasts" hardly ever commit parent-child or sibling-sibling incest. The inhibitions are already up and running in other primates, and for good evolutionary reasons. We demean the other animals when we attribute to them whatever predispositions to incest we find in ourselves. Plato feared that the animal within will incline us to "any deed of blood." But monkeys and apes and other "wild beasts" are powerfully inhibited against shedding blood, at least within the group. The established lexicon of dominance and submission, friendships, alliances, and sexual partnerships keeps real crimes of violence down to a dull roar. Mass murder is unknown. True main-force warfare has never been observed. Again, we undervalue our non-human ancestors when we blame them for our violent proclivities. Very likely, they had inhibitions in place that we routinely circumvent.

Killing an enemy with teeth and bare hands is emotionally far more demanding than pulling a trigger or pressing a button. In inventing tools and weapons, in contriving civilization, we have disinhibited the controls—sometimes thoughtlessly and inadvertently, but sometimes with cool premeditation. If the beasts who are our nearest relatives engaged recklessly in incest and mass murder, they would have rendered themselves extinct. If our non-human ancestors did, we would not be here. For the deficiencies of the human condition, we have only ourselves and our statecraft to blame—not the "wild beasts," and not our distant ancestors, who cannot defend themselves against self-serving accusations.

There is no reason for despair or timidity here. What we should be ashamed of is the counsel that urges us to avoid self-doubt even at the cost of hiding our nature from ourselves. We can solve our problems only if we know who it is we're dealing with. To balance whatever dangerous tendencies we perceive in ourselves is the knowledge that in our ancestors and close relatives, violence is inhibited, controlled, and, in encounters within the species at least, devoted mainly to symbolic ends; that we are gifted in making alliances and friendships, that politics is our business, that we are capable of self-knowledge and new forms of social organization; and that we are able, better than any species that ever lived on Earth, to figure things out and to build things that never were.

Even in the fossil remains of the earliest lifeforms, there is unmistakable evidence of communal living arrangements and mutual cooperation. We humans have been able to design effective cultures that for hundreds of thousands of years have fostered one set of inborn characteristics and discouraged another. From brain anatomy, human behavior, personal introspection, the annals of recorded history, the fossil record, DNA sequencing, and the behavior of our closest relatives, a clear lesson emerges: There is more than one side to human nature. If our greater intelligence is the hallmark of our species, then we should use it as all the other beings use their distinctive advantages—to help ensure that their offspring prosper and their heredity is passed on. It is our business to understand that some predilections we bear as remnants of our evolutionary history, when coupled with our intelligence—especially with intelligence in the subordinate role—might threaten our future. Our intelligence is imperfect, surely, and newly arisen; the ease with which it can be sweet-talked, overwhelmed, or subverted by other hardwired propensities—sometimes themselves disguised as the cool light of reason—is worrisome. But if intelligence is our only edge, we must learn to use it better, to sharpen it, to understand its limitations and deficiencies—to use it as cats use stealth, as walking sticks use camouflage, to make it the tool of our survival.

ON IMPERMANENCE

Death, like a hidden Tiger, lies in wait to slay the unsuspecting.

ASHVAGHOSHA,
Saundaranandakavya, ca. A.D. 1165[38]

Chapter 21

SHADOWS OF FORGOTTEN ANCESTORS

Already have I once been a boy and a girl, and
a bush and a bird, and a silent fish in the sea.

EMPEDOCLES

Purifications[1]

The evolutionary process has made the Earth brim over with life. There are beings that walk, jump, hop, fly, glide, float, slither, burrow, stride on the water's surface, canter, waddle, brachiate, swim, tumble, and patiently wait. Damsel flies molt, deciduous trees bud, great cats stalk, antelopes take fright, birds chatter, nematodes worry a grain of humus, perfect insect imitations of leaves and twigs rest incognito on a branch, earthworms entwine themselves in passionate bisexual embrace, algae and fungi are comfortable roommates in the lichen partnership, great whales sing their plaintive songs as they traverse the world ocean, willows suck moisture from unseen underground aquifers, and a universe of microbes swarms in every thimbleful of muck. There is hardly a clod of soil, a drop of water, a breath of air that is not teeming with life. It fills every nook and cranny of our planet's surface. There are bacteria in the upper air, jumping spiders at the tops of the highest mountains, sulfur-metabolizing worms in the deep ocean trenches, and heat-loving microbes kilometers below the surface of the land. Almost all these beings are in intimate contact. They eat and drink one another, breathe each other's waste gases, inhabit one another's bodies, disguise themselves to look like one another, construct intricate networks of mutual cooperation, and gratuitously fiddle with each other's genetic instructions. They have generated a web of mutual dependence and interaction that embraces the planet.

By 3 billion years ago, life had changed the color of the inland seas; by 2 billion years ago, the gross composition of the atmosphere; by 1 billion years ago, the weather and the climate; by a third of a billion years ago, the geology of the soil; and in the past few hundred million years the close-up appearance of the planet. These profound changes, all brought about by forms of life we tend to consider "primitive," and of course by processes we describe as natural, mock the concerns of those who hold that humans, through their technology, have now achieved "the end of Nature." We are rendering many species extinct; we may even succeed in destroying ourselves. But this is nothing new for the Earth. Humans would then be just the latest in a long sequence of upstart species that arrive on-stage, make some alterations in the scenery, kill off some of the cast, and then themselves exit stage-left forever. New players appear in the next act. The Earth abides. It has seen all this before.

Life has penetrated only a thin surface layer, bounded by the heavens above and something very much like hell below. The planet itself —rotating once a day, revolving about the Sun once a year, circumnavigating the center of the Milky Way Galaxy once every quarter billion years, this world of rock and metal with its deep convection currents that make and destroy continents and generate the planet's magnetic field—the planet knows nothing of life. The Earth would continue on its way as readily without life as with it. The Earth is indifferent, and all but that shallow clement zone at its very surface is impervious to anything life has been able to serve up.

———

Our family tree was rooted when the Earth was just emerging from a time of massive, obliterating impacts, molten red-hot landscapes, and pitch-black skies; when the oceans and the stuff of life were still falling in from space; when our connection with the Universe around us was manifest. The orphan's file began in epic style.

The family tree of a few rare individuals of our species, we've argued, can be traced back perhaps as much as two or three dozen generations. Most of us, in contrast, are able to penetrate only three or four generations into the past before the record fades and is lost. With a rare exception here and there, all earlier ancestors are the merest phantoms. But hundreds of generations link us to the time that civilization was invented, thousands of generations run to the origin of our species, and a hundred thousand generations lie between us and the first member of the genus *Homo*. How many generations link us back through our non-human primate, mammal, reptile, amphibian, fish, and still earlier ancestors to the microbes of the primeval sea, and how many generations before that to the first organic molecules able to make crude copies of themselves is unfathomed—but it might approach 100 billion. The family tree of each of us is graced by all those great inventors: the beings who first tried out self-replication, the manufacture of protein machine tools, the cell, cooperation, predation, symbiosis, photosynthesis, breathing oxygen, sex, hormones, brains, and all the rest—inventions we use, some of them, minute-by-minute without ever wondering who devised them and how much we owe to these unknown benefactors, in a chain 100 billion links long.

Many have construed our clear kinship with the other animals as an affront to human dignity. But any one of us is much more closely related to Einstein and Stalin, to Gandhi and Hitler than to any member of another species. Shall we think more or less of ourselves in consequence? The discovery of a deep connection between human nature, *all* of human nature, and the other living things on Earth comes not a moment too soon. We are helped to know ourselves.

In acknowledging our ties of kinship, we are forced to reconsider the morality (as well as the prudence) of our conduct: wiping out another species every few minutes, night and day, all over the planet. Over the last few decades we have caused the extinction of something like a million species—some providing potential new foods, some desperately needed medicines, but all unique DNA sequences, tortuously evolved over 4 billion years of the evolution of life and all now lost forever. We have been faithless heirs, squandering the family inheritance with little thought for the generations to come.

We must stop pretending we're something we are not. Somewhere between romantic, uncritical anthropomorphizing of the animals and an anxious, obdurate refusal to recognize our kinship with them—the latter made tellingly clear in the still-widespread notion of "special" creation—there is a broad middle ground on which we humans can take our stand.

If the Universe really were made for us, if there really is a benevolent, omnipotent, and omniscient God, then science has done something cruel and heartless, whose chief virtue would perhaps be a testing of our ancient faiths. But if the Universe is heedless of our aspirations and our destiny, science provides the greatest possible service by awakening us to our true circumstances. In accord with the unforgiving principle of natural selection, we are charged with our own preservation—under penalty of extinction.

And yet we go from massacre to massacre; and as our technology becomes more powerful, the magnitude of the potential tragedy grows. The many sorrows of our recent history suggest that we humans have a learning disability. We might have thought that the horrors of World War II and the Holocaust were enough to innoculate us against the toxins there revealed and unleashed. But our resistance quickly fades. A new generation gladly abandons its critical and skeptical faculties. Old slogans and hatreds are dusted off. What was only recently muttered guiltily is now offered as political axiom and agenda. There are renewed appeals to ethnocentrism, xenophobia, homophobia, racism, sexism, and territoriality. And with a sigh of relief we are apt to surrender to the will of the alpha, or long for an alpha we can surrender to.

Ten thousand generations ago, when we were divided into many small groups, these propensities may have served our species well. We can understand why they are almost reflexive, why they should be so easy to evoke, why they are the stock in trade of every demagogue and hack politician. But we cannot wait for natural selection to further mitigate these ancient primate algorithms. That would take too long. We must work with what tools we have—to understand who we are, how we got to be that way, and how to transcend our deficiencies. Then we can begin to create a society less apt to bring out the worst in us.

Still, from the perspective of the last ten thousand years extraordinary transformations have lately been playing themselves out. Consider how we humans organize ourselves. Dominance hierarchies requiring debasing submission and obedience to the alpha male, as well as hereditary alphahood, were once the global standard of human political structure, justified as right and proper and divinely ordained by our greatest philosophers and religious leaders. These institutions have now almost vanished from the Earth. Chattel slavery—likewise long defended by revered thinkers as preordained and deeply consonant with human nature—has been nearly abolished worldwide. Just a minute ago, all over the planet, with only a few exceptions, women were subordinate to men and denied equal status and power; this also was thought predetermined and inevitable. Here too, clear signs of change are now evident nearly everywhere. A common appreciation of democracy and what are called human rights is, with some backsliding, sweeping the planet.

Taken together, these dramatic societal shifts—often in ten generations or less—provide a compelling refutation of the claim that we are condemned, without hope of reprieve, to live out our lives in a barely disguised chimpanzee social order. Moreover, the shifts are occurring so swiftly that they cannot possibly be due to natural selection. Instead, our culture must be drawing forth propensities and predispositions that already reside deep within us.

We humans hold at least 99.9% of our DNA sequences in common. We are far more closely related to one another than we are to any other animal. By the similarity standards we use in other matters, humans—even of the most disparate cultures and ethnic origins—are essentially identical in our heredities. Of the immense number of possible beings, realized and unrealized, we all are cut from the same cloth, made on the same pattern, granted the same strengths and weaknesses, and will ultimately share the same fate. Given the reality of our mutual interdependence, our intelligence, and what is at stake, are we really unable to break out of behavior patterns evolved to benefit our ancestors of long ago?

We have been dismantling ancient institutions that no longer serve, and are tentatively trying out others. Our species is becoming an intercommunicating whole, with powerful economic and cultural bonds linking up the planet. Our problems, increasingly, are global in venue, admitting only global solutions. We have been uncovering the mysteries of our past and the nature of the Universe around us. We have invented tools of awesome power. We have explored the nearby worlds and have set sail for the stars. Granted, prophecy is a lost art and we are not vouchsafed an unclouded view of our future. Indeed, we are almost wholly ignorant of what is coming. But by what right, what argument can pessimism be justified? Whatever else may be hidden in those shadows, our ancestors have bequeathed us—within certain limits, to be sure—the ability to change our institutions and ourselves. Nothing is preordained.

We achieve some measure of adulthood when we recognize our parents as they really were, without sentimentalizing or mythologizing, but also without blaming them unfairly for our imperfections. Maturity entails a readiness, painful and wrenching though it may be, to look squarely into the long dark places, into the fearsome shadows. In this act of ancestral remembrance and acceptance may be found a light by which to see our children safely home.

EPILOGUE

It is not possible to be ignorant of the end of things if we know their beginning.

THOMAS AQUINAS
Summa Theologica[1]

We have described the Earth before humans set foot upon it. We have tried to understand something about our ancestors, using as our guide the fossil record and the gorgeous panorama of life that now graces our planet. While there are still vast numbers of missing pages in our orphan's file, the progress of science has enabled us to glimpse a few of the lost or forgotten entries—perhaps even many of the important items. But we have explored only the early chapters of the file. Its key central section—chronicling the dawn of our species and its evolution up to the invention of civilization—is the subject of the next book in this series.

Notes

Prologue
THE ORPHAN'S FILE

1. Attributed to Empedocles by Sextus Empiricus, in *Against the Mathematicians*, VII, 122–125, in Jonathan Barnes, editor and translator, *Early Greek Philosophy* (Harmondsworth, Middlesex, England: Penguin Books, 1987), p. 163.
2. *Science and Humanism* (Cambridge: Cambridge University Press, 1951). Schrödinger was one of the discoverers of quantum mechanics.
3. In many scientific accounts of the origin of the human species, there is a story something like this. (Cf., e.g., Misia Landau, *Narratives of Human Evolution* [New Haven and London: Yale University Press, 1991].) But rather than being imposed on the evidence, we hold that it flows naturally out of the evidence. Human origins have in fact been very humble. We have in fact, by many standards, become the dominant species on the planet, and done it partly by dint of our own efforts. We are in fact profoundly ignorant of many of the details of our origins. It is natural to represent ourselves in metaphor as a favored child brought up in obscure circumstances, and then as hero venturing forth into the world to seek our identity. The principal danger of the metaphor would be if we thought our success due to one generation or people or nation; or if our success were to blind us to the danger we have placed ourselves in.
4. Robert Redfield, *The Primitive World and Its Transformations* (Ithaca, NY: Cornell University Press, 1953), p. 108.
5. Fyodor Dostoyevsky, *Brothers Karamazov* (1880), translated by Richard Pevear and Larissa Volokhonsky (San Francisco: North Point Press, 1990), Book Six, Chapter 3, p. 318.
6. Mary Midgley, *Beast and Man: The Roots of Human Nature* (Ithaca, NY: Cornell University Press, 1978), pp. 4, 5.

7. A similar metaphor was employed in *The Origin of Species*, Chapter 10, where Charles Darwin compared the geological record to "a history of the world imperfectly kept, and written in a changing dialect; of this history, we possess the last volume alone. . . Only here and there a short chapter has been preserved; and of each page only here and there a few lines."

Chapter 1
ON EARTH AS IT IS IN HEAVEN

1. In Lucien Stryk and Takashi Ikemoto, translators, *Zen Poems of China and Japan: The Crane's Bill* (New York: Grove Press, 1973), p. 20.
2. Translated by Dennis Tedlock (New York: Simon and Schuster/ Touchstone, 1985, 1986), p. 73.
3. What we are describing here is the origin of our Solar System— not the origin of the Universe, or at least its latest incarnation, which is most often described as the Big Bang.
4. The Second Law of Thermodynamics specifies that in any process, the net orderliness of the Universe must decrease. Some places may get more orderly as long as others get more chaotic. There is plenty of order to draw on in the Universe, and nothing in the Second Law is inconsistent with the origin of the planets or the beginnings of life.
5. Except for a tiny fraction generated by the radioactive decay of atoms hailing originally from elsewhere in the Galaxy.
6. Two millennia after his last worshipper died, the name of this god was given to a newly discovered planet.

Chapter 2
SNOWFLAKES FALLEN ON THE HEARTH

1. Translated by Dennis Tedlock (New York: Simon and Schuster/ Touchstone, 1985, 1986), p. 72.
2. In *Just So Stories* (New York: Doubleday, Page & Company, 1902), p. 171.
3. The image of an hour's drive up or down is, so far as we know, originally due to the astronomer Fred Hoyle.
4. Suppose, for the sake of argument, that the primeval sea had just

the same size and depth as our present ocean. Suppose also that the organic molecules on the primitive Earth, in the absence of any life to eat them up, lasted about 10 million years before they fell to pieces from molecular old age, or were carried down toward the Earth's molten interior. Then, in the best case, the primitive oceans would have been about a 0.1% solution of organic matter (about the consistency of a very thin beef broth). For the whole world ocean. Some lakes, bays, and inlets may have been a much more concentrated solution of organic molucules. (Christopher Chyba and Carl Sagan, "Endogenous Production, Exogenous Delivery, and Impact-Shock Synthesis of Organic Molecules: An Inventory for the Origins of Life," *Nature* 355 [1992], pp. 125–132.)

5. D. H. Erwin, "The End-Permian Mass Extinction," *Annual Review of Ecology and Systematics* 21 (1990), pp. 69–91.

6. The end-Permian catastrophe was far more severe than the end-Cretaceous catastrophe some 200 million years later in which all the dinosaurs died.

7. Marcus Aurelius, *Marcus Aurelius: Meditations*, IV, 48, translated by Maxwell Staniforth (Harmondsworth, UK: Penguin Books, 1964), quoted in Michael Grant, ed., *Greek Literature: An Anthology* (London and Harmondsworth, Middlesex, England: Penguin Books, 1977), p. 430.

8. The Venerable Bede, *The Ecclesiastical History of the English Nation (Historia Ecclesiastica)* (London: J. M. Dent, 1910, 1935) (written in 732), Book II, Chapter XIII, p. 91.

Chapter 3
"WHAT MAKEST THOU?"

1. And still it burns. On the day we write this, the authors received yet another expression of outrage from a viewer offended by the endorsement of evolution in our *Cosmos* television series. "We teach our children that they are descended from monkeys, and then are surprised when they act accordingly," he writes. "Throw out an absolute standard of morality, make all behavior relative, and the result must be moral chaos." He offers no critique of the evidence for evolution, but only of its imagined social consequences.

Even today, some American high school biology curricula are still giving equal time to special creation (and to a subject oxymoronically called "scientific creationism"). Should time also be devoted in school geography curricula to the evidence for the proposition that the Earth is flat?—a view clearly held by the authors of the Bible and still supported by fringe advocacy groups. Both special creation and the flat Earth hypothesis were reasonable scientific guesses in the sixth century B.C., when Genesis was compiled. They are no longer.

Standard works defending creationism include D. T. Gish, *Evolution? The Fossils Say No!* (San Diego: Creation Life Publishers, 1979), and H. M. Morris, *Scientific Creationism* (ibid, 1974). Among the many refutations by scientists are A. N. Strahler, *Science and Earth History* (Buffalo, N.Y.: Prometheus, 1987); D. J. Futuyama, *Science on Trial: The Case for Evolution* (New York: Pantheon, 1983); G. B. Dalrymple, *The Age of the Earth* (Stanford, CA: Stanford University Press, 1991); Tim M. Berra, *Evolution and the Myth of Creationism* (ibid, 1990); and a forthright pamphlet by the National Academy of Sciences, *Science and Creationism* (Washington, D.C.: National Academy Press, 1984) that describes special creation as "an invalidated hypothesis," and concludes: "No body of beliefs that has its origin in doctrinal material [such as the Bible] rather than scientific observation should be admissible as science . . . Incorporating the teaching of such doctrines into a science curriculum stifles the development of critical thinking . . . and seriously compromises the best interests of public education." Among the many virtues of Berra's book is its dedication ("For my mother, who allowed me to read during meals").

In a 1982 Gallup poll, 44% of American respondents supported the statement "God created man pretty much in his present form at one time within the last ten thousand years." Only 9% supported the statement "Man has developed over millions of years from less advanced forms of life. God had no part in this process." (*Creation/Evolution*, No. 10 [Fall 1982], p. 38.)

In a 1988 survey of 43 members of the U.S. Congress who chose to answer a questionnaire, 88% felt that "modern evolutionary theory has a valid scientific foundation," but less than half could say, even roughly, what the basic idea of evolution might be. Only one in three strongly agreed with the statement that the Earth was 4 to 5 billion years old. In an identical survey of a quarter of the members of the Ohio legislature, the corresponding numbers were 74%, 23%, and 23%. (Michael Zimmerman, "A Survey of Pseudoscientific Sentiments of Elected Officials," *Creation/Evolution*, No. 29 [Winter 1991/1992], pp. 26–45.)

2. Erasmus Darwin, *The Botanic Garden*, Part II, *The Loves of the Plants* (1789), Canto III, line 456; in Desmond King-Hele, editor, *The Essential Writings of Erasmus Darwin* (London: MacGibbon & Kee, 1968), p. 149.

3. Dumas Malone, *Jefferson and His Time*, Volume One, *Jefferson the Virginian* (Boston: Little, Brown, 1948), p. 52.

4. Gerhard Wichler, *Charles Darwin: The Founder of the Theory of Evolution and Natural Selection* (Oxford: Pergamon Press, 1961), p. 23.

5. London, 1803 (published posthumously). Quoted in Howard E. Gruber, *Darwin on Man: A Psychological Study of Scientific Creativity* (Chicago: The University of Chicago Press, 1974), p. 50.

6. This example is from J. B. S. Haldane, *The Causes of Evolution* (New York: Harper, 1932), p. 130.

7. And in August Weismann's late-nineteenth-century experiment, five successive generations of mice had their tails cut off with no effect on the progeny. George Bernard Shaw dismissed such examples as missing Lamarck's point: The mice do not *aspire* to be tailless, as the giraffes are purported to strive for long necks (*Back to Methuselah: A Metabiological Pentateuch* [New York: Brentano's, 1929]). This is magical thinking. Surviving incarnations of Lamarck's hypothesis include the idea that the disobedience of Adam in the Garden of Eden caused an "original sin" genetically propagated to future generations (accepted by the Catholic Church at the Council of Trent and reaffirmed in a 1950 papal encyclical of Pius XII); and the fraudulent agricultural genetics of Stalin's favorite pseudoscientist, Trofim Lysenko. Nevertheless, the inheritance of acquired characteristics—while apparently wrong at the level of the organism—may be right at the level of the gene: A mutation is a chemical accident slightly changing the

structure of a gene. Descendent genes inherit the accident. But the knife of August Weismann was too blunt to reach into the genes.

8. Sir Francis Darwin, editor, *Charles Darwin's Autobiography, with His Notes and Letters Depicting the Growth of the ORIGIN OF SPECIES* (New York: Henry Schuman, 1950), pp. 29, 30.

9. *Ibid.*, pp. 34, 35.

10. John Bowlby, *Charles Darwin: A New Life* (New York: W. W. Norton, 1990), p. 110.

11. *Ibid.*, p. 118.

12. *Charles Darwin's Autobiography*, p. 33.

13. *Ibid.*, p. 37.

14. Stephen Jay Gould, *Ever Since Darwin* (New York: Norton, 1977), p. 33.

15. Charles Darwin, *The Voyage of the Beagle* (London: J. M. Dent & Sons Ltd., 1906), p. 18.

16. Frank H. T. Rhodes, "Darwin's Search for a Theory of the Earth: Symmetry, Simplicity and Speculation," *British Journal of the History of Science* 24 (1991), pp. 193–229.

17. *The Autobiography of Charles Darwin* (unexpurgated edition edited by Nora Barlow, his granddaughter) (New York: Harcourt Brace, 1958), p. 95.

18. Bowlby, *op. cit.*, p. 233.

19. Francis Darwin, editor, *The Life and Letters of Charles Darwin* (London: John Murray, 1888), Volume II, p. 16.

20. Ronald W. Clark, *The Survival of Charles Darwin: A Biography of a Man and an Idea* (New York: Random House, 1984), p. 90.

21. *Ibid.*, pp. 90, 91.

22. *Ibid.*, p. 105.

23. An excerpt from Wallace's article:

"Wild cats are prolific and have few enemies; why then are they never as abundant as rabbits? The only intelligible answer is, that their supply of food is more precarious. It appears evident, therefore, that so long as a country remains physically unchanged, the numbers of its animal population cannot materially increase. If one species does so, some others requiring the same kind of food must diminish in proportion. The numbers that die annually must be immense; and as the individual existence of each animal depends upon itself, those that die must be the weakest—the very young, the aged, and the diseased,—while those that prolong their existence can only be the most perfect in health and vigour —those who are best able to obtain food regularly, and avoid their numerous enemies. It is, as we commenced by remarking, 'a struggle for existence,' in which the weakest and least perfectly organized must always succumb. . ." (Alfred Russel Wallace, "On the Tendency of Varieties to Depart Indefinitely from the Original Type" [Wallace's contribution to Darwin and Wallace, "On the Tendency of Species to Form Varieties; and on the Perpetuation of Varieties and Species by Natural Means of Selection"], in *Journal of the Proceedings of the Linnean Society: Zoology*, Volume III [London: Longman, Brown, Green, Longmans & Roberts, and Williams and Norgate, 1859], pp. 56, 57.)

24. In subsequent editions, the sentence was amended to read "*Much* light will be thrown on the origin of man and his history" (our emphasis).

Chapter 4
A GOSPEL OF DIRT

1. In *Philosophical Works, with Notes and Supplementary Dissertations by Sir William Hamilton*, with an Introduction by Harry M. Bracken, 2 volumes (Hildesheim: Georg Olms Verlagsbuchhandlung, 1967), Vol. 1, p. 52.

2. Charles Darwin, *The Origin of Species by Means of Natural Selection or the Preservation of Favored Races in the Struggle for Life* (New York: The Modern Library, n.d.) (originally published in 1859) (Modern Library edition also contains *The Descent of Man and Selection in Relation to Sex*), Chapter XV, "Recapitulation and Conclusion," p. 371.

3. Of course, the traditional religious understanding of adaptation has been God's will. However, this is not an explication of process.

4. Unattributed quotations in this chapter are excerpted from Charles Darwin, *op. cit.*, pp. 29, 31, 33, 34, 64–67, 359, and 370; and from Charles Darwin and Alfred R. Wallace, "On the Tendency of Species to Form Varieties; and on the Perpetuation of Varieties and Species by Natural Means of Selection," *Journal of*

the Proceedings of the Linnean Society: Zoology, Volume III (London: Longman, Brown, Green, Longmans & Roberts, and Williams and Norgate, 1859), p. 51.

5. Francis Darwin, editor, *The Life and Letters of Charles Darwin* (John Murray: London, 1888), Volume III, p. 18.

6. *The Westminster Review* 143 (January 1860), pp. 165–168.

7. *The Edinburgh Review* 226 (April 1860), pp. 251–275.

8. John A. Endler's *Natural Selection in the Wild* (Princeton: Princeton University Press, 1986) provides a useful modern summary of what natural selection is and isn't, its role in evolution, and how to test that it operates. His Table 5.1, culled from the recent scientific literature, summarizes over 160 "direct demonstrations" of natural selection in the wild.

9. *The North American Review* 90 (April 1860), pp. 487 and 504.

10. *The London Quarterly Review* 215 (July 1860), pp. 118–138.

11. *The North British Review* 64 (May 1860), pp. 245–263.

12. *The London Quarterly Review* 36 (July 1871), pp. 266–309.

13. George Bernard Shaw, *Back to Methusaleh: A Metabiological Pentateuch* (New York: Brentano's, 1929), p. xlvi. The last sentence is in fact the modern evolutionary point of view.

14. James Watt, U.S. Secretary of the Interior in the first Reagan term, justified despoiling public lands on the grounds that he was unsure how much time we had "until the Lord comes." Manuel Lujan, U.S. Secretary of the Interior under President Bush, argued against protecting endangered species because "[M]an is at the top of the pecking order. I think that God gave us dominion over these creatures . . . consider the human being on a higher scale. Maybe that's because a chicken doesn't talk . . . God created Adam and Eve, and from there all of us came. God created us pretty much as we look today." (Ted Gup, "The Stealth Secretary," *Time*, May 25, 1992, pp. 57–59.) Genesis urges us to "subdue" Nature, and predicts that "fear" and "dread" of us is to be upon "every beast." These religious precepts have practical consequences in the human assault on the environment (cf. John Passmore, *Man's Responsibility for Nature: Ecological Problems and Western Traditions* [New York: Scribner's, 1974]). Leaders of a wide variety of religions have nevertheless taken forthright stands and political action to protect the environment (e.g., Carl

Sagan, "To Avert a Common Danger: Science and Religion Forge an Alliance," *Parade*, March 1, 1992, pp. 10–15).

15. Alfred Russel Wallace, the co-discoverer with Darwin of evolution by natural selection—a generous and self-effacing man who described himself as "shy, awkward and unused to good society"—differed with him on one crucial matter. He was willing to accept that every beast and vegetable had so evolved, but not humans. Some divine (and self-reproducing) spark had to be injected at a comparatively recent date in the evolutionary process, he held. Wallace's evidence?

Unlike the racists of his time, Wallace was struck that the brain size and anatomy of all humans are sensibly the same: "The more I see of uncivilized people, the better I think of human nature, and the essential differences between civilized and savage men seem to disappear . . . We [also] find many broad statements as to the low state of morality and of intellect in all prehistoric men which the facts hardly warrant." (Quoted in Loren Eiseley, *Darwin's Century* [New York: Doubleday, 1958], p. 303.) But pretechnological peoples, he thought, had no need of a brain able, say, to invent steam engines. So the human brain must somehow have been contrived early *in order to* perform complex adaptive functions much later. Such foresight, he well understood, was inconsistent with the fortuitous and short-term nature of natural selection. Thus, "some higher intelligence may have directed the process by which the human race was developed." (*Ibid.*, p. 312.)

However, Wallace greatly underestimated the complexity of pre-industrial societies. There has never been a pretechnological human culture. Fashioning stone tools and hunting large animals are by no means easy. Big brains were an advantage to us from the start.

Wallace was also transfixed by the spate of spiritualist demonstrations so popular in late Victorian England, including spirit rapping, seances, conversations with the dead, materializations of "ectoplasm," and the like. These seemed to reveal a hidden spirit component of humans, but of no other living things. So far as we know, this heady brew was concocted out of equal parts skillful charlatans and credulous upper-class audiences. The magician Harry Houdini played an important role in later exposing some of these impostures. Wallace was hardly the only eminent Victorian to be taken in.

When, toward the end of this book, we explore the extraordinary cognitive talents of chimpanzees as revealed in laboratory tests, a similar question occurs to us: How can they be preadapted to solve such complex problems? And the answer, or at least part of it, may be the same as for Wallace's conundrum: In their everyday lives in the wild, chimps need a broad-gauge, multi-purpose intelligence—not nearly as advanced as what humans have, but much more than we might think.

16. Nora Barlow, editor, *The Autobiography of Charles Darwin* (New York: Harcourt Brace, 1958), p. 95.

17. James H. Jandl, *Blood: Textbook of Hematology* (Boston: Little Brown, 1987), pp. 319 *et seq.* See also David G. Nathan and Frank A. Oski, *Hematology of Infancy and Childhood*, 3rd ed. (Philadelphia: W. B. Saunders, 1987), Chapter 22.

18. A. C. Allison, "Abnormal Haemoglobin and Erythrocyte Enzyme Deficiency Traits," in D. F. Roberts, editor, *Human Variation and Natural Selection, Symposium of the Society for the Study of Human Biology* 13 (1975), pp. 101–122.

19. Nora Barlow, *op. cit.*, p. 93.

20. An influential modern assessment from the Darwinian perspective of the behavior of animals in groups is E. O. Wilson's *Sociobiology: The New Synthesis* (Cambridge, MA: Harvard University Press, 1975). The book in general excited little controversy, but the closing chapter—in which natural selection was applied to humans—elicited a storm of criticism, including the pouring of a pitcher of water over the author's head at a scientific meeting. Wilson has taken care to stress that human behavior is the product of both hereditary and environmental influences, and has generally made his claims modestly and cautiously: "I might easily be wrong—in any particular conclusion, in the grander hopes for the role of the natural sciences, and in the trust gambled on scientific materialism . . . The uncompromising application of evolutionary theory to all aspects of human existence will come to nothing if the scientific spirit itself falters, if ideas are not constructed so as to be submitted to objective testing and hence made mortal." (E. O. Wilson, *On Human Nature* [Cambridge, MA: Harvard University Press, 1978], pp. x–xi.)

We can glimpse something of the fervor of this debate in the following, perhaps intemperate, remarks: "American social scientists fear and despise biology, although few of them have troubled to learn any . . . Again and again in the writings of social scientists, we find 'biological' equated with 'invariant'. . . This usage betrays an incomprehension of the domain of biology." (Martin Daly and Margo Wilson, *Homicide* [New York: Aldine de Gruyter, 1988], p. 154.)

Excellent recent books on evolution for the general reader include those by Richard Dawkins (e.g., *The Selfish Gene* [Oxford: Oxford University Press, 1976]; *The Extended Phenotype* [Oxford: Oxford University Press, 1982]; *The Blind Watchmaker* [New York: Norton, 1986]) and by Stephen J. Gould (e.g., *Ever Since Darwin* [New York: Norton, 1977]; *The Panda's Thumb* [New York: Norton, 1980]; *Wonderful Life* [New York: Norton, 1990]). By comparing these books, we can glimpse the healthy and vigorous scientific debate that thrives under the aegis of modern evolutionary biology.

21. John Bowlby, *Charles Darwin: A New Life* (New York: W. W. Norton, 1990), p. 381.

22. Francis Darwin, *op. cit.*, Volume I, pp. 134, 135.

23. *Ibid.*, Volume III, p. 358.

24. See, e.g., Leonard Huxley, *Thomas Henry Huxley* (Freeport, NY: Books for Libraries, 1969); Cyril Bibby, *Scientist Extraordinary* (Oxford: Pergamon, 1972).

25. Cyril Bibby, *T. H. Huxley: Scientist, Humanist and Educator* (London: Watts, 1959), pp. 35, 36.

26. Thomas H. Huxley, "On the Hypothesis that Animals Are Automata, and its History" (1874), in *Collected Essays*, Volume I, *Method and Results: Essays* (London: Macmillan, 1901), p. 243.

27. Francis Darwin, editor, *The Life and Letters of Charles Darwin* (London: John Murray, 1888), Volume III, p. 358.

28. Bibby, 1959, *op. cit.*, p. 259.

29. All quotations except that attributed to Emma Darwin at the end are taken from eyewitness accounts, although most were written down years and even decades after the event. A memorable essay on the debate, "Knight Takes Bishop?" is in Steven J. Gould's *Bully for Brontosaurus* (New York: W. W. Norton, 1991). Our version of Huxley's response to Wilberforce is from the recollec-

tions of G. Johnstone Stoney, who was present. (Stoney did pioneering work on the escape to space of planetary atmospheres, and was the first to understand why the Moon is airless.) It differs from Huxley's own later recollection, which went like this: "If then, said I, the question is put to me would I rather have a miserable ape for a grandfather or a man highly endowed by nature and possessed of great means of influence, & yet who employs those faculties & that influence for the mere purpose of introducing ridicule into a grave scientific discussion—I unhesitatingly affirm my preference for the ape." (Bibby, 1959, *op. cit.*, p. 69.)

Chapter 5
LIFE IS JUST A THREE-LETTER WORD

1. *The Bhagavad Gita*, translated by Juan Mascaró (London: Penguin, 1962), Introduction, p. 14.
2. Lucien Stryk and Takashi Ikemoto, translators, *Zen Poems of China and Japan: The Crane's Bill* (New York: Grove Press, 1973), p. 87.
3. Even in our language there remains the idea that motion requires a soul. But if there is a dusty soul that decides for every mote how and when it is to move, what animates that soul? Does it have a still smaller soul—a soul's soul—and so on, in an infinite regress of microscopic immaterial motivators? No one believes this. And if the soul of the dust mote *doesn't* need its own smaller soul to tell it what to do, why does the dust mote itself need a soul? Might it move on its own, without spiritual direction?
4. The discovery of discrete units of heredity, the genes, dates back to experiments first published in 1866 by the plant breeder Gregor Mendel. His work was essentially unread until his laws of genetics were independently rediscovered at the beginning of the twentieth century. Charles Darwin knew nothing about Mendel's work; it would have made his task much easier if he had. While nucleic acids were discovered in cells in 1868, their central importance for heredity was first suspected only in the 1940s. The remarkable structure of DNA—with long chains of nucleotides like the letters in a book, and two intertwined strands suggesting a ready means

of replication—was first understood in 1953 by James Watson and Francis Crick. Classical genetics had been wholly innocent of the chemistry of the gene.

5. How reading the genetic instructions of different organisms might unlock the evolutionary record was first stated by Emile Zuckerkandl and Linus Pauling, "Molecules as Documents of Evolutionary History," *Journal of Theoretical Biology* 9 (1965), pp. 357–366.

6. Loren Eiseley, *The Immense Journey* (New York: Vintage, 1957).

7. Wen-Hsiung Li and Dan Graur, *Fundamentals of Molecular Evolution* (Sunderland, MA: Sinauer Associates, 1991), Figure 21, p. 135. The sequences shown are from the DNA encoding the 5S ribosomal-RNA [r-RNA] sequences.

8. *Ibid.*, pp. 6, 10.

9. Cf. Edward N. Trifonov and Volker Brendel, *Gnomic: A Dictionary of Genetic Codes* (New York: Balaban Publishers, 1986), p. 8.

10. Natalie Angier, "Repair Kit for DNA Saves Cells from Chaos," *New York Times*, June 4, 1991, pp. C1, C11.

11. Daniel E. Dykhuizen, "Experimental Studies of Natural Selection in Bacteria," *Annual Review of Ecology and Systematics* 21 (1990), pp. 373–398.

12. Quoted in Monroe W. Strickberger, *Evolution* (Boston: Jones and Bartlett, 1990), p. 34.

13. A semi-popular early exposition by Lord Kelvin of his argument (he was then merely "W. Thomson" of the University of Glasgow) appeared as "On the Age of the Sun's Heat" in the March 1862 number of *Macmillan's Magazine*.

14. Thomas Henry Huxley, "On a Piece of Chalk," in *Collected Essays*, Volume VIII, *Discourses: Biological and Geological* (London and New York: Macmillan, 1902), p. 31.

15. Niles Eldredge, *Time Frames: The Rethinking of Darwinian Evolution and the Theory of Punctuated Equilibria* (New York: Simon and Schuster, 1985). There are several different kinds of "punctuation" possible. Those stressed (and for good reason) by Eldredge and Gould are consistent with the prevailing views of evolutionary biologists since World War II (e.g., George Gaylord Simpson, *Tempo and Mode in Evolution* [New York: Columbia University Press, 1944]), or, indeed, with the views of Darwin

himself (e.g., Richard Dawkins, *The Blind Watchmaker* [New York: Norton, 1986], Chapter 9). Contrary to the claims of creationists, the debate about punctuated equilibrium poses no challenge to evolution or natural selection. Gould has been especially effective in defending the teaching of Darwinian evolution in the schools.

16. More exactly, each strand manufactures a complementary strand, in which As are substituted for Ts, Gs for Cs, and vice versa. When, in due time, the complement reproduces, the original strand is duplicated, and so on. But the same genetic *information* is copied every generation.

17. RNA is the messenger by which DNA conveys what proteins are to be made by the cell. It is also the catalyst that presides over the linking up of amino acids into the proteins specified by the DNA. (M. Mitchell Waldrop, "Finding RNA Makes Proteins Gives 'RNA World' a Big Boost," *Science* 256 [1992], pp. 1396–1397, and other articles in the June 5, 1992 issue of *Science*.) To an increasing number of molecular biologists, these facts suggest an early form of life in which RNA did the information storage, replication, and catalysis all by itself, with DNA and proteins taking over later.

18. Jong-In Jong, Qing Feng, Vincent Rotello, and Julius Rebek, Jr., "Competition, Cooperation, and Mutation: Improvement of a Synthetic Replicator by Light Irradiation," *Science* 255 (1992), pp. 848–850; J. Rebek, Jr., private communication, 1992. A survey of the present state of knowledge is Leslie Orgel, "Molecular Replication," *Nature* 358 (1992), pp. 203–209.

19. In Lucien Stryk and Takashi Ikemoto, translators, *Zen Poems of China and Japan: The Crane's Bill* (New York: Grove Press, 1973), p. xlii.

Chapter 6
US AND THEM

1. Book XXII, line 262.
2. Lynn Margulis, *Symbiosis in Cell Evolution* (San Francisco: W. H. Freeman, 1981).
3. Andrew H. Knoll, "The Early Evolution of Eukaryotes: A Geological Perspective," *Science* 256 (1992), pp. 622–627.

4. Margulis, *op. cit.*

5. L. L. Woodruff, "Eleven Thousand Generations of *Paramecium*," *Quarterly Review of Biology* 1 (1926), pp. 436–438.

6. Z. Y. Kuo, "The Genesis of the Cat's Response to the Rat," *Journal of Comparative Psychology* 11 (1930), pp. 1–30.

7. Benjamin L. Hart, "Behavioral Adaptations to Pathogens and Parasites: Five Strategies," *Neuroscience and Biobehavioral Reviews* 14 (1990), pp. 273–294.

8. George C. Williams and Randolph M. Nesse, "The Dawn of Darwinian Medicine," *Quarterly Review of Biology* 66 (1991), pp. 1–22.

9. Harry J. Jerison, "The Evolution of Biological Intelligence," Chapter 12 of Robert J. Sternberg, editor, *Handbook of Human Intelligence* (Cambridge: Cambridge University Press, 1982), Figure 12–11, p. 774.

10. A view championed in recent times by the neurophysiologist Paul D. MacLean and described in Carl Sagan's *The Dragons of Eden: Speculations on the Evolution of Human Intelligence* (New York: Random House, 1977). MacLean sets forth a comprehensive summary of his views in *The Triune Brain in Evolution: Role in Paleocerebral Functions* (New York and London: Plenum Press, 1990).

11. This approach is made most accessible to the general reader in Richard Dawkins's book *The Selfish Gene*, revised edition (Oxford: Oxford University Press, 1989). In a vivid passage (pp. 19–20), he describes the genes as swarming "in huge colonies, safe inside gigantic lumbering robots, sealed off from the outside world, communicating with it by tortuous indirect routes, manipulating it by remote control. They are in you and me; they created us, body and mind; and their preservation is the ultimate rationale for our existence . . . [W]e are their survival machines."

12. A related and even more heated controversy—on whether the mother bird has any notion of what she's doing or is merely some carbon-based automaton—is addressed later in this book. Reciprocal altruism, an exchange of present for future favors, is also admitted by those who deny group selection per se.

13. Martin Daly and Margo Wilson, *Homicide* (New York: Aldine de Gruyter, 1988), pp. 88, 89.

14. W. D. Hamilton, "The Genetical Evolution of Social Behavior,"

Journal of Theoretical Biology 7 (1964), pp. 1–51; John Maynard Smith, "Kin Selection and Group Selection," *Nature* 201 (1964), pp. 1145–1147.

15. Imagine that the huddled group (of, say, insects) is in the shape of a sphere. The heat generated by the group is proportional to its volume (to the cube of its size), but the heat radiatively lost by the group is proportional to its area (to the square of its size). Thus the bigger the group is, the more heat it retains. In a large group, only a small proportion of members are on the surface of the sphere, where an individual is exposed to the cold; the remainder are satisfyingly surrounded by warm bodies on all sides. The smaller the group is, the greater the proportion of individuals on the chilly periphery.

16. Up to some limit, when the individuals doing the mobbing get in each other's way.

17. Dawkins, *op. cit.*, p. 171, citing the work of Amotz Zahavi.

18. *Ibid.*, Preface to 1989 edition. For an opposing, now minority, point of view, see V. C. Wynne-Edwards, *Evolution Through Group Selection* (Oxford: Blackwell, 1986): "The view, widely held, that group selection can be dismissed as an effective evolutionary force is based on assumptions, not on evidence . . . It is an argument uncritically derived from human experience, of cheaters, criminals and oppressors who live at other people's expense; and it ignores the fact that all viable kinds of exploiters in the animal world must be able when necessary to limit their own numbers" (p. 313).

    It seems strange that, in the real world as well as in contrived optical illusions, two completely different interpretations can give equivalent results. But this is a commonplace in physics—in quantum mechanics, say, or in the study of elementary particles—where two approaches with different starting assumptions and different mathematical apparatus turn out to give identical quantitative answers, and are therefore understood to be equivalent formulations of the solution to the problem.

19. K. Aoki and K. Nozawa, "Average Coefficient of Relationship Within Troops of the Japanese Monkey and Other Primate Species with Reference to the Possibility of Group Selection," *Primates* 25 (1984), pp. 171–184; J. F. Crow and Kenichi Aoki,

"Group Selection for a Polygenic Behavioral Trait: Estimating the Degree of Population Subdivision," *Proceedings, National Academy of Sciences 81* (1984), pp. 6073–6077.

20. Aoki and Nozawa, *op. cit.*

21. Jules H. Masserman, S. Wechkin, and W. Terris, " 'Altruistic' Behavior in Rhesus Monkeys," *American Journal of Psychiatry* 121 (1964), pp. 584, 585; Stanley Wechkin, J. H. Masserman, and W. Terris, "Shock to a Conspecific as an Aversive Stimulus," *Psychonomic Science* 1 (1964), pp. 47, 48.

22. Especially when there is an authority figure urging us to administer the electric shocks, we humans seem disturbingly willing to cause pain—and for a reward much more paltry than food is for a starving macaque (cf. Stanley Milgram, *Obedience to Authority: An Experimental View* [New York: Harper & Row, 1974]).

23. Translated by Richmond Lattimore (Chicago: The University of Chicago Press, 1951), Book XXI, lines 463–466, p. 430.

Chapter 7
WHEN FIRE WAS NEW

1. Fragment 118 in *Herakleitos and Diogenes*, Guy Davenport, translator (Bolinas, CA: Grey Fox Press, 1979).

2. Jonathan Barnes, editor, *Early Greek Philosophy* (Harmondsworth, UK: Penguin Books, 1987), p. 104.

3. Wen-Hsiung Li and Dan Graur, *Fundamentals of Molecular Evolution* (Sunderland, MA: Sinauer Associates, 1991), pp. 10–12.

4. B. Widegren, U. Arnason, and G. Akusjarvi, "Characteristics of Conserved 1, 579-bp High Repetitive Component in the Killer Whale, *Orcinus orca*," *Molecular Biology and Evolution* 2 (1985), pp. 411–419 (bp is an abbrevation for nucleotide basepairs, the letters in the genetic sequences).

5. It can be very serious on the human level. For example, on Chromosome 19 most people have a sequence of nucleotides that goes CTGCTGCTGCTGCTG, a five-fold repeat. But some have hundreds or even thousands of consecutive CTG sequences, and they suffer in consequence from a grave disease called myotonic dystrophy. Some other genetic diseases may have a similar cause.

6. M. Herdman, "The Evolution of Bacterial Genomes," In *The*

Evolution of Genome Size, T. Cavalier-Smith, ed. (New York: Wiley, 1985), pp. 37–68.

7. Richard Dawkins, *The Blind Watchmaker* (New York: Norton, 1986), pp. 46–49.

8. J. W. Schopf, private communication, 1991; Andrew W. Knoll, "The Early Evolution of Eukaryotes: A Geological Perspective," *Science* 256 (1992), pp. 622–627.

9. Philip W. Signor, "The Geologic History of Diversity," *Annual Review of Ecology and Systematics* 21 (1990), pp. 509–539.

10. Sewall Wright, *Evolution and the Genetics of Populations: A Treatise in Four Volumes*, Volume 4, *Variability Within and Among Natural Populations* (Chicago: The University of Chicago Press, 1978), p. 525.

11. Sewall Wright, "Surfaces of Selective Value Revisited," *The American Naturalist* 131 (1) (January 1988), p. 122. This article was written when the pioneering population geneticist was ninety-eight.

12. Cf. Ilkka Hanski and Yves Cambefort, editors, *Dung Beetle Ecology* (Princeton: Princeton University Press, 1991); Natalie Angier, "In Recycling Waste, the Noble Scarab Is Peerless," *New York Times*, December 19, 1991.

13. Charles Darwin, *Origin of Species*, quoted in John L. Harper, "A Darwinian Plant Ecology," in D. S. Bendall, editor, *Evolution from Molecules to Men* (Cambridge: Cambridge University Press, 1983), p. 323.

14. Clair Folsome, "Microbes," in T. P. Snyder, editor, *The Biosphere Catalogue* (Fort Worth, TX: Synergetic Press, 1985), quoted in Dorion Sagan, *Biospheres: Metamorphosis of Planet Earth* (New York: McGraw-Hill, 1990), p. 69.

Chapter 8
SEX AND DEATH

1. George Santayana, *The Works of George Santayana*, Volume II, *The Sense of Beauty: Being the Outlines of Æesthetic Theory*, edited by William G. Holzberger and Herman J. Saatkamp, Jr. (Cambridge: The MIT Press, 1988), Part II, §13, p. 41.

2. Richard Taylor, editor, quoted in George Seldes, *The Great Thoughts* (New York: Random House, 1985), p. 373.

3. The first clear explanations of sex both as a means of rapid evolution and as an escape of populations—especially small populations—from the cumulative impact of deleterious mutations were made by the geneticist H. J. Muller (e.g., "Some Genetic Aspects of Sex," *American Naturalist* 66 [1932], pp. 118–138; "The Relation of Recombination to Mutational Advance," *Mutation Research* 1 [1964], pp. 2–9). There is theoretical and experimental support for his proposals (e.g., Joseph Felsenstein, "The Evolutionary Advantage of Recombination," *Genetics* 78 [1974], pp. 737–756; Graham Bell, *Sex and Death in Protozoa: The History of an Obsession* [Cambridge: Cambridge University Press, 1988]; Lin Chao, Thutrang Than, and Crystal Matthews, "Muller's Ratchet and the Advantage of Sex in the RNA Virus φ6," *Evolution* 46 [1922], pp. 289–299).

   Muller stressed that sexual reproduction was hardly necessary for survival, but that "lack of recomination would greatly handicap a species, in long-term evolutionary advancement, in keeping pace with sexually reproducing competitors." The idea of sex providing a long-term benefit for the species certainly seems to be an example of group selection, as was explicitly noted, without undue alarm, by one of the founders of modern population genetics, R. A. Fisher (*The Genetical Theory of Natural Selection* [Oxford: Clarendon Press, 1930]). Fisher was one of the first to suggest that, in other cases, what superficially looks like group selection may in fact be kin selection.

4. D. Crews, "Courtship in Unisexual Lizards: A Model for Brain Evolution," *Scientific American* 259 (June 1987), pp. 116–121.

5. Raoul E. Benveniste, "The Contributions of Retroviruses to the Study of Mammalian Evolution," Chapter 6 in R. I. MacIntyre, editor, *Molecular Evolutionary Genetics* (New York: Plenum, 1985), pp. 359–417.

6. We have scarcely touched on the complexity and diversity of the sexual machinery, both on the molecular level and the level of individual organisms. Nor have we given a full flavor of the debate on what sex is good for. An excellent short summary is in James L. Gould and Carol Grant Gould, *Sexual Selection* (New York: W. H. Freeman, 1989). See also the influential book by John Maynard Smith, *The Evolution of Sex* (Cambridge: Cambridge University Press, 1978); H. O. Halvorson and A. Monroy, editors,

The Origin and Evolution of Sex (New York: A. R. Liss, 1985); Lynn Margulis and Dorion Sagan, *Origins of Sex* (New Haven: Yale University Press, 1986); R. E. Michod and B. R. Levin, *The Evolution of Sex* (Sunderland, MA: Sinauer, 1988); Alun Anderson, "The Evolution of Sexes," *Science* 257 (1992), pp. 324–326; and Bell, *op. cit.* in Note 3.

7. D. J. Roberts, A. B. Craig, A. R. Berendt, R. Pinches, G. Nash, K. Marsh and C. I. Newbold, "Rapid Switching to Multiple Antigenic and Adhesive Phenotypes in Malaria," *Nature* 357 (1992), pp. 689–692.

8. W. D. Hamilton, R. Axelrod, and R. Tanese, "Sexual Reproduction as an Adaptation to Resist Parasites (A Review)," *Proceedings of the National Academy of Sciences* 87 (1990), pp. 3566–3573.

9. Helen Fisher, "Monogamy, Adultery, and Divorce in Cross-Species Perspective," in Michael H. Robinson and Lionel Tiger, editors, *Man and Beast Revisited* (Washington and London: Smithsonian Institution Press, 1991), p. 97.

10. E. A. Armstrong, *Bird Display and Bird Behaviour: An Introduction to the Study of Bird Psychology* (New York: Dover, 1965), p. 305.

11. W. D. Hamilton and M. Zuk, "Heritable True Fitness and Bright Birds: A Role for Parasites?" *Science 218* (1982), pp. 384–387.

12. The same bargain is made in the common, sexually repressive version of the story of the Garden of Eden—in which it is sexual activity between Adam and Eve that excites God's wrath and makes them mortal.

13. This wonderfully vivid image is Frans de Waal's, in *Peacemaking Among Primates* (Cambridge: Harvard University Press, 1989), p. 11.

14. Translated by Edward Kissam and Michael Schmidt (Tempe, AZ: Bilingual Press/Editorial Bilingüe, 1983), p. 47.

Chapter 9
WHAT THIN PARTITIONS...

1. Alexander Pope, *An Essay on Man*, Frank Brady, editor (Indianapolis: Bobbs-Merrill, 1965) (originally published in 1733–1734), Epistle I, "Argument of the Nature and State of Man, with Respect to the Universe," p. 13, lines 221–226.

2. An updating after Jakob von Uexküll, "A Stroll Through the Worlds of Animals and Men: A Picture Book of Invisible Worlds" (1934), reprinted in Claire H. Schiller, translator and editor, *Instinctive Behavior: The Development of a Modern Concept* (New York: International Universities Press, 1957), pp. 6 ff.

3. Six carbon atoms make up the ring in this molecule. Chemists number them in sequence from 1 to 6. The chlorine atoms are attached in the 2 and 6 positions. If instead they were attached in, say, the 2 and 5 positions, the tick of the opposite sex would not be interested.

4. Ticks are arachnids with eight legs, like spiders, tarantulas, and scorpions. They're a matter of practical concern because they are the vectors for the spread of Rocky Mountain spotted fever, Lyme disease, and other illnesses—of livestock as well as of humans. We've described many of the essential sensory skills of a particular species, but other strategies and capabilities appear on closer examination or in other species. Some species have not one but three different mammalian hosts at different stages of their life cycles. Those ticks that live in caves may wait years for an appropriate host. Ticks chemically interfere with fibrinogen and other machinery that works to staunch the flow of their host's blood, permitting some species to stuff themselves with a hundred times their unfed body weight in blood. Not only butyric acid is sensed in their quest for mammalian blood, but also lactic acid ($CH_3HCOHCOOH$) and ammonia (NH_3). Ticks use pheromones for purposes other than attracting the opposite sex—an assembly pheromone, for example, for a gathering of the tribes in cracks and crevices, or in caves. (See Daniel E. Sonenshine, *Biology of Ticks*, Volume 1 [New York: Oxford University Press, 1991]). Nevertheless, the basic sensory armamentarium of tick life still seems, as it did in the 1930s, very simple.

5. J. L. Gould and C. G. Gould, "The Insect Mind: Physics or Metaphysics?" in D. R. Griffin, editor, *Animal Mind–Human Mind* (Report of the Dahlem Workshop on Animal Mind–Human Mind, Berlin, March 22–27, 1981) (Berlin: Springer-Verlag, 1982), p. 283.

6. Thomas H. Huxley, "On the Hypothesis that Animals Are Automata, and its History" (1874), in *Collected Essays*, Volume I, *Method and Results: Essays* (London: Macmillan, 1901), p. 218.

7. von Uexküll, *op. cit.*, pp. 43, 46.

8. Karl von Frisch, *The Dancing Bees* (New York: Harcourt, Brace, 1953).

9. A provocative modern discussion, informed by neurophysiology and computer science, is Daniel C. Dennett's *Consciousness Explained* (Boston: Little, Brown, 1991). Optimistic assessments of the near future of artificial intelligence and artificial life include Hans Moravec, *Mind Children* (Cambridge: Harvard University Press, 1988) and Maureen Caudill, *In Our Own Image: Building an Artificial Person* (New York: Oxford University Press, 1992). A more pessimistic assessment is Roger Penrose, *The Emperor's New Mind* (New York: Oxford University Press, 1990).

10. Quoted in Konrad Lorenz, "Companionship in Bird Life: Fellow Members of the Species as Releasers of Social Behavior," in Schiller, *op. cit.*, p. 126.

11. René Descartes, letter to the Marquis of Newcastle, quoted in Mortimer J. Adler and Charles Van Doren, *Great Treasury of Western Thought: A Compendium of Important Statements on Man and His Institutions by the Great Thinkers in Western History* (New York and London: R. R. Bowker Company, 1977), p. 12.

12. Aristotle, *History of Animals*, Book VIII, 1, 588ᵃ, in *The Works of Aristotle*, Great Books edition, Volume II, translated into English under the editorship of W. D. Ross (Chicago: Encyclopaedia Britannica, 1952) p. 114.

13. Charles Darwin, *The Descent of Man and Selection in Relation to Sex* (New York: The Modern Library, n.d.) (originally published in 1871) (Modern Library edition also contains *The Origin of Species by Means of Natural Selection or the Preservation of Favored Races in the Struggle for Life*), Chapters 1 and 3.

14. René Descartes, *Traité de l'Homme*, Victor Cousin, editor, pp. 347, 427, as translated by T. H. Huxley, in Huxley, *Collected Essays*, Volume I, *Method and Results: Essays* (London: Macmillan, 1901), "On Descartes' 'Discourse Touching the Method of Using One's Reason Rightly and of Seeking Scientific Truth' " (1870).

15. Voltaire, "Animals," *Philosophical Dictionary* (1764), T. H. Huxley, translator.

16. Thomas H. Huxley, "On Descartes' 'Discourse Touching the Method of Using One's Reason Rightly and of Seeking Scientific Truth' " (1870), and "On the Hypothesis that Animals Are Automata, and its History" (1874), in Huxley, *Collected Essays*, Volume I, *Method and Results: Essays* (London: Macmillan, 1901), pp. 186–187, 184, 187–189, 237–238, 243–244.

17. J. L. and C. J. Gould, "The Insect Mind: Physics or Metaphysics?" in D. R. Griffin, editor, *Animal Mind–Human Mind* (Report of the Dahlem Workshop on Animal Mind–Human Mind, Berlin, March 22–27, 1981) (Berlin: Springer-Verlag, 1982), pp. 288, 289, 292.

Chapter 10
THE NEXT-TO-LAST REMEDY

1. Thomas Hobbes, *Leviathan, or the Matter, Forme and Power of a Commonwealth Ecclesiasticall and Civil*, Michael Oakeshott, editor (Oxford: Basil Blackwell, 1960), Part 2, Chapter 30, p. 227.

2. Charles Darwin and Alfred R. Wallace, "On the Tendency of Species to Form Varieties; and on the Perpetuation of Varieties and Species by Natural Means of Selection," *Journal of the Proceedings of the Linnean Society: Zoology*, Volume III (London: Longman, Brown, Green, Longmans & Roberts, and Williams and Norgate, 1859), p. 50. Here Darwin also describes sexual selection in which the males compete for the favors of the female, or she selects from among several males on the basis of some quality she finds attractive: "This kind of selection, however, is less rigorous than the other," Darwin said; "it does not require the death of the less successful, but gives to them fewer descendants."

3. Curt P. Richter, "Rats, Man, and the Welfare State," *The American Psychologist* 14 (1959), pp. 18–28.

4. John B. Calhoun, "Population Density and Social Pathology," *Scientific American* 206 (2) (February 1962), pp. 139–146, 148; and references cited there.

5. Frans de Waals, *Peacemaking Among Primates* (Cambridge, MA: Harvard University Press, 1989).

6. Richard Dawkins argues that lowered birth rates in response to overcrowding are explained equally well (not better) by individual

as by group selection (*The Selfish Gene* [Oxford: Oxford University Press, 1989], p. 119).

7. John F. Eisenberg, "Mammalian Social Organization and the Case of *Alouatta*," in Michael H. Robinson and Lionel Tiger, editors, *Man and Beast Revisited* (Washington: Smithsonian Institution Press, 1991), p. 135.

8. Peter Marler, "*Golobus guereza*: Territoriality and Group Composition," *Science 163* (1969), pp. 93–95.

9. John F. Eisenberg and Devra G. Kleiman, "Olfactory Communication in Mammals," in *Annual Review of Ecology and Systematics 3* (1972), pp. 1–32.

10. As first pointed out by Charles Darwin (1872) in *The Expression of the Emotions in Man and the Animals* (Chicago: University of Chicago Press, 1965, 1967), p. 119.

11. C. G. Beer, "Study of Vertebrate Communication—Its Cognitive Implications," in D. R. Griffin, editor, *Animal Mind–Human Mind* (Report of the Dahlem Workshop on Animal Mind–Human Mind, Berlin, March 22–27, 1981) (Berlin: Springer-Verlag, 1982), p. 264.

12. Lorenz's translation from cranish. Konrad Lorenz, *On Aggression* (New York: Harcourt Brace, 1966), pp. 174, 175.

13. An example:

"My friend and teacher, Bill Drury, invited me to go bird-watching one day on a small island off the coast of Maine. We left bird books and binoculars behind and strode to the nearest small tree growing alone in the open. He then made a series of high-pitched bird sounds and soon the tree began to fill up with birds, themselves making a series of calls. As the tree started to fill up, it seemed to attract more and more birds, so that as if by magic all small songbirds in the area were streaking toward the tree under which we were standing. By this time Bill was down on his knees, bent over, and most of the time making a deep kind of moaning sound. The birds actually appeared to wait in line to get the closest look at Bill they could; that is, they hopped from branch to branch until they rested on a branch about eight feet off the ground and not more than two feet from my face. As each bird hopped down, Bill, as if on cue, would introduce them. 'This is a male, black-capped chickadee. You can tell because of the black along the neck and shoulders. I would guess he's about two to three years old. Can you see if there is yellow on his back between his shoulders? This is a good index of age.'

"For me the moment was utterly magical. In a matter of minutes Bill had reduced the distance between us and these birds by orders of magnitude, both physically and socially. Our relationship was so completely different that I was permitted individual introductions at a distance of a couple of feet. Obviously Bill was pulling some kind of trick and had induced some kind of trance through his bird song. . . . Bill was at first only imitating the mobbing calls of a couple of the small passerines in the area and interspersing these with occasional owl hoots. The owl is deadly at night but is vulnerable in the daytime, and groups of songbirds will mob it in order (presumably) to run it out of their area, or even harass and kill it on the spot. This drew them into the tree at an ever-increasing rate, since mobbing assemblages gain in individual safety with each new arrival (as well as gaining in power to harass the owl). Once they landed in the tree, however, they could see two four-eyed human beings but could not see the owl. Bill's bending over and hooting from the ground was meant to suggest the owl was hidden underneath him. This drew them as close as they could get for a good look, which put them two feet from my face. Unlike some magic tricks, knowing how Bill's was done did not detract from my enjoyment." (Robert Trivers, "Deceit and Self-Deception: The Relationship Between Communication and Consciousness," in Michael H. Robinson and Lionel Tiger, editors, *Man and Beast Revisited* [Washington: Smithsonian Institution Press, 1991], pp. 182, 183.)

14. Mary Jane West-Eberhard, "Sexual Selection and Social Behavior," in Robinson and Tiger, *op. cit.*, p. 165.

15. T. J. Fillion and E. M. Blass, "Infantile Experience with Suckling Odors Determines Adult Sexual Behavior in Male Rats," *Science* 231 (1986), pp. 729–731.

16. Marcus Aurelius, *Meditations*, translated with an introduction by Maxwell Staniforth (Harmondsworth, Middlesex, England: Penguin, 1964), II, 17, p. 51.

Chapter 11
DOMINANCE AND SUBMISSION

1. Charles Darwin, *The Origin of Species by Means of Natural Selection or the Preservation of Favored Races in the Struggle for Life* (New York: The Modern Library, n.d.) (originally published in 1859) Chapter XV, "Recapitulation and Conclusion," p. 371.

2. From George Seldes, *The Great Thoughts* (New York: Ballantine, 1985), p. 302.

3. E.g., Natalie Angier, "Pit Viper's Life: Bizarre, Gallant and Venomous," *New York Times*, October 15, 1991, pp. C1, C10.

4. Snakes certainly fight over territory as well—rat snakes, for example, over knotholes in trees where birds nest. The loser looks for another tree.

5. David Duvall, Stevan J. Arnold, and Gordon W. Schuett, "Pit Viper Mating Systems: Ecological Potential, Sexual Selection, and Microevolution," in *Biology of Pitvipers*, J. A. Campbell and E. D. Brodie, Jr., editors (Tyler, TX: Selva, 1992).

6. B. J. Le Boeuf, "Male-male Competition and Reproductive Success in Elephant Seals," *American Zoologist* 14 (1974), pp. 163–176.

7. C. R. Cox and B. J. Le Boeuf, "Female Incitation of Male Competition: A Mechanism in Sexual Selection," *American Naturalist* 111 (1977), pp. 317–335.

8. E.g., Peter Maxim, "Dominance: A Useful Dimension of Social Communication," *Behavioral and Brain Sciences* 4 (3) (September 1981), pp. 444, 445.

9. Charles Darwin, *The Descent of Man and Selection in Relation to Sex* (New York: The Modern Library, n.d.) (originally published in 1871) Part II, "Sexual Selection," Chapter XVIII, "Secondary Sexual Characters of Mammals—continued," p. 863.

10. Paul F. Brain and David Benton, "Conditions of Housing, Hormones, and Aggressive Behavior," in Bruce B. Svare, editor, *Hormones and Aggressive Behavior* (New York and London: Plenum Press, 1983), p. 359.

11. *Ibid.*, Table II, "Characteristics of Dominant and Subordinate Mice from Small Groups," p. 358.

12. Dominance in a one-on-one encounter and dominance rank

within a hierarchy are not necessarily the same and cannot always be predicted from one another. See Irwin S. Bernstein, "Dominance: The Baby and the Bathwater," and subsequent commentary, *Behavioral and Brain Sciences* 4 (3) (September 1981), pp. 419–457. Some animals distinguish only between those lower and those higher in rank. Others—baboons, for example—behave differently to those of very distant rank than to those nearly co-equal in rank (Robert M. Seyfarth, "Do Monkeys Rank Each Other?" *ibid.*, pp. 447–448).

13. W. C. Allee, *The Social Life of Animals* (Boston: Beacon Press paperback, 1958), especially p. 135 (originally published in 1938 by Abelard-Schuman Ltd.; this revised edition published in hardback in 1951 under the title *Cooperation Among Animals With Human Implications*).

14. V. C. Wynne-Edwards, *Evolution Through Group Selection* (Oxford: Blackwell, 1986), pp. 8–9.

15. Neil Greenberg and David Crews, "Physiological Ethology of Aggression in Amphibians and Reptiles," in Svare, *op. cit.*, pp. 483 (varanids), 481 (crocodiles), 474 (*Dendrobates* [dendratobids]), and 483 (skinks).

16. B. Hazlett, "Size Relations and Aggressive Behaviour in the Hermit Crab, *Clibanarius Vitatus*," *Zeitschrift für Tierpsychologie* 25 (1968), pp. 608–614.

17. Patricia S. Brown, Rodger D. Humm, and Robert B. Fischer, "The Influence of a Male's Dominance Status on Female Choice in Syrian Hamsters," *Hormones and Behavior* 22 (1988), pp. 143–149.

18. One of many other examples: Bart Kempenaers, Geert Verheyen, Marleen van den Broeck, Terry Burke, Christine van Broeckhoven, and Andre Dhondt, "Extra-pair Paternity Results from Female Preference for High-Quality Males in the Blue Tit," *Nature* 357 (1992), pp. 494–496.

19. Mary Jane West-Eberhard, "Sexual Selection and Social Behavior," in Michael H. Robinson and Lionel Tiger, editors, *Man and Beast Revisited* (Washington and London: Smithsonian Institution Press, 1991), p. 165.

20. In 1857, Elizabeth Cady Stanton wrote: "[H]ow perfectly [woman's dress] describes her condition. Her tight waist and long, trail-

ing skirts deprive her of all freedom of breath and motion. No wonder man prescribes her sphere. She needs his aid at every turn. He must help her up stairs and down, in the carriage and out, on the horse, up the hill, over the ditch and fence, and thus teach her the poetry of dependence." (J. C. Lauer and R. H. Lauer, "The Language of Dress: A Sociohistorical Study of the Meaning of Clothing in America," *Canadian Review of American Studies* 10 [1979], pp. 305–323.) Stunning change has occurred since 1857, although the poetry of dependence is still widely recited in the women's fashion industry.

21. Owen R. Floody, "Hormones and Aggression in Female Mammals," in Svare, *op. cit.*, pp. 51, 52.

Chapter 12
THE RAPE OF CAENIS

1. Elizabeth Wyckoff, translator (Chicago: University of Chicago Press, 1954).

2. David Grene, translator (Chicago: University of Chicago Press, 1942).

3. Ovid, *Metamorphoses*, translation by Frank Justus Miller (Cambridge: Harvard University Press/Loeb Classical Library, 1916, 1976), Book XII, pp. 192–195; Robert Graves, *The Greek Myths* (Harmondsworth, Middlesex, England: Penguin Books, 1955, 1960), Volume 1, pp. 260–262; Froma Zeitlin, "Configurations of Rape in Greek Myth," in Sylvana Tomaselli and Roy Porter, editors, *Rape: An Historical and Social Enquiry* (Oxford and New York: Basil Blackwell, 1986), pp. 133, 134.

4. Smaller amounts of androgens are produced in the cortex of the adrenal gland that surmounts each kidney, from other hormones in the body, and in the placenta.

5. R. M. Rose, I. S. Bernstein, and J. W. Holaday, "Plasma Testosterone, Dominance Rank, and Aggressive Behavior in a Group of Male Rhesus Monkeys," *Nature* 231 (1971), pp. 366–368; G. G. Eaton and J. A. Resko, "Plasma Testosterone and Male Dominance in a Japanese Macaque (*Macaca fuscata*) Troop Compared with Repeated Measures of Testosterone in Laboratory Males," *Hormones and Behavior* 5 (1974), pp. 251–259.

6. Peter Marler and William J. Hamilton III, *Mechanisms of Animal Behavior* (New York: John Wiley & Sons, 1966), p. 177.

7. D. Michael Stoddart, *The Scented Ape: The Biology and Culture of Human Odour* (Cambridge: Cambridge University Press, 1990), pp. 136, 137, 163.

8. J. Money and A. Ehrhardt, *Man and Woman, Boy and Girl: The Differentiation and Dimorphism of Gender Identity from Conception to Maturity* (Baltimore: Johns Hopkins University Press, 1972); J. Money and M. Schwartz, "Fetal Androgens in the Early Treated Adrenogenital Syndrome of 46XX Hermaphroditism: Influence on Assertive and Aggressive Types of Behavior," in *Aggressive Behavior* 2 (1976), pp. 19–30; J. Money, M. Schwartz, and V. G. Lewis, "Adult Erotosexual Status and Fetal Hormonal Masculinization and Demasculinization," *Psychoneuroendocrinology* 9 (1984), pp. 405–414; Sheri A. Berenbaum and Melissa Hines, "Early Androgens Are Related to Childhood Sex-Typed Toy Preferences," *Psychological Science* 3 (1992), pp. 203–206.

9. Aristotle, *Generation of Animals*, in *The Oxford Translation of Aristotle*, W. D. Ross, translator and editor (London: Oxford University Press, 1928), 737ª28.

10. Stefan Hansen, "Mechanisms Involved in the Control of Punished Responding in Mother Rats," *Hormones and Behavior* 24 (1990), pp. 186–197.

11. Mary Midgley, *Beast and Man* (Ithaca, NY: Cornell University Press, 1978), p. 39.

12. John Sparks with Tony Soper, *Parrots: A Natural History* (New York: Facts on File, 1990), p. 90.

13. Owen R. Floody, "Hormones and Aggression in Female Mammals," in Bruce B. Svare, editor, *Hormones and Aggressive Behavior* (New York: Plenum Press, 1983), pp. 44–46.

14. Alfred M. Dufty, Jr., "Testosterone and Survival: A Cost of Aggressiveness?" *Hormones and Behavior* 23 (1989), pp. 185–193.

15. Hansen, *op. cit.*

16. Lester Grinspoon, Harvard Medical School, private communication, 1991.

17. John C. Wingfield and M. Ramenofsky, "Testosterone and Aggressive Behaviour During the Reproductive Cycle of Male

Birds," in R. Gilles and J. Balthazart, editors, *Neurobiology* (Berlin: Springer-Verlag, 1985), pp. 92–104.

18. Stephen T. Emlen, Cornell University, private communication, 1991.

19. R. L. Sprott, "Fear Communication via Odor in Inbred Mice," *Psychological Reports* 25 (1969), pp. 263–268; John F. Eisenberg and Devra G. Kleiman, "Olfactory Communication in Mammals," in *Annual Review of Ecology and Systematics* 3 (1972), pp. 1–32.

20. These classic experiments were described by Konrad Lorenz in 1939 and by Nikko Tinbergen in 1948. Some later research suggests that the chicks and goslings become less afraid of a silhouette as they become habituated to it (and it doesn't eat anyone). Wolfgang Schleidt ("Über die Auslösung der Flucht vor Raubvögeln bei Truthühnern," *Die Naturwissenschaften* 48 [1961], pp. 141–142) suggests that birds on the ground are afraid of *any* unfamiliar flying silhouette, become used to the harmless image of a flying goose, but retain a fear of the less familiar hawk. This is not far from the toddler's shyness about strangers and fear of "monsters."

21. Peter Marler, "Communication Signals of Animals: Emotion or Reference?" Address, Centennial Conference, Department of Psychology, Cornell University, July 20, 1991.

22. Marcel Gyger, Stephen J. Karakashian, Alfred M. Dufty, Jr., and Peter Marler, "Alarm Signals in Birds: The Role of Testosterone," *Hormones and Behavior* 22 (1988), pp. 305–314.

23. Stoddart, *op. cit.*, pp. 116–119.

24. The chemicals in question are gamma aminobutyric acid and serotonin. Cf., e.g., Jon Franklin, *Molecules of the Mind* (New York: Laurel/Dell, 1987), pp. 155–157.

23. Heidi H. Swanson and Richard Schuster, "Cooperative Social Coordination and Aggression in Male Laboratory Rats: Effects of Housing and Testosterone," *Hormones and Behavior* 21 (1987), pp. 310–330.

Chapter 13
THE OCEAN OF BECOMING

1. Edward Conze, editor, *Buddhist Scriptures* (Harmondsworth, UK: Penguin, 1959), p. 241.

2. The initial rate of increase of the new mutation in the population is very slow. The thousand-generation estimate, courtesy of the population geneticist James F. Crow, is what it takes to go from gene frequencies of 0.001 (almost nobody) to 0.9 (almost everybody).

3. Sewall Wright, *Evolution and the Genetics of Populations: A Treatise in Four Volumes*, Volume 4, *Variability Within and Among Natural Populations* (Chicago: The University of Chicago Press, 1978); Wright, *Evolution: Selected Papers*, edited by William B. Provine (Chicago: The University of Chicago Press, 1986); Wright, "Surfaces of Selective Value Revisited," *The American Naturalist* 131 (January 1988), pp. 115–123; William B. Provine, *Sewall Wright and Evolutionary Biology* (Chicago: University of Chicago Press, 1986); J. F. Crow, W. R. Engels, and C. Denniston, "Phase Three of Wright's Shifting-Balance Theory," *Evolution* 44 (1990), pp. 233–247. Also, Roger Lewin, "The Uncertain Perils of an Invisible Landscape," *Science* 240 (1988), pp. 1405, 1406.

4. Carl Sagan, "Croesus and Cassandra: Policy Responses to Global Change," *American Journal of Physics* 58 (1990), pp. 721–730.

5. Plutarch, "Antony," *The Lives of the Noble Grecians and Romans*, translated by John Dryden and revised by Arthur Hugh Clough (New York: The Modern Library, 1932), p. 1119.

6. Stewart Henry Perowne, "Cleopatra," *Encyclopaedia Britannica*, 15th Edition (1974), *Macropaedia*, Volume 4, p. 712.

7. Graham Bell, *Sex and Death in Protozoa: The History of an Obsession* (Cambridge: Cambridge University Press, 1988), pp. 65–66.

8. K. Ralls, J. D. Ballou, and A. Templeton, "Estimates of Lethal Equivalents and Cost of Inbreeding in Mammals," *Conservation Biology* 2 (1988), pp. 185–193; P. H. Harvey and A. F. Read, "Copulation Genetics: When Incest Is Not Best," *Nature* 336 (1988), pp. 514–515.

9. James L. Gould and Carol Grant Gould, *Sexual Selection* (New York: W. H. Freeman, 1989), p. 64.

10. Anne E. Pusey and Craig Packer, "Dispersal and Philopatry," Chapter 21 of Barbara B. Smuts, Dorothy L. Cheney, Robert M. Seyfarth, Richard W. Wrangham, and Thomas T. Struhsaker, editors, *Primate Societies* (Chicago: University of Chicago Press, 1986), p. 263.

11. P. H. Harvey and K. Ralls, "Do Animals Avoid Incest?" *Nature*

320 (1986), pp. 575, 576; D. Charlesworth and B. Charlesworth, "Inbreeding Depression and Its Evolutionary Consequences," *Annual Review of Ecology and Systematics* 18 (1987), pp. 237–268. The latter reference contains a good summary of the means by which the incest taboo is enforced in plants.

12. John Paul Scott and John L. Fuller, *Genetics and the Social Behavior of the Dog* (Chicago: University of Chicago Press, 1965), pp. 406, 407.

13. William J. Schull and James V. Neel, *The Effects of Inbreeding on Japanese Children* (New York: Harper and Row, 1965).

14. Morton S. Adams and James V. Neel, "Children of Incest," *Pediatrics* 40 (1967), pp. 55–62.

15. Theodosius Dobzhansky was a leading twentieth-century geneticist. He gives this example in his *Mankind Evolving* (New Haven: Yale University Press, 1962), p. 281.

16. Over long enough intervals, isolation—even in large populations—generates diversity. When, for example, the Pangaea supercontinent broke up, the populations on adjacent land masses were no longer able (or at least not much able) to interbreed, and gene combinations established on one continent would by no means automatically be transferred to another; no longer did outbreeding link up the gene pools of widely separated populations. The unique biology of such isolated regions as Australia, New Zealand, Madagascar, or the Galapagos Islands is due to tectonic or other kinds of geographical isolation.

17. George Gaylord Simpson, *Tempo and Mode in Evolution* (New York: Columbia University Press, 1944), p. 119.

18. We recognize with Wright that we are close to postulating group selection here. But any argument for optimum gene *frequencies* in a population must, it seems to us, do so.

19. John Tyler Bonner, *The Evolution of Culture in Animals* (Princeton, NJ: Princeton University Press, 1980): "We can see the seeds, the origins, of everything we know about our culture in the distant past. This means that every aspect of our culture can benefit from some understanding of the biology from which it sprang" (p. 186).

Chapter 14
GANGLAND

1. (London and Edinburgh: Williams and Norgate, 1863), p. 59.

Chapter 15
MORTIFYING REFLECTIONS

1. Translated by E. Gurney Salter (London: J. M. Dent and Co., 1904), Chapter VIII, p. 85.
2. Book III, Chapter 30 (added as a footnote to the edition of 1781); translated by Arthur O. Lovejoy in *The Great Chain of Being: A Study of the History of an Idea* (Cambridge, MA: Harvard University Press, 1953), p. 235.
3. For Hanno's expedition, see Jacques Ramin, "The Periplus of Hanno," *British Archaeological Reports*, Supplementary Series 3 (Oxford: 1976). For scholarly debate on which kind of primates Hanno and his men slaughtered, see William Coffmann McDermott, *The Ape in Antiquity* (Baltimore: Johns Hopkins Press, 1938), pp. 51–55.
4. Aristotle, *History of Animals*, Book II, 8–9, 502^a-502^b, in *The Works of Aristotle*, Great Books edition, Volume II, translated into English under the editorship of W. D. Ross (Chicago: Encyclopaedia Britannica, 1952) (originally published by Oxford University Press), pp. 24, 25.
5. H. W. Janson, *Apes and Ape Lore in the Middle Ages and the Renaissance* (London: University of London, 1952).
6. Paul H. Barrett *et al.*, editors, *Charles Darwin's Notebooks, 1836-1844* (Ithaca, N.Y.: Cornell University Press, 1987), p. 539.
7. Thomas N. Savage and Jeffries Wyman, "Observations on the External Characters and Habits of the Troglodytes niger, by Thomas N. Savage, M.D., and on its Organization, by Jeffries Wyman, M.D.," *Boston Journal of Natural History*, Volume IV, 1843–4; quoted in Thomas Henry Huxley, *Man's Place in Nature and Other Anthropological Essays* (London and New York: Macmillan, 1901).
8. Quoted in Keith Thomas, *Man and the Natural World: A History*

of the Modern Sensibility (New York: Pantheon Books, 1983), p. 66.

9. William Congreve, *The Way of the World*, edited by Brian Gibbons (New York: W. W. Norton, 1971), pp. 37, 42, 44.

10. Letter of July 10, 1695; in William Congreve, *Letters and Documents*, John C. Hodges, editor (New York: Harcourt, Brace and World, 1964), p. 178.

11. Jeremy Collier, *A Short View of the Immorality and Profaneness of the English Stage*, edited by Benjamin Hellinger (New York: Garland Publishing, 1987) (originally published in London in 1698), p. 13.

12. G. L. Prestige, *The Life of Charles Gore: A Great Englishman* (London: William Heinemann, 1935), pp. 431, 432.

13. Aelian, quoted by McDermott, *op. cit.*, p. 76.

14. The Linnaean Society of London was named after Linnaeus. It was in this Society's journal that the world first learned, from the pens of Darwin and Wallace, about natural selection.

15. Arthur O. Lovejoy, *The Great Chain of Being: A Study of the History of an Idea* (Cambridge: Harvard University Press, 1953), p. 235.

16. Letter to J. G. Gmelin, February 14, 1747, quoted in George Seldes, *The Great Thoughts* (New York: Ballantine, 1985), p. 247.

17. Thomas Henry Huxley, *Evidence as to Man's Place in Nature* (London and Edinburgh: Williams and Norgate, 1863), pp. 69, 70.

18. *Ibid.*, p. 102.

19. Quoted in Monroe W. Strickberger, *Evolution* (Boston: Jones and Bartlett, 1990), p. 57.

20. Michael M. Miyamoto and Morris Goodman, "DNA Systematics and Evolution of Primates," *Annual Review of Ecology and Systematics* 21 (1990), pp. 197–220. In humans the genes coding for beta-globins are on Chromosome 11.

21. M. Goodman, B. F. Koop, J. Czelusniak, D. H. A. Fitch, D. A. Tagle, and J. L. Slightom, "Molecular Phylogeny of the Family of Apes and Humans," *Genome* 31 (1989), pp. 316–335; and Morris Goodman, private communication, 1992. Similar results are found from DNA hybridization studies: C. G. Sibley, J. A. Comstock and J. E. Ahlquist, "DNA Hybridization Evidence of Hominoid Phylogeny: A Reanalysis of the Data," *Journal of Molecular Evolution* 30 (1990) pp. 202–236.

22. Based on data in Strickberger, *op cit.*, pp. 227, 228.

23. E.g., Richard C. Lewontin, "The Dream of the Human Genome," *New York Review of Books*, May 28, 1992, pp. 31–40. (This is, incidentally, an engaging critical review of the justifications offered for the project to map all of the roughly 4 billion nucleotides in human DNA, and is at variance with the views of many prominent molecular biologists). Also ref. 21.

24. Donald R. Griffin, "Prospects for a Cognitive Ethology," *Behavioral and Brain Sciences* 1 (4) (December 1978), pp. 527–538.

25. Jane Goodall, *The Chimpanzees of Gombe: Patterns of Behavior* (Cambridge, MA: The Belknap Press of Harvard University Press, 1986); Goodall, *Through a Window: My Thirty Years with the Chimpanzees of Gombe* (Boston: Houghton Mifflin, 1990); Toshisada Nishida and Mariko Hiraiwa-Hasegawa, "Chimpanzees and Bonobos: Cooperative Relationships among Males," Chapter 15 in Barbara B. Smuts, Dorothy L. Cheney, Robert M. Seyfarth, Richard W. Wrangham, and Thomas T. Struhsaker, editors, *Primate Societies* (Chicago: University of Chicago Press, 1986); Nishida, "Local Traditions and Cultural Transmission," Chapter 38 in Smuts *et al.*, eds., *op. cit.*; Nishida, editor, *The Chimpanzees of the Mahale Mountains: Sexual and Life History Strategies* (Tokyo: University of Tokyo Press, 1990); Frans de Waal, *Chimpanzee Politics: Power and Sex among Apes* (New York: Harper & Row, 1982); de Waal, *Peacemaking among Primates* (Cambridge, MA: Harvard University Press, 1989).

26. B. M. F. Galdikas, "Orangutan Reproduction in the Wild," in C. E. Graham, editor, *Reproductive Biology of the Great Apes* (New York: Academic Press, 1981), pp. 281–300.

27. Anne C. Zeller, "Communication by Sight and Smell," Chapter 35 of Barbara B. Smuts, Dorothy L. Cheney, Robert M. Seyfarth, Richard W. Wrangham, and Thomas T. Struhsaker, editors, *Primate Societies* (Chicago: University of Chicago Press, 1986), p. 438.

28. Jane Goodall, *The Chimpanzees of Gombe: Patterns of Behavior*, (Cambridge, MA: The Belknap Press of Harvard University Press, 1986), p. 368.

29. Very much the vengeance that—in the horrifying close to one of the most beautiful of the Psalms—the Israelites during the Babylonian exile proposed visiting on the children of their captors:

> O daughter of Babylon, who art to be destroyed; happy shall he be, that rewardeth thee as thou hast served us.
> Happy shall he be, that taketh and dasheth thy little ones against the stones.
>
> —Psalm 137, verses 8 and 9

30. Janis Carter, "A Journey to Freedom," *Smithsonian* 12 (April 1981), pp. 90–101.
31. Goodall, *The Chimpanzees of Gombe*, pp. 490, 491.
32. Thomas, *op. cit.* (ref. 8), p. 22.
33. Euripides, *The Trojan Women*, in *The Medea*, Gilbert Murray, translator (New York: Oxford University Press, 1906), p. 59.

Chapter 16
LIVES OF THE APES

1. In Greg Whincup, editor and translator, *The Heart of Chinese Poetry* (New York: Anchor Press/Doubleday, 1987), p. 48.
2. The principal sources for unattributed details on chimpanzee life in Chapters 14, 15, and 16 are Goodall, Nishida, and de Waal: Jane Goodall, *The Chimpanzees of Gombe: Patterns of Behavior* (Cambridge, MA: The Belknap Press of Harvard University Press, 1986); Goodall, *Through a Window: My Thirty Years with the Chimpanzees of Gombe* (Boston: Houghton Mifflin, 1990); Toshisada Nishida and Mariko Hiraiwa-Hasegawa, "Chimpanzees and Bonobos: Cooperative Relationships among Males," Chapter 15 in Barbara B. Smuts, Dorothy L. Cheney, Robert M. Seyfarth, Richard W. Wrangham, and Thomas T. Struhsaker, editors, *Primate Societies* (Chicago: University of Chicago Press, 1986); Nishida, "Local Traditions and Cultural Transmission," Chapter 38 in Smuts *et al.*, eds., *op. cit.*; Nishida, editor, *The Chimpanzees of the Mahale Mountains: Sexual and Life History Strategies* (Tokyo: University of Tokyo Press, 1990); Frans de Waal, *Chimpanzee Politics: Power and Sex among Apes* (New York: Harper & Row, 1982); de Waal, *Peacemaking among Primates* (Cambridge, MA: Harvard University Press, 1989). Also other chapters of Smuts, *et al.*
3. Chapter III, verse 1.

4. Frans de Waal, *Peacemaking among Primates* (Cambridge, MA: Harvard University Press, 1989), p. 49.
5. Frans de Waal, *Chimpanzee Politics: Power and Sex among Apes* (New York: Harper & Row, 1982), pp. 37, 38.
6. Here is Darwin's argument about pink bottoms in the season of love:

"In the discussion on Sexual Selection in my 'Descent of Man,' no case interested and perplexed me so much as the brightly-coloured hinder ends and adjoining parts of certain monkeys. As these parts are more brightly coloured in one sex than the other, and as they become more brilliant during the season of love, I concluded that the colours had been gained as a sexual attraction. I was well aware that I thus laid myself open to ridicule; though in fact it is not more surprising that a monkey should display his bright-red hinder end than that a peacock should display his magnificent tail. I had, however, at that time no evidence of monkeys exhibiting this part of their bodies during their courtship; and such display in the case of birds affords the best evidence that the ornaments of the males are of service to them by attracting or exciting the females. . . . Joh. von Fischer, of Gotha . . . finds that not only the mandrill but the drill and three other kinds of baboons, also *Cynopithecus niger*, and *Macacus rhesus* and *nemestrinus*, turn this part of their bodies, which in all these species is more or less brightly coloured, to him when they are pleased, and to other persons as a sort of greeting. He took pains to cure a *Macacus rhesus*, which he had kept for five years, of this indecorous habit, and at last succeeded. These monkeys are particularly apt to act in this manner, grinning at the same time, when first introduced to a new monkey, but often also to their old monkey friends; and after this mutual display they begin to play together . . .

"[T]he habit with adult animals is connected to a certain extent with sexual feelings, for Von Fischer watched through a glass door a female *Cynopithecus niger*, and she during several days, 'turned and displayed her intensely reddened backside while making gurgling sounds—something I had never before observed in this animal. Seeing this, the male grew agitated; he vigorously rattled the bars of the cage, likewise emitting gurgling noises' [this quotation was cautiously rendered by Darwin in the original German and is translated here]. As all the monkeys which have the hinder parts of their bodies more or less brightly coloured live, according to Von Fischer, in open rocky places, he thinks that these colours serve to render one sex conspicuous at a distance to the other; but, as monkeys are such gregarious animals, I should have thought there was no need for the sexes to recognise each other at a distance. It seems to me more probable that the bright colours, whether on the face or hinder end, or, as in the mandrill, on both, serve as a sexual ornament and attraction." (Charles Darwin, "Supplemental Note on Sexual Selection in Relation to Monkeys," *Nature*, November 2, 1876, p. 18.)

7. R. M. Yerkes and J. H. Elder, "Oestrus, Receptivity and Mating in the Chimpanzee," *Comparative Psychology Monographs* 13 (1936), pp. 1–39.

8. Helen Fisher, "Monogamy, Adultery, and Divorce in Cross-Species Perspective," in Michael H. Robinson and Lionel Tiger, editors, *Man and Beast Revisited* (Washington and London: Smithsonian Institution Press, 1991), p. 98.

9. de Waal, *Peacemaking among Primates*, p. 82.

10. Sarah Blaffer Hrdy, "The Primate Origins of Human Sexuality," in Robert Bellig and George Stevens, eds., *Nobel Conference XXIII: The Evolution of Sex* (San Francisco: Harper & Row, 1988), pp. 112 ff.

11. Kelly J. Stewart and Alexander H. Harcourt, "Gorillas: Variation in Female Relationships," Chapter 14 of Barbara B. Smuts, Dorothy L. Cheney, Robert M. Seyfarth, Richard W. Wrangham, and Thomas T. Struhsaker, editors, *Primate Societies* (Chicago: University of Chicago Press, 1986), p. 163.

12. Work of Nicholas Davies in the U.K., described by Stephen Emlen, private communication, 1991.

13. Emily Martin, "The Egg and the Sperm: How Science Has Constructed a Romance Based on Stereotypical Male-Female Roles," *Signs: Journal of Women in Culture and Society*, Spring 1991.

14. This is less true to the extent that the attributes of the sperm cells are determined by the *father's* genes, and not the DNA instructions for making the next generation that the sperm cell itself is carrying. Sperm competition will in any case be very important in

those animals—primates prominent among them—where more than one male ejaculates in rapid succession into a given female.

15. Goodall, *The Chimpanzees of Gombe*, p. 366.
16. H[ippolyte] A. Taine, *History of English Literature*, translated by H. van Laun, second edition (Edinburgh: Edmonston and Douglas, 1872), Volume I, p. 340.
17. Jacqueline Goodchilds and Gail Zellman, "Sexual Signaling and Sexual Aggression in Adolescent Relationships," in *Pornography and Sexual Aggression*, Neil Malamuth and Edward Donnerstein, editors (New York: Academic Press, 1984).
18. Neil Malamuth, "Rape Proclivity among Males," *Journal of Social Issues* 37 (1981), pp. 138–157; Malamuth, "Aggression against Women: Cultural and Individual Causes," in Malamuth and Donnerstein, editors, *op. cit.*
19. The most comprehensive national survey was sponsored by the National Victim Center and the Crime Victims Research and Treatment Center of the Medical University of South Carolina, with financial support from the U.S. Department of Health and Human Services. See David Johnston, "Survey Shows Number of Rapes Far Higher than Official Figures," *New York Times*, April 24, 1992, p. A14.
20. Bondage and rape are popular themes in pornography designed for male audiences in, e.g., Britain, France, Germany, South America, and Japan, as well as the United States. A recurrent subject of Japanese pornographic movies is the rape of a high school girl (Paul Abramson and Haruo Hayashi, "Pornography in Japan," in Malamuth and Donnerstein, editors, *op. cit.*).
21. Robert A. Prentky and Vernon L. Quinsey, *Human Sexual Aggression: Current Perspectives*, Volume 528 of the Annals of the New York Academy of Sciences (New York: New York Academy of Sciences, 1988); Howard E. Barbaree and William L. Marshall, "The Role of Male Sexual Arousal in Rape: Six Models," *Journal of Consulting and Clinical Psychology* 59 (1991), pp. 621–630; Gene Abel, J. Rouleau, and J. Cunningham-Rather, "Sexually Aggressive Behavior," in *Modern Legal Psychiatry and Psychology*, A. L. McGarry and S. A. Shah, editors (Philadelphia: Davis, 1985); Gene Abel, quoted in Faye Knopp, *Retraining Adult Sex*

Offenders: Methods and Models (Syracuse, NY: Safer Society Press, 1984), p. 9.

22. E.g., Lee Ellis, "A Synthesized (Biosocial) Theory of Rape," *Journal of Consulting and Clinical Psychology* 59 (1991), pp. 631–642.

23. E.g., Susan Brownmiller, *Against Our Will: Men, Women and Rape* (New York: Simon & Schuster, 1975); Judith Lewis Herman, "Considering Sex Offenders: A Model of Addiction," *Signs: Journal of Women in Culture and Society* 13 (1988), pp. 695–724.

24. Lee Ellis, *Theories of Rape* (New York: Hemisphere, 1989).

25. Peggy Reeves Sanday, "The Socio-Cultural Context of Rape: A Cross-Cultural Study," *Journal of Social Issues* 37 (1981), pp. 5–27.

Chapter 17

ADMONISHING THE CONQUEROR

1. (London and Edinburgh: Williams and Norgate, 1863), p. 105.

2. Sarah Blaffer Hrdy, "Raising Darwin's Consciousness: Females and Evolutionary Theory," in Robert Bellig and George Stevens, editors, *Nobel Conference XXIII: The Evolution of Sex* (San Francisco: Harper & Row, 1988), p. 161.

3. John Paul Scott, "Agonistic Behavior of Primates: A Comparative Perspective," in Ralph L. Holloway, editor, *Primate Aggression, Territoriality, and Xenophobia: A Comparative Perspective* (New York: Academic Press, 1974), especially p. 427; Shirley C. Strum, *Almost Human: A Journey into the World of Baboons* (New York: Random House, 1987).

4. Dorothy L. Cheney, "Interactions and Relationships Between Groups," Chapter 22 in Barbara B. Smuts, Dorothy L. Cheney, Robert M. Seyfarth, Richard W. Wrangham, and Thomas T. Struhsaker, editors, *Primate Societies* (Chicago: University of Chicago Press, 1986), p. 281.

5. Solly Zuckerman, *The Social Life of Monkeys and Apes* (New York: Harcourt, Brace, 1932), pp. 49, 50.

6. Solly Zuckerman, *From Apes to Warlords* (New York: Harper & Row, 1978), p. 39.

7. *Ibid.*, p. 12.

8. F. W. Fitzsimons, *The Natural History of South Africa*, Volume

1, *Mammals* (London: Longmans, Green, 1919), quoted in Zuckerman, *The Social Life of Monkeys and Apes*, p. 293.

9. Zuckerman, *From Apes to Warlords*, pp. 220, 219, and footnote, p. 220.

10. Zuckerman, *The Social Life of Monkeys and Apes*, pp. 228, 229.

11. *Ibid.*, p. 237.

12. Scott, *op. cit.*; H. Kummer, *Social Origin of Hamadryas Baboons* (Chicago: University of Chicago Press, 1968).

13. Zuckerman, *From Apes to Warlords*, p. 41.

14. *Ibid.*, p. 42.

15. Zuckerman, *The Social Life of Monkeys and Apes*, p. 148.

16. Hrdy, *op. cit.* (ref. 2), p. 163.

17. Donna Robbins Leighton, "Gibbons: Territoriality and Monogamy," Chapter 12 in Smuts *et al.*, eds., *op. cit.*, pp. 135–145.

18. Randall Susman, editor, *The Pygmy Chimpanzee: Evolutionary Biology and Behavior* (New York: Plenum, 1984).

19. Frans de Waal, *Peacemaking among Primates* (Cambridge, MA: Harvard University Press, 1989), p. 181.

20. Toshisada Nishida and Mariko Hiraiwa-Hasegawa, "Chimpanzees and Bonobos: Cooperative Relationships among Males," Chapter 15 in Smuts et al., *op. cit.*, p. 167.

21. Charles Darwin, *The Descent of Man and Selection in Relation to Sex* (New York: The Modern Library, n.d.) (originally published in 1871) pp. 396, 397. Both Pliny and Aelian wrote about wine-imbibing apes who could be captured when drunk.

22. Edward O. Wilson, *Sociobiology: The New Synthesis* (Cambridge, MA: The Belknap Press of Harvard University Press, 1975), p. 538.

23. Irenäus Eibl-Eibesfeldt, *The Biology of Peace and War: Men, Animals, and Aggression*, translated by Eric Mosbacher (New York: The Viking Press, 1979) (originally published in 1975 as *Krieg und Frieden* by R. Piper, München), p. 108.

24. Paul D. MacLean, "Special Award Lecture: New Findings on Brain Function and Sociosexual Behavior," Chapter 4 in Joseph Zubin and John Money, editors, *Contemporary Sexual Behavior: Critical Issues in the 1970s* (Baltimore: The Johns Hopkins University Press, 1973), p. 65.

25. Barbara B. Smuts, "Sexual Competition and Mate Choice," Chapter 31 in Barbara B. Smuts, Dorothy L. Cheney, Robert M.

Seyfarth, Richard W. Wrangham, and Thomas T. Struhsaker, editors, *Primate Societies* (Chicago: University of Chicago Press, 1986), p. 392.

26. Sarah Blaffer Hrdy, "The Primate Origins of Human Sexuality," in Robert Bellig and George Stevens, editors, *Nobel Conference XXIII: The Evolution of Sex* (San Francisco: Harper & Row, 1988).

27. Alison F. Richard, "Malagasy Prosimians: Female Dominance," Chapter 3 in Smuts *et al.*, eds., *op. cit.*, p. 32. Reference for quotation within passage: A. Jolly, "The Puzzle of Female Feeding Priority," in M. Small, ed., *Female Primates: Studies by Women Primatologists* (New York: Alan R. Liss, 1984), p. 198.

28. Toshisada Nishida and Mariko Hiraiwa-Hasegawa, "Chimpanzees and Bonobos: Cooperative Relationships among Males," Chapter 15 in Smuts *et al.*, eds., *op. cit.*, p. 174.

29. Mireille Bertrand, Bibliotheca Primatologica, Number 11, *The Behavioral Repertoire of the Stumptail Macaque: A Descriptive and Comparative Study* (Basel: S. Karger, 1969), p. 191.

30. Frans de Waal, *Peacemaking among Primates* (Cambridge, MA: Harvard University Press, 1989), pp. 153, 154.

31. Frank E. Poirier, "Colobine Aggression: A Review," in Ralph L. Holloway, editor, *Primate Aggression, Territoriality, and Xenophobia: A Comparative Perspective* (New York and London: Academic Press, 1974), pp. 146–147, 130–131, 140–141.

32. Sherwood L. Washburn, "The Evolution of Human Behavior," in John D. Roslansky, editor, *The Uniqueness of Man* (Amsterdam: North-Holland, 1969), p. 170.

33. Robert M. Seyfarth, "Vocal Communication and Its Relation to Language," Chapter 36 in Smuts *et al.*, eds., *op. cit.*, pp. 444, 450, 445.

34. P. D. MacClean, "New Findings on Brain Function and Sociosexual Behavior," in *Contemporary Sexual Behavior*, Zubin and Money, eds., *op. cit.*

35. Solly Zuckerman, *The Social Life of Monkeys and Apes* (New York: Harcourt, Brace, 1932), p. 259.

36. Darwin, *op. cit.*, p. 449.

37. Zuckerman, *op. cit.*, p. 474.

38. Patricia L. Whitten, "Infants and Adult Males," Chapter 28 in Smuts *et al.*, eds., *op. cit.*, pp. 343, 344.

Chapter 18
THE ARCHIMEDES OF THE MACAQUES

1. Translated by John Dryden and revised by Arthur Hugh Clough (New York: The Modern Library, 1932), pp. 378, 379.
2. Work of Wendy Bailey and Morris Goodman; private communication from Morris Goodman, 1992. See also ref. 12.
3. Michael M. Miyamoto and Morris Goodman, "DNA Systematics and Evolution of Primates," *Annual Review of Ecology and Systematics* 21 (1990), pp. 197–220.
4. Marc Godinot and Mohamed Mahboubi, "Earliest Known Simian Primate Found in Algeria," *Nature* 357 (1992), pp. 324–326.
5. Leonard Krishtalka, Richard K. Stucky, and K. Christopher Beard, "The Earliest Fossil Evidence for Sexual Dimorphism in Primates," *Proceedings of the National Academy of Sciences of the United States of America* 87 (13) (July 1990), pp. 5223–5226.
6. Almost 9% of the volume of the brain of insectivores ("insect-eaters," small mammals that may resemble the ancestors of primates) is concerned with the analysis of odors. For prosimians, the number is down to 1.8%; for monkeys, around 0.15%; and for great apes, 0.07%. The fraction for humans is only 0.01%: Only one part in ten thousand of the volume of our brain is devoted to the understanding of smell. (H. Stephan, R. Bauchot, and O. J. Andy, "Data on Size of the Brain and of Various Brain Parts in Insectivores and Primates," in *The Primate Brain*, C. Noback and W. Montagna, editors [New York: Appleton-Century-Crofts, 1970], pp. 289–297.) For insectivores, smell is a major part of what the brain does. For humans, it is an almost insignificant part of our perception of the world—as everyday experience confirms. Humans require 10 million times more butyric acid in the air than dogs do in order to smell it reliably. For acetic acid the factor is 200 million; for caproic acid, 100 million; and for ethyl mercaptan, which is not involved in sexual signaling, two thousand times. (R. H. Wright, *The Sense of Smell* [London: George Allen & Unwin, 1964]; D. Michael Stoddart, *The Scented Ape: The Biology and Culture of Human Odour* [Cambridge: Cambridge University Press, 1990], Table 9.1, p. 235.)
7. J. Terborgh, "The Social Systems of the New World Primates: An

Adaptationist View," in J. G. Else and P. C. Lee, eds., *Primate Ecology and Conservation* (Cambridge: Cambridge University Press, 1986), pp. 199–211.

8. H. Sigg, "Differentiation of Female Positions in Hamadryas One-Male-Units," *Zeitschrift für Tierpsychologie* 53 (1980), pp. 265–302.

9. Connie M. Anderson, "Female Age: Male Preference and Reproductive Success in Primates," *International Journal of Primatology* 7 (1986), pp. 305–326.

10. Dorothy L. Cheney and Richard W. Wrangham, "Predation," Chapter 19 in Barbara B. Smuts, Dorothy L. Cheney, Robert M. Seyfarth, Richard W. Wrangham, and Thomas T. Struhsaker, editors, *Primate Societies* (Chicago: University of Chicago Press, 1986), pp. 227–239.

11. Susan Mineka, Richard Keir, and Veda Price, "Fear of Snakes in Wild- and Laboratory-reared Rhesus Monkeys (*Macaca mulatta*)," *Animal Learning and Behavior* 8 (4) (1980), pp. 653–663.

12. Wendy J. Bailey, Kenji Hayasaka, Christopher G. Skinner, Susanne Kehoe, Leang C. Sien, Jerry L. Slighton and Morris Goodman, "Re-examination of the African Hominoid Trichotomy with Additional Sequences from the Primate β-Globin Gene Cluster," *Molecular Phylogenetics and Evolution*, in press, 1993. See also, C. G. Sibley, J. A. Comstock and J. E. Ahlquist, "DNA Hybridization Evidence of Hominid Phylogeny: a Reanalysis of the Data," *Journal of Molecular Evolution* 30 (1990), pp. 202–236.

13. Toshisada Nishida, "Local Traditions and Cultural Transmission," Chapter 38 in Smuts *et al.*, eds., *op. cit.*, pp. 467, 468. One of the original discussions is by S. Kawamura, "The Process of Subculture Propagation Among Japanese Macaques," *Journal of Primatology* 2 (1959), pp. 43–60. See also Kawamura, "Subcultural Propagation Among Japanese Macaques," in *Primate Social Behavior*, C. A. Southwick, ed. (New York: van Nostrand, 1963); and A. Tsumori, "Newly Acquired Behavior and Social Interaction of Japanese Monkeys," in *Social Communication Among Primates*, S. Altman, ed. (Chicago: University of Chicago Press, 1982).

14. Masao Kawai, "On the Newly-Acquired Pre-Cultural Behavior of the Natural Troop of Japanese Monkeys on Koshima Islet," *Primates* 6 (1965), pp. 1–30.

15. These findings have led to a widely accepted, but wholly unsub-
stantiated myth sometimes called the hundredth-monkey phenomenon (Lyall Watson, *Lifetide* [New York: Simon and
Schuster, 1979]; Ken Keyes, Jr., *The Hundredth Monkey* [Coos
Bay, OR: Vision, 1982]). Potato washing spread slowly through
the macaque colony, it is said, until some critical threshold was
reached; as soon as the hundredth monkey learned the technique,
this knowledge was achieved by everyone, "overnight"—a kind of
paranormal collective consciousness. Various edifying lessons for
human society are then drawn. Unfortunately, there is no evidence at all in support of this heartwarming account (Ron
Amundson, "The Hundredth Monkey Phenomenon," in *The
Hundredth Monkey and Other Paradigms of the Paranormal*, Ken-
drick Frazier, editor [Buffalo, N.Y.: Prometheus, 1991], pp. 171–
181.) It seems to have been invented out of whole cloth.

16. The pioneering physicist Max Planck remarked, after encountering enormous resistance to his new quantum theory, that it takes
a generation for physicists to accept radically new ideas, no matter
how much they explain.

17. William Coffmann McDermott, *The Ape in Antiquity* (Baltimore:
Johns Hopkins Press, 1938).

18. Julian Huxley, *The Uniqueness of Man* (London: Chatto and Windus, 1943), p. 3.

19. B. T. Gardner and R. A. Gardner, "Comparing the Early Utter-
ances of Child and Chimpanzee," in A. Pick, editor, *Minnesota
Symposium in Child Psychology* (Minneapolis, MN: University of
Minnesota Press, 1974), volume 8, pp. 3–23.

20. H. S. Terrace, L. A. Pettito, R. J. Sanders, and T. G. Bever, "Can
an Ape Create a Sentence?" *Science* 206 (1979), pp. 891–902; C.
A. Ristau and D. Robbins, "Cognitive Aspects of Ape Language
Experiments," in D. R. Griffin, editor, *Animal Mind–Human
Mind* (Report of the Dahlem Workshop on Animal Mind–Human
Mind, Berlin, March 22–27, 1981) (Berlin: Springer-Verlag, 1982),
p. 317.

21. Herbert S. Terrace, *Nim* (New York: Knopf, 1979); H. S. Terrace,
L. A. Pettito, R. J. Sanders, and T. G. Bever, "Can an Ape Create
a Sentence?" *Science* 206 (1979), pp. 891–902; Robert M. Seyfarth,
"Vocal Communication and Its Relation to Language," Chapter
36 in Smuts *et al.*, eds., *op. cit.*

22. Roger S. Fouts, Deborah H. Fouts, and Thomas E. Van Cantfort, "The Infant Loulis Learns Signs from Cross-fostered Chimpanzees," in R. A. Gardner, B. T. Gardner, and T. E. Van Cantfort, eds., *Teaching Sign Language to Chimpanzees* (New York: State University of New York Press, 1989).

23. *The Great Ideas: A Syntopicon of Great Books of the Western World*, Volume II, Mortimer J. Adler, editor in chief, William Gorman, general editor, Volume 3 of *Great Books of the Western World*, Robert Maynard Hutchins, editor in chief (Chicago: William Benton/Encyclopaedia Britannica, 1952, 1977), Introduction to Chapter 51, "Man."

24. E. S. Savage-Rumbaugh, D. M. Savage-Rumbaugh, S. T. Smith, and J. Lawson, "Reference—the Linguistic Essential," *Science* 210 (1980), pp. 922–925.

25. Patricia Marks Greenfield and E. Sue Savage-Rumbaugh, "Grammatical Combination in *Pan paniscus*: Processes of Learning and Invention in the Evolution and Development of Language," in *"Language" and Intelligence in Monkeys and Apes*, Sue Taylor Parker and Kathleen Gibson, editors (Cambridge: Cambridge University Press, 1990); *idem*, "Imitation, Grammatical Development, and the Invention of Protogrammar by an Ape," in *Biological and Behavioral Determinants of Language Development*, Norman Krasnegor, D. M. Rumbaugh, R. L. Schiefelbusch and M. Studdert-Kennedy, editors (Hillsdale, NJ: Erlbaum, 1991).

26. These experiments by Sue Savage-Rumbaugh and Duane Rumbaugh are briefly described in D. S. Rumbaugh, "Comparative Psychology and the Great Apes: Their Competence in Learning, Language and Numbers," *The Psychological Record* 40 (1990), pp. 15–39. A detailed description is in E. Sue Savage-Rumbaugh, Jeannine Murphy, Rose Sevcik, S. Williams, K. Brakke, and Duane M. Rumbaugh, "Language Comprehension in Ape and Child," *Monographs of the Society for Research in Child Development*, in press, 1993.

27. D. M. Rumbaugh, W. D. Hopkins, D. A. Washburn, and E. Sue Savage-Rumbaugh, "Comparative Perspectives of Brain, Cognition and Language," In N. A. Krasnegor, *et al.*, editors, *op. cit.* (ref. 22).

28. David Premack, *Intelligence in Ape and Man* (Hillsdale, NJ: Erlbaum, 1976).

29. D. J. Gillan, D. Premack, and G. Woodruff, "Reasoning in the Chimpanzee: I. Analogical Reasoning," *Journal of Experimental Psychology and Animal Behavior* 7 (1981), pp. 1–17; D. J. Gillan, "Reasoning in the Chimpanzee: II. Transitive Inference," *ibid.*, pp. 150–164.

30. David Premack and G. Woodruff, "Chimpanzee Problem-solving: A Test for Comprehension," *Science* 202 (1978), pp. 532–535; Premack and Woodruff, "Does the Chimpanzee Have a Theory of Mind?" *Behavior and Brain Sciences* 4 (1978), pp. 515–526.

31. An early, although limited attempt: Duane M. Rumbaugh, Timothy V. Gill and E. C. von Glasersfeld, "Reading and Sentence Completion by a Chimpanzee (Pan)," *Science* 182 (1973), pp. 731–733; James L. Pate and Duane M. Rumbaugh, "The Language-Like Behavior of Lana Chimpanzee," *Animal Learning and Behavior* 11 (1983), pp. 134–138.

32. This quotation and the basis for its supporting paragraph is from Derek Bickerton's stimulating *Language and Species* (Chicago: University of Chicago Press, 1990).

33. E. Sue Savage-Rumbaugh *et al.*, *op. cit.* (Note 24).

34. Eugene Linden, *Silent Partners: The Legacy of the Ape Language Experiments* (New York: Times Books, 1986), pp. 144, 145.

35. Jane Goodall, *Through a Window* (Boston: Houghton Mifflin, 1990), p. 13.

36. Linden, *op. cit.*, pp. 79, 81.

37. Janis Carter, "Survival Training for Chimps: Freed from Keepers and Cages, Chimps Come of Age on Baboon Island," *The Smithsonian* 19 (1) (June 1988), pp. 36–49.

38. The total number of chimps left on Earth is now about fifty thousand. They are very much an endangered species.

39. II, 17, translated by Maxwell Staniforth (Harmondsworth, UK: Penguin Books, 1964); in Michael Grant, editor, *Greek Literature: An Anthology* (Harmondsworth, UK: Penguin Books, 1977) (first published in Pelican Books as *Greek Literature in Translation*, 1973), p. 427.

Chapter 19
WHAT IS HUMAN?

1. Quoted in Gavin Rylands de Beer, editor, "Darwin's Notebooks on Transmutation of Species, Part IV: Fourth Notebook (October 1838–10 July 1839)," *Bulletin of the British Museum (Natural History), Historical Series* (London) 2 (5) (1960), pp. 151–183; quotation (from notebook entry 47) appears on p. 163.

2. Frank Roper, *The Missing Link: Consul the Remarkable Chimpanzee* (Manchester: Abel Heywood, 1904). A now-extinct primate of some 30 million years ago, perhaps ancestral to both apes and humans, has been named Proconsul, in honor of the Victorian sophisticate.

3. Mortimer J. Adler, *The Difference of Man and the Difference It Makes* (New York: Holt, Rinehart and Winston, 1967), p. 84.

4. Theodosius Dobzhansky, *Mankind Evolving* (New Haven: Yale University Press, 1962), p. 339.

5. George Gaylord Simpson, *The Meaning of Evolution* (New Haven: Yale University Press, 1949), p. 284.

6. Adler, *op. cit.*, p. 136.

7. This answer was first proposed in a lecture to the Yale Divinity School in 1880 by Darwin's friend, the botanist and evolutionary biologist Asa Gray (*Natural Science and Religion* [New York: Scribner's, 1880]).

8. *Metaphysics, Materialism and the Evolution of Mind: Early Writings of Charles Darwin*, transcribed and annotated by Paul H. Barrett, commentary by Howard E. Gruber (Chicago: University of Chicago Press, 1974), p. 187.

9. Especially in *The Descent of Man*.

10. Adam Smith, *An Inquiry into the Nature and Causes of the Wealth of Nations*, Edwin Cannan, editor (New York: Modern Library/Random House, 1937), Chapter II, "Of the Principle Which Gives Occasion to the Division of Labour," p. 13.

11. Keith Thomas, *Man and the Natural World: A History of the Modern Sensibility* (New York: Pantheon, 1983), p. 31.

12. Frans de Waal, *Peacemaking Among Primates* (Cambridge, MA: Harvard University Press, 1989), p. 82.

13. Smith, *op. cit.*, p. 14.

14. Tacitus, *The Histories*, translated by Alfred John Church and William Jackson Brodribb, in Volume 15 of *Great Books of the Western World*, Robert Maynard Hutchins, editor in chief (Chicago: William Benton/Encyclopaedia Britannica, 1952, 1977), Book IV, 13, 17, pp. 269, 271.

15. Another purported distinction of humans based solely on bodily form: "Man is, I believe the only animal that has a marked projection in the middle of the face," an opinion of the eighteenth-century aesthete Uvedale Price. (Quoted in Keith Thomas, *op. cit.*, p. 32.) He may have been ignorant of tapirs and proboscis monkeys, but elephants?

16. Thomas Aquinas, *Summa Theologica*, Volume I, translated by Fathers of the English Dominican Province, revised by Daniel J. Sullivan, Volume 19 of *Great Books of the Western World* (Chicago: Encyclopaedia Britannica, 1952), Second Part, Part I, I. "Treatise on the Last End," Question I, "On Man's Last End" (p. 610); Part I, II. "Treatise on Human Acts," Question XIII, "Of Choice" (pp. 673, 674); and Question XVII, "Of the Acts Commanded by the Will" (p. 688).

17. Jakob von Uexküll, "A Stroll Through the Worlds of Animals and Men: A Picture Book of Invisible Worlds" (1934), Part I of Claire H. Schiller, translator and editor, *Instinctive Behavior: The Development of a Modern Concept* (New York: International Universities Press, 1957), p. 42.

18. John Dewey, *Reconstruction in Philosophy* (New York: Henry Holt, 1920), p. 1.

19. Hugh Morris, *The Art of Kissing* (1946), forty-seven pages; no publisher is given in this demure little pamphlet.

20. Desmond Morris, *The Naked Ape* (New York: Dell, 1984) (originally published in 1967 by McGraw Hill; revised edition published in 1983), p. 62.

21. Donald Symons, *The Evolution of Human Sexuality* (New York: Oxford University Press, 1979), pp. 78, 79.

22. Gerritt S. Miller, "Some Elements of Sexual Behavior in Primates, and Their Possible Influence on the Beginnings of Human Social Development," *Journal of Mammalogy* 9 (1928), pp. 273–293.

23. Gordon D. Jensen, "Human Sexual Behavior in Primate Perspec-

tive," Chapter 2 in Joseph Zubin and John Money, editors, *Contemporary Sexual Behavior: Critical Issues in the 1970s* (Baltimore: The Johns Hopkins University Press, 1973), p. 20.

24. Cf. *ibid.*, p. 22.

25. For example, K. Imanishi, "The Origin of the Human Family: A Primatological Approach," *Japanese Journal of Ethnology* 25 (1961), pp. 110–130 (in Japanese); discussed in Toshisada Nishida, editor, *The Chimpanzees of the Mahale Mountains: Sexual and Life History Strategies* (Tokyo: University of Tokyo Press, 1990), p. 10.

26. By the philosopher Johan Huizinga, *Homo Ludens* (Boston: Beacon, 1955).

27. Epictetus, *The Discourses of Epictetus*, translated by George Long, pp. 105–252 of Volume 12, *Great Books of the Western World* (Chicago: Encyclopaedia Britannica, 1952), Book IV, Chapter 11, "About Purity," pp. 240, 241. (In Book III, Chapter 7, Epictetus proposes another "unique" quality: shame and blushing.)

28. E.g., Jane Goodall, *Through a Window: My Thirty Years with the Chimpanzees of Gombe* (Boston: Houghton-Mifflin, 1990).

29. Plato, *The Dialogues of Plato*, translated by Benjamin Jowett (in Volume 7 of *Great Books of the Western World*), *Laws*, Book VII, p. 715.

30. Goodall, *op. cit.*

31. Charles Darwin, *The Descent of Man and Selection in Relation to Sex* (New York: The Modern Library, n.d.) (originally published in 1871) p. 449.

32. Leo K. Bustad, "Man and Beast Interface: An Overview of Our Interrelationships," in Michael H. Robinson and Lionel Tiger, editors, *Man and Beast Revisited* (Washington and London: Smithsonian Institution Press, 1991), p. 250.

33. Toshisada Nishida, "Local Traditions and Cultural Transmission," Chapter 38 of Barbara B. Smuts, Dorothy L. Cheney, Robert M. Seyfarth, Richard W. Wrangham, and Thomas T. Struhsaker, editors, *Primate Societies* (Chicago: University of Chicago Press, 1986), p. 473.

34. Martin Daly and Margo Wilson, *Homicide* (New York: Aldine de Gruyter, 1988), p. 187.

35. Owen Chadwick, *The Secularization of the European Mind in the 19th Century* (Cambridge: Cambridge University Press, 1975), p. 269.

36. Solly Zuckerman, *The Social Life of Monkeys and Apes* (New York: Harcourt, Brace, 1932), p. 313.

37. Leslie A. White, "Human Culture," *Encyclopaedia Britannica,Macropaedia* (1978), Volume 8, p. 1152.

38. Toshisada Nishida, "A Quarter Century of Research in the Mahale Mountains: An Overview," Chapter 1 of Nishida, editor, *The Chimpanzees of the Mahale Mountains*, p. 34.

39. Henri Bergson, *The Two Sources of Morality and Religion* (New York: Holt, 1935).

40. Nishida, *op. cit.* (Note 38), p. 24. Chimpanzee folk medicine seems to have been independently rediscovered by other primatologists (Ann Gibbons, "Plants of the Apes," *Science* 255 [1992], p. 921). Among pre-industrial humans, most plants are used for something. The botanist Gillian Prance and his colleagues found (private communication, 1992) that 95 percent of the rainforest trees accessible to a group of Bolivian indigenous peoples are employed—for example, the sap of a tree in the nutmeg family as a potent fungicide.

41. E.g., Raymond Firth, *Elements of Social Organisation* (London: Watts and Co., 1951), pp. 183, 184; D. Michael Stoddart, *The Scented Ape: The Biology and Culture of Human Odour* (Cambridge: Cambridge University Press, 1990), p. 126.

42. Napoleon A. Chagnon, *Yanomamo: The Fierce People* (New York: Holt, Rinehart, Winston, 1968), p. 65.

43. Desmond Morris, *The Biology of Art* (London: Methuen, 1962); R. A. Gardner and B. T. Gardner, "Comparative Psychology and Language Acquisition," in K. Salzinger and F. E. Denmarks, editors, *Psychology: The State of the Art* (New York: Annals of New York Academy of Sciences, 1978), pp. 37–76; K. Beach, R. S. Fouts, and D. H. Fouts, "Representational Art in Chimpanzees," *Friends of Washoe*, 3:2–4, 4:1–4. Oil paintings by a chimp named Congo, which today hang in several private collections, exhibit a gaudy abstract expressionism and are considered the best of the chimp *oeuvres*.

44. Birds, for example, recognize and mob a novel predator (or even

a milk bottle) that frightened their ancestors four generations earlier. And speaking of milk bottles, soon after one blue tit pierced the metal foil cap of a milk bottle left on a doorstep and drank the cream, blue tits all over England are said to have begun drinking cream. (John Tyler Bonner, *The Evolution of Culture in Animals* [Princeton, NJ: Princeton University Press, 1980].) Of course no one knows who this pioneering bird was. This may not be learning by imitation, though. An already opened milk bottle and another bird present nearby and happy may be enough to give a naive bird the idea. (D. F. Sherry and B. G. Galef, Jr., "Social Learning Without Imitation: More About Milk Bottle Opening by Birds," *Animal Behaviour* 40 [1990], pp. 987–989.)

45. Zuckerman, *op. cit.*, pp. 315, 316.

46. Nishida, "A Quarter Century of Research," p. 12.

47. So could souls have provided consciousness back then? A deity responsible on a case-by-case basis for precision injection of souls into this immense host of tiny creatures over the full range of geological time would be a very fussy as well as a very inefficient creator. Why not design it right from the beginning, and let life run by itself? Would the god responsible for the subtle, elegant, and universally applicable laws of physics do such slapdash, error-ridden, journeyman work in biology—requiring hands-on attention to every pathetic little microbe when they already know perfectly well how to reproduce themselves and vast stores of information? Instead, all the god has to do is to encode directly into the DNA of a few ancestors whatever information souls are required to know. Souls and consciousness could then pass, on their own, from generation to generation, freeing the god for other matters, perhaps some of greater urgency. But if the information in the DNA has come to be through the patient evolutionary process, why is a god needed to explain the injection of data, genes, or souls in the first place?

48. A. I. Hallowell, "Culture, Personality and Society," in *Anthropology Today*, A. L. Kroeber, editor (Chicago: University of Chicago Press, 1953), pp. 597–620; Hallowell, "Self, Society and Culture in Phylogenetic Perspective," in *Evolution After Darwin*, Volume 2, S. Tax, editor (Chicago: University of Chicago Press, 1960), pp. 309–371. The contention that only humans are self-aware can be

found in many philosophical and scientific disquisitions, e.g., Karl R. Popper and John C. Eccles, *The Self and Its Brain* (New York: Springer, 1977).

49. G. G. Gallup, Jr., "Self-Recognition in Primates: A Comparative Approach to the Bidirectional Properties of Consciousness," *American Psychologist* 32 (1977), pp. 329–338.

50. A common literary and iconographic theme in medieval Europe beginning in the thirteenth century is an alleged propensity for apes to admire themselves in mirrors. Cf. H. W. Janson, *Apes and Ape Lore in the Middle Ages and the Renaissance* (London: University of London, 1952), pp. 212 *et seq.*

51. Montaigne, *The Essays of Michel Eyquem de Montaigne*, Book II, Essay XII, "Apology for Raimond de Sebonde," translated by Charles Cotton, edited by W. Carew Hazlitt, Volume 25 of *Great Books of the Western World* (Chicago: Encyclopaedia Britannica, 1952), p. 227. In a nearby passage, Montaigne quotes the Roman epigramist Juvenal: "What stronger lion ever took the life from a weaker?" But, as we've mentioned, lions routinely kill all the cubs on taking over a pride. This saves the male the trouble of caring for young not his, and helps bring the females back into heat.

52. E.g., R. L. Trivers, *Social Evolution* (Menlo Park, CA: Benjamin/ Cummings, 1985), especially the chapter "Deceit and Self-Deception"; Joan Lockard and Delroy Paulhus, editors, *Self-Deception: An Adaptive Mechanism?* (Englewood Cliffs, NJ: Prentice-Hall, 1989).

53. C. G. Beer, "Study of Vertebrate Communication—Its Cognitive Implications," in *D. R. Griffin, editor, Animal Mind–Human Mind* (Report of the Dahlem Workshop on Animal Mind–Human Mind, Berlin, March 22–27, 1981) (Berlin: Springer-Verlag, 1982), p. 264; E. W. Menzel, "A Group of Young Chimpanzees in a One-acre Field," in A. M. Schrier and F. Stollnitz, editors, *Behavior of Nonhuman Primates* (New York: Academic Press, 1974).

54. Stuart Hampshire, *Thought and Action* (London: Chatto and Windus, 1959).

55. T. H. Huxley, *Evidence as to Man's Place in Nature* (London: Williams and Norgate, 1863), p. 132.

56. Letter of February 5, 1649, in Mortimer J. Adler and Charles Van Doren, *Great Treasury of Western Thought: A Compendium of*

Important Statements on Man and His Institutions by the Great Thinkers in Western History (New York and London: R. R. Bowker Company, 1977), p. 12.

57. See, for example, Eugene Linden, *Silent Partners: The Legacy of the Ape Language Experiments* (New York: Times Books, 1986); Roger Fouts, "Capacities for Language in the Great Apes," in *Proceedings, Ninth International Congress of Anthropological and Ethnological Sciences* (The Hague: Mouton, 1973).

58. For example, "Man is the only animal . . . that can use symbols" (Max Black, *The Labyrinth of Language* [New York: Praeger, 1968]); "Animals cannot have language . . . If they had it, they would . . . no longer be animals. They would be human beings" (K. Goldstein, "The Nature of Language," in *Language: An Enquiry into Its Meaning and Function* [New York: Harper, 1957]); "There seems to be no substance to the view that human language is simply a more complex instance of something to be found elsewhere in the animal world" (Noam Chomsky, *Language and Mind* [New York: Harcourt Brace Jovanovich, 1972]). These examples are taken from Donald R. Griffin's *The Question of Animal Awareness*, revised edition (New York: Rockefeller University Press, 1981). Only occasionally is a contrary note sounded (e.g., A. I. Hallowell, *Philosophical Theology*, Vol. 2 [Cambridge: Cambridge University Press, 1937], p. 94.)

59. Derek Bickerton, *Language and Species* (Chicago: University of Chicago Press, 1990), especially pp. 8, 15–16.

60. Bickerton, *op. cit.*, proposes that the early speech of children is a "protolanguage" fundamentally different from fully developed human languages, that this protolanguage may be accessible to apes, and that it was used by our ancestors in the transition from apes to humans.

Chapter 20
THE ANIMAL WITHIN

1. (New York: Doubleday, 1958), p. 345.

2. In the wild there are occasional female chimps who reject males under all circumstances and at great cost. They of course produce no children. Might this correlation be noticed? Might there be,

occasionally, a chimp that ponders the possible connection between sex and babies? How sure can we be that this might not be so?

3. Bolingbroke (1809), quoted in Arthur O. Lovejoy, *The Great Chain of Being: A Study of the History of an Idea* (Cambridge: Harvard University Press, 1953), p. 196.

4. Ambrose Bierce, "Reverence," in *The Enlarged Devil's Dictionary*, Ernest Jerome Hopkins, editor (Garden City, NY: Doubleday, 1967), p. 247.

5. Walt Whitman, *Leaves of Grass*, Harold W. Blodgett and Sculley Bradley, editors (New York: New York University Press, 1965), "Song of Myself," stanza 32, lines 684–691, p. 60.

6. *The Essays of Michel Eyquem de Montaigne*, translated by Charles Cotton, edited by W. Carew Hazlitt, Volume 25 of *Great Books of the Western World*, Robert Maynard Hutchins, editor in chief (Chicago: William Benton/Encyclopaedia Britannica, 1952, 1977), Book III, Essay I, "Of Profit and Honesty," p. 381.

7. C. Boesch and H. Boesch, "Possible Causes of Sex Differences in the Use of Natural Hammers by Wild Chimpanzees," *Journal of Human Evolution* 13 (1984), pp. 415–440, and references given there.

8. See, e.g., John Alcock, "The Evolution of the Use of Tools by Feeding Animals," *Evolution* 26 (1972), pp. 464–473; K. R. L. Hall and G. B. Schaller, "Tool-using Behavior of the Californian Sea Otter," *Journal of Mammalogy* 45 (1964), pp. 287–298; A. H. Chisholm, "The Use by Birds of 'Tools' or 'Instruments,' " *Ibis* 96 (1954), pp. 380–383; J. van Lawick-Goodall and H. van Lawick, "Use of Tools by Egyptian Vultures," *Nature* 12 (1966), pp. 1468-1469.

9. Anthony J. Podlecki, *The Political Background of Aeschylean Tragedy* (Ann Arbor: University of Michigan Press, 1966), pp. 1, 7, 155.

10. Mortimer J. Adler, *The Difference of Man and the Difference It Makes* (New York: Holt, Rinehart, Winston, 1967), p. 121.

11. Geza Teleki, "Chimpanzee Subsistence Technology: Materials and Skills," *Journal of Human Evolution* 3 (6) (November 1974), pp. 575–594; our quotes are from pp. 585–588 and p. 593.

12. Michael Tomasello, "Cultural Transmission in the Tool Use and

Communicatory Signalling of Chimpanzees?" in *"Language" and Intelligence in Monkeys and Apes*, Sue Taylor Parker and Kathleen Gibson, editors (Cambridge: Cambridge University Press, 1990).

13. Teleki, *op. cit.*

14. C. Jones and J. Sabater Pi, "Sticks Used by Chimpanzees in Rio Muni, West Africa," *Nature* 223 (1969), pp. 100–101; Y. Sugiyama, "The Brush-stick of Chimpanzees Found in Southwest Cameroon and Their Cultural Characteristics," *Primates* 26 (1985), pp. 361–374; W. McGrew and M. Rogers, "Chimpanzees, Tools and Termites: New Record from Gabon," *American Journal of Primatology* 5 (1983), pp. 171–174.

15. Teleki, *op. cit.*

16. E.g., Kenneth P. Oakley, *Man the Tool-Maker* (Chicago: University of Chicago Press, 1964).

17. E. Sue Savage-Rumbaugh, Jeannine Murphy, Rose Sevcik, S. Williams, K. Brakke and Duane M. Rumbaugh, "Language Comprehension in Ape and Child," *Monographs of the Society for Research in Child Development*, in press, 1993; Duane M. Rumbaugh, private communication, 1992.

18. Susan Essock-Vitale and Robert M. Seyfarth, "Intelligence and Social Cognition," Chapter 37 of Barbara B. Smuts, Dorothy L. Cheney, Robert M. Seyfarth, Richard W. Wrangham, and Thomas T. Struhsaker, editors, *Primate Societies* (Chicago: University of Chicago Press, 1986), pp. 456, 457; Wolfgang Kohler, *The Mentality of Apes*, second edition (New York: Viking, 1959) (originally published in 1925), p. 38.

19. Richard Wrangham, quoted by Ann Gibbons, "Chimps: More Diverse than a Barrel of Monkeys," *Science* 255 (1992), pp. 287, 288.

20. H. J. Jerison, *Evolution of the Brain and Intelligence* (New York: Academic Press, 1973); Carl Sagan, *The Dragons of Eden: Speculations on the Evolution of Human Intelligence* (New York: Random House, 1977), Chapter 2; William S. Cleveland, *The Elements of Graphing Data* (Monterey, CA: Wadsworth, 1985). Cleveland notes that "Happily, modern man is at the top."

21. R. E. Passingham, "Changes in the Size and Organization of the

Brain in Man and His Ancestors," *Brain and Behavioral Evolution* 11 (1980), pp. 73–90.

22. *Ibid.*

23. E.g., Sagan, *op. cit.* (note 20).

24. Gordon Thomas Frost, "Tool Behavior and the Origins of Laterality," *Journal of Human Evolution* 9 (1980), pp. 447–459.

25. E.g., Mortimer J. Adler, *op. cit.* (note 10), p. 120.

26. F. Nottebohm, "Neural Asymmetries in the Vocal Control of the Canary," in *Lateralization in the Nervous System*, S. R. Harnad and R. W. Doty, editors (New York: Academic, 1977).

27. E.g., W. D. Hopkins and R. D. Morris, "Laterality for Visual-Spatial Processing in Two Language-Trained Chimpanzees," *Behavioral Neuroscience* 103 (1989), pp. 227–234.

28. Thomas Henry Huxley, *Evidence as to Man's Place in Nature* (London and Edinburgh: Williams and Norgate, 1863), pp. 109, 110.

29. Aristotle, *Ethica Nicomachea*, in Volume IX of *The Works of Aristotle*, translated into English under the editorship of W. D. Ross (Oxford: Clarendon Press, 1925), Book X, "Pleasure; Happiness," 7, 1178ª5.

30. Mark Twain, *Letters from the Earth*, Bernard DeVoto, editor (New York and Evanston: Harper & Row, 1962), "The Damned Human Race," V, "The Lowest Animal," p. 227.

31. E.g., Carl Sagan and Richard Turco, *A Path Where No Man Thought: Nuclear Winter and the End of the Arms Race* (New York: Random House, 1990).

32. Henry D. Thoreau, *Walden*, edited by J. Lyndon Shanley (Princeton, NJ: Princeton University Press, 1971), "Higher Laws," p. 219.

33. Plato, *The Republic*, translated by Benjamin Jowett (New York: The Modern Library, 1941), IX, 571, p. 330.

34. J. Hughlings Jackson, *Evolution and Dissolution of the Nervous System* (London: John Bale, 1888), p. 38.

35. Paul D. MacLean, *The Triune Brain in Evolution: Role in Paleocerebral Functions* (New York and London: Plenum Press, 1990).

36. *Romans* 7:18 (King James translation).

37. So far as we know, the testosterone defense has not yet been tried in a court of law.

38. *Buddhist Scriptures*, Edward Conze, editor (Harmondsworth, UK:

Penguin, 1959), p. 112; *The Saundarananda of Ashvaghosha*, E. H. Johnston, editor and translator (Delhi: Motilal Banarsidass, 1928, 1975), Canto XV, "Emptying the Mind," p. 86 of English translation, verse 53.

Chapter 21
SHADOWS OF FORGOTTEN ANCESTORS

1. Attributed to Empedocles by Hippolytus, in *Refutation of All Heresies*, I, iii, 2, in Jonathan Barnes, editor, *Early Greek Philosophy* (Harmondsworth, Middlesex, England: Penguin Books, 1987), p. 196.

Epilogue

1. Thomas Aquinas, *Summa Theologica*, Volume I of *Basic Writings of Saint Thomas Aquinas*, translated by Father Laurence Shapcote, edited and translation revised by Anton C. Pegis (New York: Random House, 1945), Part I, VIII, "The Divine Government," Question 103, Article 2, p. 952.

Permissions Acknowledgments

Grateful acknowledgment is made to the following for permission to reprint previously published material:

The Belknap Press of Harvard University Press: Excerpts from *The Chimpanzees of Gombe: Patterns of Behavior* by Jane Goodall. Copyright © 1986 by the President and Fellows of Harvard College; excerpts from *The New Synthesis* by Edward O. Wilson. Copyright © 1975 by the President and Fellows of Harvard College. Reprinted by permission of The Belknap Press of Harvard University Press.

Bilingual Press and Anvil Poetry Press Ltd.: Excerpt from *Poems of the Aztec Peoples*, translated by Edward Kissam and Michael Schmidt. Copyright © 1977, 1983 by Edward Kissam and Michael Schmidt. Rights throughout the world excluding the United States are controlled by Anvil Press Poetry Ltd. Reprinted by permission of Bilingual Press and Anvil Press Poetry Ltd.

Doubleday, a division of Bantam, Doubleday, Dell Publishing Group, Inc.: Excerpts from *Darwin's Century* by Loren Eiseley. Copyright © 1958 by Loren Eiseley; excerpt from "Written for Old Friends in Yang-jou . . ." from *The Heart of Chinese Poetry* by Greg Whincup. Copyright © 1987 by Greg Whincup. Reprinted by permission of Doubleday, a division of Bantam, Doubleday, Dell Publishing Group, Inc.

Encyclopedia Britannica: Excerpts from "Human Culture" by Leslie A. White from *Encyclopedia Britannica*, 15th edition (1978), 8:1152. Reprinted by permission of Encyclopedia Britannica.

Grove Press, Inc.: Excerpts from *Zen Poems of China and Japan: The Crane's Bill* by Lucien Stryk, Takashi Ikemoto, and Taigan Takayama. Copyright © 1973 by Lucien Stryk, Takashi Ikemoto, and Taigan Takayama. Reprinted by permission of Grove Press, Inc.

Harcourt Brace Jovanovich, Inc.: Excerpts from *The Social Life of Monkeys and Apes* by Solly Zuckerman. Reprinted by permission of Harcourt Brace Jovanovich, Inc.

HarperCollins Publishers, Inc.: Excerpts from *Nobel Conference XXIII* by Sarah Blaffer Hrdy, edited by Bellig and Stevens; excerpts from *From Apes to Warlords* by Solly Zuckerman. Reprinted by permission of HarperCollins Publishers, Inc.

Harvard University Press: Excerpts from *Peacemaking Among Primates* by Frans B. M. de Waal. Copyright © 1989 by Frans B. M. de Waal. Reprinted by permission of Harvard University Press.

Houghton Mifflin Company and Weidenfeld & Nicolson Ltd: Excerpts from *Through a Window: My Thirty Years with the Chimpanzees of Gombe* by Jane Goodall. Copyright © 1990 by Soko Publications Ltd. Rights throughout the British Commonwealth are controlled by Weidenfeld & Nicolson. Reprinted by permission of Houghton Mifflin Company and Weidenfeld & Nicholson Ltd.

The Johns Hopkins University Press: Excerpts from "Special Awards Lecture" by MacLean from *Contemporary Sexual Behavior*, edited by John Money and Joseph Zubin, published by The Johns Hopkins University Press, Baltimore/London, in 1973. Reprinted by permission.

John Murray (Publishers) Ltd: Excerpts from *The Bhagavad Ghita*, translated by Juan Mascaro. Reprinted by permission of John Murray (Publishers) Ltd.

Penguin Books Ltd.: Excerpts from *Early Greek Philosophy*, translated and edited by Jonathan Barnes (Penguin Classics, 1987). Copyright © 1987 by Jonathan Barnes. Reprinted by permission of Penguin Books Ltd.

Simon and Schuster, Inc. Excerpts from *Popul Vuh*, translated by Dennis Tedlock. Copyright © 1985 by Dennis Tedlock. Reprinted by permission of Simon and Schuster, Inc.

Smithsonian Institution Press: Excerpts from "Deceit and Self-Deception: The Relationship Between Communications and Consciousness" by Robert Trivers in *Man and Beast Revisited*, edited by Michael H. Robinson and Lionel Tiger. Copyright © 1991 by Smithsonian Institution. Reprinted by permission of Smithsonian Institution Press.

University of Chicago Press: Excerpt from Williams and Nesse, *Quarterly Review of Books* 66:1 (March 1991); excerpts from *Primate Societies*, edited by Smuts et. al.; excerpts from "Hippolytus" translated by Grene from *Complete Greek Tragedies*; excerpts from *Genetics and the Social Behavior of the Dog* by Scott and Fuller. All excerpts reprinted by permission of University of Chicago Press.

Viking Penguin, a division of Penguin Books USA Inc. Excerpts from *The Biology of Peace and War* by Irenaus Eibl-Eibesfeldt, translated by Eric Mosbacher. Translation copyright © 1966 by R. Piper & Co., Verlag, Munchen. English translation copyright © 1979 by Viking Penguin, a division of Penguin Books USA, Inc. Reprinted by permission.

Index

CARL SAGAN is the David Duncan Professor of Astronomy and Space Sciences and Director of the Laboratory for Planetary Studies at Cornell University. He has played a leading role in the Mariner, Viking, and Voyager spacecraft expeditions to the planets, for which he received the NASA Medals for Exceptional Scientific Achievement and (twice) for Distinguished Public Service. Once a research assistant of the Nobel Prize–winning geneticist H. J. Muller, his continuing research on the origin of life began in the 1950s. The Masursky Award from the American Astronomical Society cites "his extraordinary contributions to the development of planetary science . . . As a scientist trained in both astronomy and biology, Dr. Sagan has made seminal contributions to the study of planetary atmospheres, planetary surfaces, the history of the Earth, and exobiology. Many of the most productive planetary scientists working today are his present and former students and associates."

His book *Cosmos* (accompanying his Emmy- and Peabody-award-winning television series of the same name) was the best-selling science book ever published in the English language. His novel *Contact* is soon to be a major motion picture (Warner Bros.). Co-founder and President of The Planetary Society, he serves as Distinguished Visiting Scientist, Jet Propulsion Laboratory, California Institute of Technology. Dr. Sagan has received the Pulitzer Prize, the Oersted Medal, and many other awards—including eighteen honorary degrees from American colleges and universities—for his contributions to science, literature, education, and the preservation of the environment.

ANN DRUYAN is the Secretary of the Washington-based Federation of American Scientists, founded in 1945 to oppose the misuse of science and high technology. As Creative Director of NASA's Voyager Interstellar Record project, she was responsible for sending rock-and-roll (and much else) on two spacecraft to the stars. She has served as writer-producer of PBS's *Nova* and of several network television specials. She was co-writer of the *Cosmos* television series and executive producer of its recent updating and reversioning. Ms. Druyan is the author of the novel A *Famous*

Broken Heart, and, with Sagan, of the best-seller *Comet*, other books, speeches, and numerous articles. She is also a Director of the New York Children's Health Project.

The authors are married and together have two children. Dr. Sagan has three grown sons. In the 1980s he and Ms. Druyan organized three of the largest demonstrations of non-violent civil disobedience at the Nevada Nuclear Test Site in protest against continued U.S. nuclear weapons testing. Recently, they have been working to bring scientists and religious leaders together to help protect the global environment.

A B O U T   T H E   T Y P E

This book was set in Electra, a typeface designed for Linotype by W. A. Dwiggins, the renowned type designer (1880–1956). Electra is a fluid typeface, avoiding the contrasts of thick and thin strokes that are prevalent in most modern typefaces.

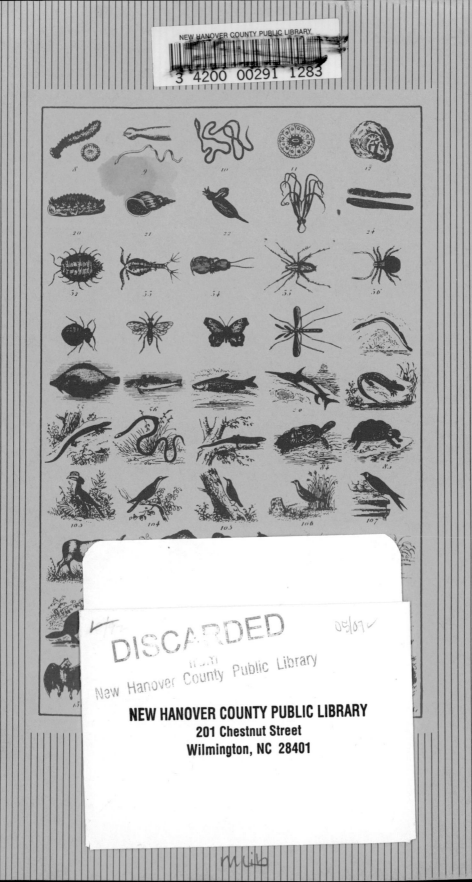